The Effects of Invasive Plants on Soil Fauna

CABI INVASIVES SERIES

Invasive species are plants, animals or microorganisms not native to an ecosystem, whose introduction has threatened biodiversity, food security, health or economic development. Many ecosystems are affected by invasive species and they pose one of the biggest threats to biodiversity worldwide. Globalization through increased trade, transport, travel and tourism will inevitably increase the intentional or accidental introduction of organisms to new environments, and it is widely predicted that climate change will further increase the threat posed by invasive species. To help control and mitigate the effects of invasive species, scientists need access to information that not only provides an overview of and background to the field, but also keeps them up to date with the latest research findings.

This series addresses all topics relating to invasive species, including biosecurity surveillance, mapping and modelling, economics of invasive species and species interactions in plant invasions. Aimed at researchers, upper-level students and policy makers, titles in the series provide international coverage of topics related to invasive species, including both a synthesis of facts and discussions of future research perspectives and possible solutions.

Titles Available

1. Invasive Alien Plants: An Ecological Appraisal for the Indian Subcontinent
Edited by J.R. Bhatt, J.S. Singh, R.S. Tripathi, S.P. Singh and R.K. Kohli

2. Invasive Plant Ecology and Management: Linking Processes to Practice
Edited by T.A. Monaco and R.L. Sheley

3. Potential Invasive Pests of Agricultural Crops
Edited by J.E. Peña

4. Invasive Species and Global Climate Change
Edited by L.H. Ziska and J.S. Dukes

5. Bioenergy and Biological Invasions: Ecological, Agronomic and Policy Perspectives on Minimizing Risk
Edited by L.D. Quinn, D.P. Matlaga and J.N. Barney

6. Biosecurity Surveillance: Quantitative Approaches
Edited by F .Jarrad S.L. Choy and K. Mengersen

7. Pest Risk Modeling and Mapping for Invasive Alien Species
Edited by Robert C. Venette

8. Invasive Alien Plants: Impacts on Development and Options for Management
Edited by Carol Ellison, K.V. Sankaran and Sean Murphy

9. Invasion Biology: Hypotheses and Evidence
Edited by Jonathan M. Jeschke and Tina Heger

10. Invasive Species and Human Health
Edited by Giuseppe Mazza and Elena Tricarico

11. Parthenium Weed: Biology, Ecology and Management
Edited by Steve Adkins, Asad Shabbir and Kunjithapatham Dhileepan

12. Biological Control In Latin America and the Caribbean: Its Rich History and Bright Future
Edited by Joop C. van Lenteren, Vanda H.P. Bueno, M. Gabriela Luna and Yelitza C. Colmenarez

13. Plant Invasions: The Role of Biotic Interactions
Edited by Anna Traveset and David M. Richardson

14. Invasive Species and Global Climate Change, 2nd Edition
Edited by L.H. Ziska

15. Parasites and Biological Invasions
Edited by Jamie Bojko, Alison Dunn and April Blakeslee

16. The Ecology And Management of Invasive Prosopis Trees in Eastern Africa
Edited by Urs Schaffner, Ketema Bekele, Albrecht Ehrensperger, Purity Rima Mbaabu and Brian van Wilgen

17. The Effects of Invasive Plants on Soil Fauna
Edited by Andrea Čerevková

The Effects of Invasive Plants on Soil Fauna

Edited by

Andrea Čerevková
Institute of Parasitology, Slovak Academy of Sciences, Hlinkova 3, Košice, 04001 Slovakia

CABI

CABI is a trading name of CAB International

CABI
Nosworthy Way
Wallingford
Oxfordshire OX10 8DE
UK

Tel: +44 (0)1491 832111
E-mail: info@cabi.org
Website: www.cabi.org

CABI
200 Portland Street
Boston
MA 02114
USA

T: +1 (617)682-9015
E-mail: cabi-nao@cabi.org

A catalogue record for this book is available from the British Library, London, UK.

Library of Congress Cataloging-in-Publication Data

Names: Čerevková, Andrea editor
Title: The effects of invasive plants on soil fauna / edited by Andrea Čerevková.
Description: Boston, MA : CAB International, 2026. | Series: Cabi invasives series | Includes bibliographical references and index. | Summary: "A comprehensive analysis of the mechanisms through which invasive flora alters soil fauna biodiversity, nutrient cycling, and the broader ecological equilibrium, from detailed case studies of specific invasive plants to broader discussions on soil fauna diversity and ecosystem functioning"– Provided by publisher.
Identifiers: LCCN 2026011908 (print) | LCCN 2026011909 (ebook) | ISBN 9781836991274 hardback | ISBN 9781836991281 ebk | ISBN 9781836991298 epub
Subjects: LCSH: Invasive plants–Environmental aspects | Soil animals | Animal ecology | Ecological disturbances
Classification: LCC SB613.5 .E45 2026 (print) | LCC SB613.5 (ebook)
LC record available at https://lccn.loc.gov/2026011908
LC ebook record available at https://lccn.loc.gov/2026011909

ISBN-13: 9781836991274 (hardback)
9781836991281 (ePDF)
9781836991298 (ePub)

DOI: 10.1079/9781836991298.0000

Commissioning Editor: David Hemming
Editorial Assistant: Emma McCann
Production Editor: Helen Elliott

Typeset by Exeter Premedia Services Pvt Ltd, Chennai, India
Printed and bound by CPI Group (UK) Ltd, Croydon, CR0 4YY

Contents

Contributors

Amir Abdullah Khan, School of Environment and Safety Engineering, Jiangsu University, Zhenjiang, 212013, China; ORCID iD: 0000-0002-4831-9032

Andrea Čerevková, Institute of Parasitology, Slovak Academy of Sciences, Hlinkova 3, Košice, 04001 Slovakia; ORCID iD: 0000-0002-9615-9870

Babar Iqbal, School of Environment and Safety Engineering, Jiangsu University, Zhenjiang, 212013, China; ORCID iD: 0000-0002-2115-8494

Barbara Simon, Hungarian University of Agriculture and Life Sciences, Gödöllő, Hungary; ORCID iD: 0000-0002-1343-8154

Dušan Gömöry, Technical University in Zvolen, Faculty of Forestry, TG Masaryka 24, 96001 Zvolen, Slovakia; ORCID iD: 0000-0002-9426-4247

Emilia Grzędzicka, Institute of Systematics and Evolution of Animals, Polish Academy of Sciences, Sławkowska Street 17, 31-016 Kraków, Poland; ORCID iD: 0000-0002-3954-4895

Erika Gömöryová, Technical University in Zvolen, Faculty of Forestry, TG Masaryka 24, 96001 Zvolen, Slovakia; ORCID iD: 0000-0002-0374-1168

Fredrick Ojija, Department of Earth Sciences, College of Science and Technical Education, Mbeya University of Science and Technology, P.O. Box 131, Mbeya, Tanzania; ORCID iD: 0000-0002-1117-5119

Gamage Kavinda Dhananjaya Gamage, Department of Agricultural Biology, Faculty of Agriculture, University of Ruhuna, Mapalana, Kamburupitiya, 81100 Sri Lanka

Gbenga Festus Akomolafe, Department of Plant Science and Biotechnology, Federal University of Lafia, PMB 146, Lafia, Nasarawa State, Nigeria; ORCID iD: 0000-0003-3886-6945

Gergely Boros, Hungarian University of Agriculture and Life Sciences, Gödöllő, Hungary; ORCID iD: 0000-0002-5067-528X

Giovanni Bacaro, Department of Life Sciences, University of Trieste, Via L. Giorgieri 10, 34127 Trieste, Italy; ORCID iD: 0000-0003-0946-4496

Gisandu Malunguja, Department of Earth Sciences, College of Science and Technical Education, Mbeya University of Science and Technology, P.O. Box 131, Mbeya, Tanzania; ORCID iD: 0000-0002-1853-5068

Hudson Laizer, School of Life Sciences and Bioengineering, Nelson Mandela African Institution of Science and Technology, P.O. Box 447, Arusha, Tanzania; ORCID iD: 0000-0003-4849-9835

Ioan Tăuşan, Lucian Blaga University of Sibiu, Faculty of Sciences, Biology and Ecology Research Centre, Sibiu, Romania; ORCID iD: 0000-0002-9599-1761

Iwona Gruss, Department of Plant Protection, Wrocław University of Environmental and Life Sciences, Grunwaldzki Sq 24a, 50-363 Wrocław, Poland; ORCID iD: 0000-0002-3562-5962

Jayawanage Oshani Kaushalya, Department of Agricultural Biology, Faculty of Agriculture, University of Ruhuna, Mapalana, Kamburupitiya, 81100 Sri Lanka

Marek Renčo, Laboratory of Plant Nematology, Institute of Parasitology, Slovak Academy of Sciences, Hlinkova 3, Košice, 04001 Slovakia; ORCID iD: 0000-0001-9771-4189

Marko Černe, Institute of Agriculture and Tourism, Poreč, 52440 Poreč, Croatia; ORCID iD: 0000-0002-0949-0638

Matthew Chidozie Ogwu, Goodnight Family Department of Sustainable Development, Appalachian State University, Boone, NC, USA; ORCID iD: 0000-0001-6054-1667

Mirjana Šipek, Department of Biology, Faculty of Natural Sciences and Mathematics, University of Maribor, Koroška c. 160, SI-2000 Maribor, Slovenia; ORCID iD: 0000-0002-4787-902X

Mohammad Danish, Section of Botany, School of Sciences, Maulana Azad National Urdu University, Hyderabad, India; ORCID iD: 0000-0002-5955-1697

Mohammad Shahid, Marwadi University Research Centre, Department of Agriculture, Faculty of Science, Marwadi University, Rajkot-360003, Gujrat, India; ORCID iD: 0000-0002-2443-9451

Monika Jakubus, Department of Soil Science and Microbiology, Poznan University of Life Sciences, Poland; ORCID iD: 0000-0001-8485-5425

Muhammad Bouhadi, Institute of Agriculture and Tourism, Poreč, 52440 Poreč, Croatia; ORCID iD: 0000-0002-1887-9768

Nina Šajna, Department of Biology, Faculty of Natural Sciences and Mathematics, University of Maribor, Koroška c. 160, SI-2000 Maribor, Slovenia; ORCID iD: 0000-0002-7830-5424

Peliyagodage Chathura Dineth Perera, Department of Agricultural Biology, Faculty of Agriculture, University of Ruhuna, Mapalana, Kamburupitiya, 81100 Sri Lanka; ORCID iD: 0000-0001-7415-7236

Philipina Shayo, Department of Earth Sciences, College of Science and Technical Education, Mbeya University of Science and Technology, P.O. Box 131, Mbeya, Tanzania; ORCID iD: 0000-0001-6331-3053

Qaiser Javed, Institute of Agriculture and Tourism, Poreč, 52440 Poreč, Croatia; ORCID iD: 0000-0003-2825-8217

Renáta Petrikovszki, Hungarian University of Agriculture and Life Sciences, Gödöllő, Hungary; ORCID iD: 0000-0003-3858-9562

S. Maqbool Ahmed, Section of Botany, School of Sciences, Maulana Azad National Urdu University, Hyderabad, India

Viktor V. Brygadyrenko, Department of Biodiversity and Ecology, Faculty of Biology and Ecology, Oles Honchar Dnipro National University, Dnipro, Ukraine; ORCID iD: 0000-0002-9448-8232

Viktoriia O. Komlyk, Department of Biodiversity and Ecology, Faculty of Biology and Ecology, Oles Honchar Dnipro National University, Dnipro, Ukraine; ORCID iD: 0000-0002-3727-5847

Preface

Invasive plant species are one of the greatest challenges to the conservation of biodiversity and the sustainable management of ecosystems in the 21st century. Their expansion beyond native ranges because of global trade, transport and land-use changes has profoundly altered the structure and functioning of terrestrial ecosystems. Invasive plants affect native vegetation, soil properties and faunal communities, leading to cascading effects on ecosystem services, agricultural productivity and climatic regulation.

The study of biological invasions has a long and distinguished history. Charles Darwin's hypothesis of naturalization, presented in his work *On the Origin of Species* (1859), provided the first theoretical insight into species' invasion beyond their native ranges, and Charles Sutherland Elton's *The Ecology of Invasions by Animals and Plants* (1958) formally defined invasion ecology as a field of study. Invasion biology has since expanded rapidly, integrating advances in genetics, community ecology and soil science to explain how alien species establish, spread and transform ecosystems.

The role of soil biota, particularly soil fauna, has gained increasing attention in recent decades as a fundamental component of invasion processes. Soil organisms, including micro- and macrofauna, play critical roles in soil structure, the decomposition of organic matter and the cycling of nutrients. Soil organisms can regulate plant growth and community dynamics, creating both barriers and opportunities for invasive species. Invasive plants often alter soil conditions and food-web structures, facilitating feedbacks that enhance their own success while suppressing native species. These belowground mechanisms are key to understanding the persistence and ecological impact of invasions.

Despite considerable progress, the interactions between invasive plants and soil fauna remain only partially understood, largely due to the complexity and heterogeneity of soil ecosystems. New methodological approaches using artificial intelligence and metadata in studies of soil biodiversity, from molecular techniques to isotopic and microcosmic studies, are now providing insights into these intricate relationships. A deeper understanding of these processes is vital for developing strategies of sustainable management and for predicting how global environmental change will influence the dynamics of invasion.

This book, *The Effects of Invasive Plants on Soil Fauna*, synthesizes current research on the interactions between invasive plants and soil organisms. It explores how soil fauna respond to biological invasions, how invasive species modify soil ecosystems and how these interactions feed back to influence plant community composition, nutrient cycling and ecosystem health. The book brings together contributions from leading experts and early-career perspective scientists, integrating findings from ecology, soil biology and environmental management.

The chapters are organized into three thematic sections that collectively build a comprehensive understanding of belowground processes in invasion ecology.

1. Foundations of soil ecology and biodiversity (Chapters 1 and 2)

This section introduces the fundamental roles of soil fauna in the functioning and resilience of ecosystems. These chapters provide the theoretical background necessary for interpreting how changes in soil biodiversity affect ecological stability and the capacity of ecosystems to resist invasions.

2. Mechanisms and processes of plant–soil interactions (Chapters 3–10)

This section examines how invasive plants influence soil properties, faunal communities and nutrient dynamics. Topics include the engineering roles of invasive vegetation, alterations to microbial and faunal structures, allelopathic interactions and feedback loops that sustain invasions. This section provides mechanistic insight into how belowground processes drive aboveground ecological change.

3. Ecological consequences and restoration perspectives (Chapters 11 and 12)

This section focuses on case studies and applied approaches to restoring ecosystems. It includes species-specific analyses, such as the impact of Solidago canadensis on arthropod communities, and explores how soil fauna can facilitate the recovery and resilience of ecosystems following invasion.

By integrating recent findings and diverse methodological perspectives, this publication highlights the central role of soil organisms in shaping plant invasions. Understanding these belowground interactions is essential not only for advancing ecological theory, but also for informing practical strategies to mitigate the impacts of invasive species. Global environmental change accelerates biological invasions, so this synthesis offers valuable insights for researchers, conservationists and policy makers striving to protect native biodiversity and maintain the integrity of the world's ecosystems.

Andrea Čerevková

1 Soil Fauna and Ecosystem Interactions

Emilia Grzędzicka*
Institute of Systematics and Evolution of Animals, Polish Academy of Sciences, Kraków, Poland

Introduction

The biodiversity of soil animals is enormous, and soil fauna influence soil processes by feeding on and decomposing plant material and organic substrates. There is extensive research on the functioning of soil fauna. Soil invertebrates (i.e. soil fauna) are important drivers of soil aggregate formation, water retention and soil organic matter transformation. Many soil fauna groups participate in litter consumption (Heděnec *et al.*, 2022). Despite the increasing frequency of invasions by exotic plants, which have global ecological and economic consequences, it remains unclear how the belowground fauna groups are impacted (Abgrall *et al.*, 2019). There are about four times fewer studies on the impact of invasions on soil fauna than on the impact of land-use intensification, pollution or nutrient enrichment (Phillips *et al.*, 2024). This is a serious oversight, as the composition of the soil fauna significantly influences the success of exotic organisms that enter a given ecosystem. It is known that plant invasions can alter soil biota and soil biological activities through rhizosphere inputs, which are essential for organic matter decomposition and nutrient cycling in ecosystems (Negesse *et al.*, 2025). Pathogen and pest outbreaks in ecosystems can also be classified as invasions. However, the belowground communities function differently from plant communities, as they are characterized by additive invasions and a lack of interspecific competition. This concept is referred to as the neutral community model (Wanner and Xylander, 2005).

This chapter aims to summarize the relationships between soil fauna and invasions to identify the roles of soil fauna in invasion processes and ecosystem interactions. The division of organisms into taxonomic groups used to search for articles in this field was selected based on the following classifications proposed by other authors. One study summarized five vector types by which soil biota influence exotic plant invasions: (1) soil microorganisms (pathogenic microorganisms, symbiotic microorganisms, saprophytic microorganisms, microbial diversity); (2) soil fauna (herbivorous insects, nematodes and protozoa, mites and collembolans, earthworms); (3) soil food webs (micro-food webs, entire food webs); (4) the relationship between plant–soil feedback and plant invasions and (5) plant above- and belowground biotic interactions (Zhao *et al.*, 2024). Another study proposed three levels of soil fauna: (1) ecosystem engineers (earthworms, termites and ants); (2) litter transformers (microarthropods: decomposers); and (3) grazers or predators (nematodes, protists, mites, etc.) (Coleman *et al.*, 2024). The articles in this review were selected according to the PRISMA framework (Page *et al.*, 2021) (Fig. 1.1). Only the Scopus database was used to search for articles.

*Corresponding author: emgrzedzicka@gmail.com

DOI: 10.1079/9781836991298.0001

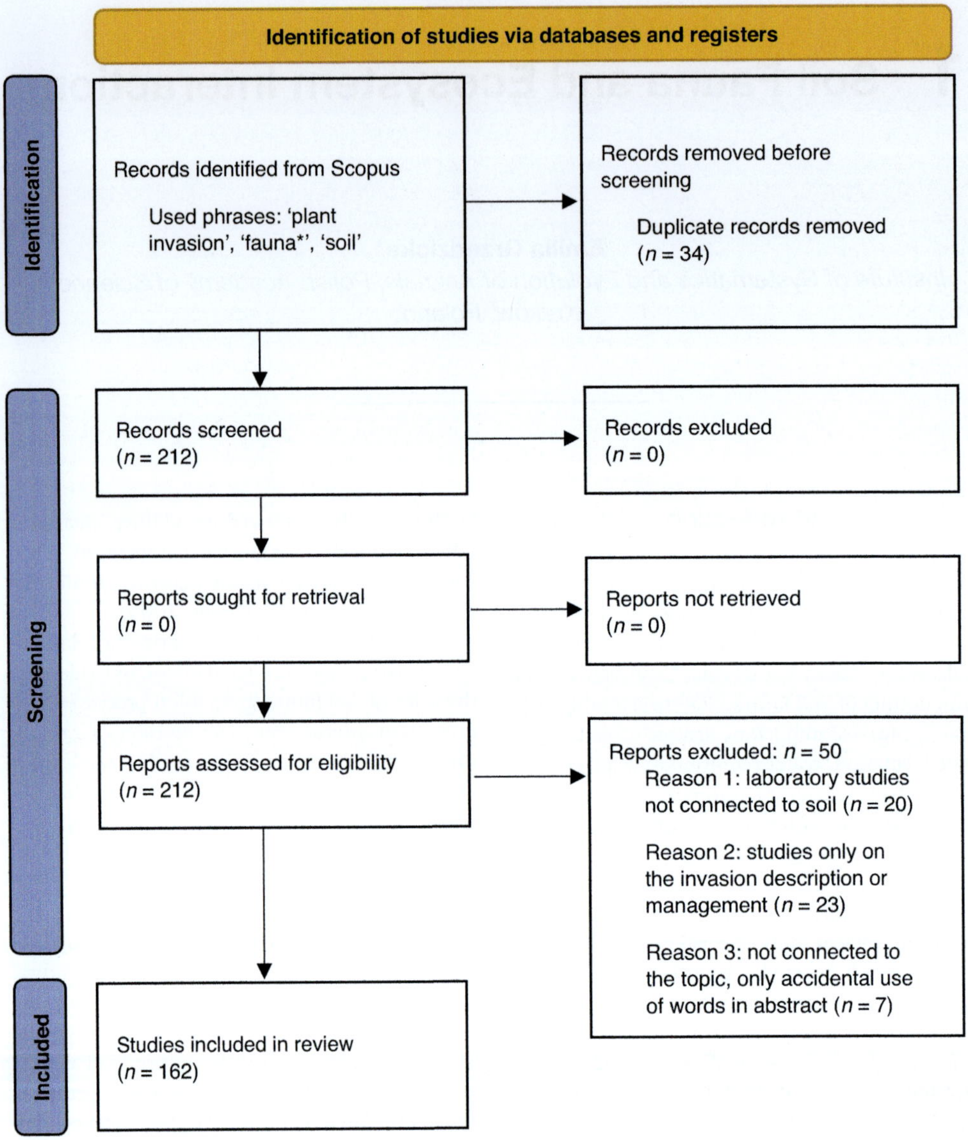

Fig. 1.1. Scheme of selecting articles for review on the relationships between soil fauna and invasions. Source of PRISMA scheme: Page *et al.* (2021), reproduced under the CC BY 4.0 licence.
Note: Groups of fauna: Protists, Macro-arthropods, Rotifera, Tardigrada, Nematoda, Micro-arthropods, mites, Collembola, enchytraeids, Oligochaeta, Formicidae, Isoptera, Isopoda, Arachnida, Mollusca and Vertebrata.

Microbes, Protists, Fungi and the Plant Rhizosphere

Microbes, protists and microfungi constitute the smallest soil organisms. Soil protozoa may interact with soil microorganisms, such as bacteria and fungi, forming a barrier and playing a role in defending plant roots against the invasion of soil-borne pathogens (Wei *et al.*, 2021b). Some studies have shown that the rhizosphere – the area of contact between plants and soil – serves as a barrier, preventing soil-borne microorganisms from penetrating plant roots (Wei *et al.*, 2021a; Ren *et al.*, 2023). Soil and plant contact zones affected by the fungus *Fusarium* spp. have been frequently studied (Li *et al.*, 2023; Ren *et al.*, 2023; Gao *et al.*, 2024). Differences in the composition of soil protists were found between uninfected and *Fusarium*-infected sites (Li *et al.*, 2023; Ren *et al.*, 2023). Other studies have examined the relationships among microorganisms. For example, the rhizosphere of Chinese cabbage plants colonized by the pathogen *Plasmodiophora brassicae* was invaded more often by Chytridiomycota fungi, illustrating a mutualistic relationship between these two microorganisms that facilitates invasion (Lebreton *et al.*, 2019). It was also found that newly emerging soil protists, surprisingly, do not directly occupy the space of existing protists, as they do not reduce the density of the 'residents', although the soil community pattern (abundances, biomasses, dominances) becomes altered by those protists (Wanner and Xylander, 2005).

Studies on the effects of plant invasions on soil microbes have identified the riparian zone – that is, the ecotone between the water body and the upland area (Fig. 1.2) – as a habitat that is particularly vulnerable to the development of invasions via soil processes. Litter decomposition and the structure and activity of microbial communities have been found to be crucial in facilitating invasions (Castro-Díez and Alonso, 2017). Riparian zones are exposed to high pressure from exotic plant propagation, and their biotic resistance, derived from native vegetation, is being eroded by natural (e.g. flooding) and anthropogenic factors. These factors explain why riparian zones are often heavily invaded (Castro-Díez and Alonso, 2017). For example, the invasive weed bitter vine (*Mikania micrantha*) spreads particularly in wet areas and stream banks, altering the soil microbial community structure (Quan *et al.*, 2011).

Nematodes and Mechanisms of Plant Invasions

Soil nematodes, the most diverse and abundant soil faunal group, play a crucial role in the mechanisms of relationships between plants and soil (Tóth and Korponai, 2025). To examine the importance of nematodes for plants, the occurrence of lepidopteran and nematode pests was examined in sugarcane fields, and crop loss from nematodes was greater than that caused by the African sugarcane borer (*Eldana saccharina*) (Lepidoptera: Pyralidae) (Berry *et al.*, 2010). Most studies on plant–nematode interactions have been connected with agriculture; such interactions in nature are less known (Wilschut and Geisen, 2021). Some nematodes are considered to be crop pests that need to be managed. For example, nematodes of the genus *Meloidogyne* spp. have been studied due to their effect on tomato plants (Khan *et al.*, 1997; Wu *et al.*, 1997). Cyst nematodes are an especially often studied group of pathogens

Fig. 1.2. Invasion of Sosnowsky's hogweed (*Heracleum sosnowskyi*), one of the biggest herbaceous plants in Europe (reaching 4–5 m in height). (A) In the riparian zone along a river in southern Poland, 15 June 2020; (B) at a meadow site in eastern Poland, 10 June 2020.

with a limited host range that cause major losses to crops, including potato, soybean and sugar beet (Wang *et al.*, 2011; Green *et al.*, 2012; Escudero Martinez *et al.*, 2019). Soil nematodes not only affect species of agricultural importance. For example, pine wilt disease caused by pinewood nematode (*Bursaphelenchus xylophilus*) alters bacterial and fungal communities (Wang *et al.*, 2014; Zhang *et al.*, 2024).

While the occurrence and distribution of nematodes in transformed and simplified ecosystems are largely dependent on humans and may lead to unpredictable invasions at any time, the responses of soil communities to plant invasions are relatively understudied (Reid and Emery, 2018). In natural habitats, the influence of nematodes on the invasion process may be described by the classical hypotheses of invasion ecology. Firstly, it is assumed that outside their native range, invasive plant species may be released from pressure from their natural enemies, which include nematodes. For example, the exotic sand ryegrass (*Leymus arenarius*) was significantly more resistant to soil nematodes than the native American beachgrass (*Ammophila breviligulata*), and gained more benefits from fungi (Reid and Emery, 2017). Similarly, populations of spotted knapweed (*Centaurea stoebe*), spreading outside of its natural range, accumulated the fewer root-feeding nematodes in soil than native brown knapweed (*Centaurea jacea*); so, *C. stoebe* experienced belowground enemy release (Wilschut *et al.*, 2020). Nematode abundance and species diversity were found to be lower in sites with invasive weeds such as Sosnowsky's hogweed (*Heracleum sosnowskyi*) (Fig. 1.2) or giant hogweed (*Heracleum mantegazzianum*) (Renčo and Baležentiené, 2015; Renčo *et al.*, 2021). Another study investigated the interactive effects between a native natural enemy, the root-knot nematode (*Meloidogyne incognita*), and the alligatorweed flea beetle (*Agasicles hygrophila*), an introduced biological control agent of the invasive alligator weed (*Alternanthera philoxeroides*). The performance of beetles on invasive *A. philoxeroides* was inhibited by nematodes, increasing the success of invasive plants (Qin *et al.*, 2021).

Secondly, the capacity of communities to control introduced plants is referred to as biotic resistance. Biotic resistance has been tested for plant–plant competition and aboveground herbivores and pathogens, while neglecting root herbivores and soil pathogens (Bezemer *et al.*, 2004). This is an oversight because the establishment and impacts of invasive plant species often involve interactions with the belowground subsystem (Dehlin *et al.*, 2008). The presence of plants that are hosts of soil pathogens may negatively affect plant invaders (Knevel *et al.*, 2004). Both the enemy release and biotic resistance hypotheses were examined in relation to the invasiveness of marram grass (*Ammophila arenaria*). This exotic plant species may experience both enemy release and biotic resistance simultaneously, with the balance between enemy escape versus biotic resistance supporting the species' invasiveness in a new habitat (Knevel *et al.*, 2004). Sometimes native plant species may even benefit from the presence of an invasive competitor mediated by changes in the soil biota community. However, this is not a rule, as some plants are not affected by the history of the soil content, or at least by the fact that an invasive plant, Canadian goldenrod (*Solidago canadensis*), previously grew in it (Schittko and Wurst, 2014).

Thirdly, invasive plants may secrete specific substances into the soil that facilitate their success, based on the novel weapon hypothesis. There is increasing evidence that allelopathy often plays an important role in explaining both invasion success and the impact on native taxa. Allelopathic effects are known in the case of the influence of Japanese knotweed (*Fallopia japonica*) (synonym: *Reynoutria japonica*; Fig. 1.3) on soil mesofauna (Nematoda, Collembola and Acari). Invasive allelopathic plants such as knotweeds can affect soil fauna at different trophic levels through addition of nutrients and secondary metabolites to the soil (Abgrall *et al.*, 2018).

Regardless of the invasion mechanisms, nematodes are a common target in studies on the effects of plant invasions on soil organisms. In the root zone of invasive creeping thistle (*Cirsium arvense*), plenty of nematodes were found, and a high proportion of those nematodes were parasites of various plants (Bezemer *et al.*, 2004). Nematodes under the invasive tree, broad-leaved paperbark (*Melaleuca quinquenervia*), were dominated by bacterivores, while the abundance of parasitic nematodes decreased under this tree (Porazinska *et al.*, 2007). A pair of invasive and

Fig. 1.3. An invasive plant, Japanese knotweed (*Fallopia japonica*; synonym: *Reynoutria japonica*), a common exotic weed at a site in eastern Poland, 20 July 2025.

native plant species, smooth cordgrass (*Spartina alterniflora*) and common reed (*Phragmites australis*), respectively, was compared to determine the impacts of their living roots or shoots and root litter on soil nematode communities. The plant-feeding nematodes were less abundant when living roots of invasive *S. alterniflora* were present (Zhang *et al.*, 2018, 2020a). Another study subjected the invasive *S. alterniflora* and native *P. australis* to herbivory by caterpillars of the native moth, reed tussock (*Laelia coenosa*), and assessed soil microbes, nematodes and other soil community parameters such as plant biomass and nitrogen content. The greater soil responses associated with *S. alterniflora* suggested stronger positive soil feedback than those with the native *P. australis* (Zhou *et al.*, 2017).

In another study, nematode populations were not negatively affected by the common milkweed (*Asclepias syriaca*) (Jurová *et al.*, 2020). However, the invasion of baby's breath (*Gypsophila paniculata*), a large invasive forb, reduced plant and nematode heterogeneity (Reid and Emery, 2018). In Canadian goldenrod (*S. canadensis*) stands, nematode taxonomic diversity declined, and herbivores and fungivores contributed most to this decline (Tóth and Korponai, 2025). *S. canadensis* invasion changed the soil's physicochemical properties, which then changed the soil microfaunal (Nematoda) community (Chen *et al.*, 2012). In contrast, polyploids of *S. canadensis* did not significantly affect bacterivores or omnivores, but significantly increased the number of herbivores (Wu *et al.*, 2019). Invasion of the cheatgrass (*Bromus tectorum*) into stands of the native grass, James' galleta (*Hilaria jamesii*), decreased richness and caused a species shift in plants, microarthropods, fungi and nematodes (Belnap *et al.*, 2005). Invasion of *H. sosnowskyi* caused significant shifts in plant species composition, which modified nematode assemblages (Renčo *et al.*, 2019). The abundance and diversity of microbe-consuming nematodes were highest under native trees when compared with nematodes under invasive trees, Douglas fir (*Pseudotsuga menziesii*) and shore pine (*Pinus contorta*) (Dehlin *et al.*, 2008). Abundance and richness of nematodes decreased in stands invaded by the tree, black locust (*Robinia pseudoacacia*) (Lazzaro *et al.*, 2018). Plots invaded by *P. contorta* contained only half as many nematode taxa as uninvaded plots. Invaded communities were simplified compared to uninvaded ones, and had a higher proportion of nematodes with short life cycles, characteristic of disturbed environments (Peralta *et al.*, 2020). The invasion of *S. alterniflora* caused biotic homogenization of soil communities (Zhang *et al.*, 2019).

Other studies explored the potential of nematodes to mediate microbial responses to invasions. It turned out that invasive plants may enhance microbial–microfaunal interactions, stimulating nutrient release (Sun *et al.*, 2019). It has been found that high concentrations of *Meloidogyne javanica* eggs favour fungal infections (Siddiqui and Shaukat, 2003). The nematode invasion rates on *Crotalaria* spp. were variable, as *Crotalaria* are highly mycorrhiza-dependent (Germani and Plenchette, 2005). Leaf litter microbial decomposition near the invasive giant ragweed (*Ambrosia trifida*) facilitated nematode proliferation and altered its community structure (Lu *et al.*, 2016). The invasion of *S. alterniflora* stimulated the growth of bacterial nematodes by producing a higher quality of litter by microbes than near the native *P. australis* (Chen *et al.*, 2007). *S. alterniflora* also has a great potential to increase soil carbon (Zhang *et al.*, 2017). Soil-borne nematodes and fungi together reduced the invasive marram grass (*A. arenaria*) growth (Beckstead and Parker, 2003).

Habitat development and management influence relationships between nematodes and invasive plants. This is important because the expansion or removal of invasive plants triggers subsequent stages of habitat modification and succession. Invasion of restored prairies by exotic grasses was connected with a decline in the diversity of plant-root feeders (Coffey and Otfinowski, 2019). At the early stages of degraded soil recovery, nematode species that are resistant to unfavourable environmental conditions dominated, and subsequently the abundance of nematodes connected with plants increased (Gruzdeva and Sushchuk, 2010). Nematode abundance increased from moorland to birch woodland, with moorland soils being dominated by a few taxa (e.g. fungal feeders; Keith *et al.*, 2006).

Nematode communities reflected resource enrichment with increasing white tea-tree (*Kunzea ericoides*) cover, whereas communities became more simplified (homogenized) with increasing exotic black pine (*Pinus nigra*) cover (Peralta *et al.*, 2019). Woody succession under non-native *P. nigra* was associated with losses from the mineral soil carbon pool affecting the belowground taxonomic richness of plants, nematodes, mites, microbes, and fungi (Dickie *et al.*, 2011). However, nematode abundance and species richness were higher in sites invaded by early goldenrod (*Solidago gigantea*) than in uninvaded sites, regardless of whether the ecosystem was forest or meadow (Čerevková *et al.*, 2020). Interestingly, soil communities of bacteria and nematodes in early successional dune systems were influenced more by climatic factors (temperature, precipitation), rather than the presence of the exotic plant, sand ryegrass (*L. arenarius*) (Reid *et al.*, 2019). Fungal biomass, microbial diversity and nematode abundance (except omnivores/predators) increased following invasion by the giant African land snail (*Lissachatina fulica*) in the dry season but decreased in the wet season (Wu *et al.*, 2024).

Some nematode species can also be invasive. Regarding the importance of habitat management for the nematode invasion, it was stated that the absence of cultivation and mammalian grazing reduce possibilities for invasion by new nematode taxa (Yeates and King, 1997). On the importance of the habitat succession stage: the older the plant, the less the infestation by nematodes (Rahman and Evans, 1987).

Microarthropods: Collembola and Acari

Microarthropods have been the subject of studies comparing their assemblages between uninvaded and invaded sites by various plants, pointing to the negative impacts of the invasions. It was found that Uropodina (belonging to Acari, order Mesostigmata and class Arachnida) differed in species abundance between sites with invasive plant steeplebush (*Spiraea tomentosa*) and uninvaded ones (Wiatrowska *et al.*, 2024). The invasive giant knotweed (*Fallopia sachalinensis*, synonym: *Reynoutria sachalinensis*) had a profound negative effect on the soil oribatid mite (belonging to order Oribatida and class Arachnida) community composition (Skubała, 2012). A low proportion of Oribatida specific to deciduous forests significantly distinguished the sites with old alien northern red oaks (*Quercus rubra*) from those with old native oaks (Kohyt and Skubała, 2020). Soil Collembola community composition differed between sites invaded by *S. gigantea* and uninvaded ones, indicating a reduction in species diversity in invaded meadows (Sterzyńska *et al.*, 2017). Most detritus-based arthropod taxa declined in areas with high annual grass invasion (Wolkovich *et al.*, 2009). Bamboo invasion negatively influenced soil Collembola communities (Long *et al.*, 2023). The invasion by the exotic palm tree *Pinanga coronata* led to a strong and significant decrease in native plants and Collembola taxonomic diversity (Forey *et al.*, 2021). The invasion of dog-strangling vine (*Vincetoxicum rossicum*) enhanced the negative effect of urbanization on microarthropod richness and abundance (Malloch *et al.*, 2020). A significant increase in Collembola abundance and a decrease in abundance, body size and diversity of macroarthropods at order and family levels in invaded plots with giant reed (*Arundo donax*) compared to native stands. Arthropods were smaller in *A. donax* soils like due to food-web alterations in invaded areas (Maceda-Veiga *et al.*, 2016). Densities and number of microarthropods (i.e. Collembola, Isotomidae, Onychiuridae, Oribatida, Gamasida, Astigmata and Prostigmata, Oppiidae and Brachychthoniidae) were reduced when the invasive earthworms *Lumbricus terrestris* and *Octolasion tyrtaeum* were present due to resource competition and habitat degradation (Eisenhauer *et al.*, 2007).

Microarthropods are often treated together with nematodes and microorganisms, which sometimes makes it difficult to distinguish studies strictly related to particular groups of microarthropods. For example, it is known that invasive Asian knotweeds (*Reynoutria* spp.) cause various, including negative, responses of two major soil mesofauna (Collembola) and microfauna (Nematodes) communities, according to trophic position or life form (De Almeida *et al.*, 2022).

Several positive effects of invasive plants on soil fauna assemblages have also been detected. The abundance of Nematoda and Collembola increased significantly in plots invaded by common ragweed (*Ambrosia artemisiifolia*) (Xie *et al.*, 2011; Qin *et al.*, 2019). Cheatgrass and perennial grasses supported distinct communities of soil bacteria, fungi, nematodes and other non-fungal, non-nematode microbial eukaryotes. Abundances of bacterial-feeding nematodes, collembolans and prostigmatid mites were higher beneath cheatgrass than native perennial grasses (Porazinska *et al.*, 2022). More macro-invertebrate and more mite species were found under invasive sprawling mat-forming grass (*Agrostis stolonifera*) (Gremmen *et al.*, 1998). Invaded areas with garlic mustard (*Alliaria petiolata*) contained about three times more springtails (Collembola) than uninvaded areas (Alerding and Hunter, 2013). This may result from the fact that springtail species assumed to be exotic have the broadest thermal tolerances and greatest tolerance to environmental stress (Janion-Scheepers *et al.*, 2018). Based on ectoparasite burdens, invasion sites harboured more chiggers (larval stages of mites) and ticks (arachnids) than uninvaded habitats (Wei *et al.*, 2020).

Some studies indicated a neutral effect of invasion on microarthropods. For example, soil properties and soil fauna assemblages, including the important decomposer group – oribatid mites – did not cluster by invasive status in sites with invasive plants such as common heather (*Calluna vulgaris*) and Scotch broom (*Cytisus scoparius*) (Pearson *et al.*, 2024). There was also no significant difference in soil decomposition rates between non-invasive and invasive plots containing leafy spurge (*Euphorbia esula*) (Pritekel *et al.*, 2006).

Other studies concerning microarthropods tested predictions of the classical ideal weed and phenotypic plasticity hypotheses of invasion biology. The ideal weed hypothesis assumes that invasive plants are, in comparison with natives, less reliant on ecological conditions and interactions like mutualisms, which enables them to grow even in disturbed environments. The phenotypic plasticity hypothesis suggests that invasive species are more phenotypically flexible than native ones. Soil fauna might help native communities to resist alien plant invasions, but this effect might be weakened under drought due to the greater tolerance (phenotypic plasticity) of exotic soil mites (Jin *et al.*, 2022). Surprisingly, there is also a study that has not confirmed the second hypothesis in the case of the introduced species of Collembola, as the extent of phenotypic plasticity did not differ between the native and invasive species, and these differences became weaker at higher temperatures (Chown *et al.*, 2023). In the first hypothesis, for example, the invasion of rush skeletonweed (*Chondrilla juncea*) was successful due to decreased competition for environmental resources. The environmental treatments designed to reduce invasive annual grasses affected the abundance of biocontrol agent (gall mite), indicating that the disruption of natural relationships can promote plant invasion despite the reduction of exotic plant populations (Lazarus and Germino, 2021), although it might be expected that invasions are independent of relationships with other organisms. Another study examined latitudinal variation in the soil biota communities associated with the invasive alligator weed (*A. philoxeroides*) and its native congener, sissoo spinach (*Alternanthera sessilis*). Confirming the ideal weed hypothesis, there were no latitudinal patterns for the invasive plant (Lu *et al.*, 2018), although both the vertical distribution of *A. philoxeroides* and the soil fauna presented surface segregation with the increased soil depth (Zhang *et al.*, 2020b).

As in the case of Nematoda, there is no shortage of research on the importance of microarthropods in soil processes influencing invasions, although there are few studies on them. It was found that the increasing density of the invasive ailanthus tree (*Ailanthus altissima*) was associated with lower soil microbial activity and a decreasing abundance of litter detritivores (i.e. Acari and Collembola; Motard *et al.*, 2015). Another study investigated how herbaceous plants expanding into alpine tundra affected soil litter and soil mesofauna, focusing on the existing feedback mechanisms between these components. The increasing herbaceous plant expansion resulted in a significant increase in litter biomass (Zhang *et al.*, 2022). Japanese stiltgrass (*Microstegium vimineum*) invasion induced soil chemical changes that altered litter microarthropod communities in acidic, nutrient-poor upland forests. Microarthropod community diversity was lower in *M. vimineum* litter (McGrath and Binkley, 2009).

Other studies have examined relationships between microfaunal groups (i.e. Acari and Collembola) and environmental conditions, which influenced the effects of various invasions and infections, including those affecting microarthropods. Grass mites (*Siteroptes* spp.) played the role of inoculum transmission vector for the pathogen *Dactylobotrys graminicola* (Tong *et al.*, 2025). Another study examined the interaction between host trees and fungi associated with a tree-killing bark beetle (*Dendroctonus frontalis*). Two of these fungi were mutualists with *D. frontalis*, and the third one was a blue-staining fungus (*Ophiostoma minus*), commonly introduced by beetles and phoretic mites (Hofstetter *et al.*, 2005). One of the most invasive moss species worldwide is the heath star moss (*Campylopus introflexus*). For Collembola, a negative effect of this moss was assumed due to lower field layer cover and soil moisture (Pehle and Schirmel, 2015). Similarly, the spatial extent of the allelopathic garlic mustard (*A. petiolata*) invasion mediated its impact on oribatid mite communities (Flaherty *et al.*, 2024). The abundance and species richness of both native and exotic species at the soil fauna (i.e. Collembola) metacommunity scale responded strongly to declining environmental favourability (Terauds *et al.*, 2011). The mesostigmatid mites and oribatid mites were assessed in patches of birch woodland dominated by moor birch (*Betula pubescens*) and heather moorland dominated by *C. vulgaris*. The community's composition differed between habitats within sites (Nielsen *et al.*, 2010).

Oligochaeta: First of All Invaders

The Collembola and Oligochaeta are considered two major groups of soil invertebrates that encompass both indigenous and invasive species. DNA barcoding was used to investigate the genetic structure in their native and introduced ranges. Most of these species exhibited a similar genetic structure, characterized by large and stable populations in North America and Europe, supporting a scenario of multiple invasions (Porco *et al.*, 2013). Ecosystems of North America existed without native earthworm fauna until European settlers arrived and introduced European species (Eisenhauer *et al.*, 2007). In the case of earthworms (Oligochaeta), it is known that community dynamics are shaped by the interplay of selection, dispersal and ecological drift (Szlavecz *et al.*, 2018). European earthworms (*Lumbricus* spp.) are spreading into previously earthworm-free forests in the USA and Canada, causing substantial changes, including habitat and soil structure homogenization, and reduction in arthropod abundance and plant species richness (Loss and Blair, 2011). In temperate grasslands, earthworms serve as crucial ecosystem engineers, making up the majority of soil faunal biomass (Zaller *et al.*, 2013). It is also known that exotic earthworm species dominate the earthworm community in nutrient-enriched soils characteristic of urbanized and rural areas, while native species persist in less disturbed habitats. Thus, exotic species may displace native species through competitive interactions, but only under conditions of high resource availability (Winsome *et al.*, 2006). Moreover, the likelihood of encountering earthworms increases close to roads and streams and in high-moisture conditions. Climatic variables and potential sources of earthworm introduction and dispersal (e.g. roads and streams) should also be used for predicting current and future sites of earthworm invasions (Holdsworth *et al.*, 2007a; Paudel *et al.*, 2016b). Long-term earthworm activities in soil pose such a great threat to biodiversity that they are now being monitored using soundscape measures and acoustic indices. Such analysis of soil soundscapes may enable larger-scale monitoring of earthworm invasions identified as a threat to the resilience of ecosystems (Keen *et al.*, 2022).

Even though, for larger invertebrates or small vertebrates, invasive earthworms may become important prey with the potential to increase resource availability, in the longer term the activity of invading earthworms can have a strong negative impact on indigenous faunal groups across multiple trophic levels (Migge-Kleian *et al.*, 2006). Herbivore behaviour has been also shown to be particularly affected by earthworms. After adding earthworms (in this case, Annelida: Lumbricidae) to planted grassland communities with different plant species composition, Spanish slugs (*Arion vulgaris*; formerly known as *A. lusitanicus*; Gastropoda: Arionidae) damaged 60% fewer leaves, regardless of the species composition of the plant communities (Zaller *et al.*, 2013). This suggests competition for space and resources between earthworms and other herbivores. Moreover, invasive earthworms substantially alter microbial communities, including mycorrhizal fungi (Paudel *et al.*, 2016a). The composition of the native Cercozoa (a phylum that includes amoeboid and flagellate organisms) soil community also shifted in response to earthworm invasion (Dumack *et al.*, 2022). There are multiple direct effects of non-native earthworms on plants (Paudel *et al.*, 2016a). Exotic earthworms have already increased the density of plant-feeding nematode taxa that are less detrimental to plant growth (*r*-strategists), although they reduced the density of harmful plant-feeding nematodes (*K*-strategists) (Shao *et al.*, 2017).

Consequently, earthworms decreased the stability of grassland communities, facilitating plant invasions (Eisenhauer *et al.*, 2008). Even removal of the invasive shrub amur honeysuckle (*Lonicera maackii*) had little effect on earthworm density and litter biomass (Mahon and Crist, 2019). Earthworms (i.e. *Lumbricus* and *Aporrectodea* spp.) reduced soil moisture and plant community productivity in tundra communities (Jonsson *et al.*, 2024). Invasion by *Lumbricus* spp. was a strong indicator of reduced vascular plant richness in forests (Holdsworth *et al.*, 2007b). A study investigated the effects of invasion by three European earthworm species – octagonal-tailed worm (*Dendrobaena octaedra*), red earthworm (*Lumbricus rubellus*) and common earthworm (*Lumbricus terrestris*) – on the mortality and biomass of native understorey plants. All three species combined caused larger impacts than any single earthworm species (Hale *et al.*, 2008). The survival and growth of native seedlings – white baneberry (*Actaea pachypoda*), Canadian columbine (*Aquilegia canadensis*), northern swamp dogwood (*Cornus racemosa*), northern red oak (*Q. rubra*) and white rattlesnake-root (*Prenanthes alba*) – in response to the presence/absence of earthworms was assessed. *Actaea*, *Cornus*, *Prenanthes* and *Quercus* seedlings had a lower proportion of fine roots in earthworm-invaded plots, while fine roots in *Aquilegia* made up a higher proportion of the root system (Dobson *et al.*, 2020).

Earthworm impacts on plant foraging can occur indirectly via physical and chemical changes to the soil and directly via root consumption or abrasion, and thus may be a factor influencing plant growth and community change following earthworm invasion (Cameron *et al.*, 2014). Earthworms accelerate litter disappearance rates and may reduce the differences in decomposition rates that result from different litter qualities at later stages of decay (Suárez *et al.*, 2006). The action of earthworms may provide a pathway through which common buckthorn (*Rhamnus cathartica*) invades forests (Roth *et al.*, 2015). Extracts from invasive plants caused more negative effects on the manure worm (*Eisenia fetida*) than did extracts from native plant species, indicating allelopathy from invasive plants may inhibit earthworm physiological functioning (Liu *et al.*, 2020). Another study aimed to distinguish between the effects of earthworm-induced changes in soils and the physical presence of earthworms on oxalate production. A two-way factorial greenhouse trial was conducted using uninvaded soils for two invasive earthworm species, Asian jumping worm (*Amynthas agrestis*) and red earthworm (*Lumbricus rubellus*). Earthworms affected oxalate concentrations, and the sequence of invasion mattered for oxalate production by jack-in-the-pulpit (*Arisaema triphyllum*) (Melnichuk *et al.*, 2022). Surprisingly, even though the herbaceous plant *Parthenium hysterophorus* is known to possess strong negative allelopathy, as well as plant/animal toxicity, its vermicompost produced by earthworms showed none of these attributes (Hussain *et al.*, 2016).

Enchytraeidae is a family of soil-inhabiting small- to medium-sized oligochaete worms using degradable plant material as a food source. This group is segregated into two genera: *Enchytraeus* and *Lumbricillus* (Christensen and Glenner, 2010). Elevated densities of enchytraeids in stands of invasive non-indigenous saltmeadow cordgrass (*Spartina patens*), and their incorporation of carbon derived from this plant, suggested that dense patches of *S. patens* may facilitate detritivore populations. *S. patens* invasion may influence soil processes and pathways mediated by detritivore activity (e.g. soil respiration rates, nutrient retention and transformation, and energy flow) (Page *et al.*, 2010).

It is worth highlighting that one study presented the New Zealand flatworm (*Arthurdendyus triangulatus*; synonym: *Artioposthia triangulata*) as an example of an invasive organism that, by reducing lumbricid earthworm populations, could have a major impact on soil ecosystems. How it was introduced into the British Isles is not known, but like many invasive species it is suspected to have been introduced by humans (Boag and Yeates, 2001).

Formicidae and Isoptera: Socialized Ecosystem Engineers

Arthropod food webs can be indirectly impacted by invasions, with cascading consequences for higher trophic levels (Clark *et al.*, 2019). Invasive ants provide some of the most spectacular examples of invasive insect-induced cascades destabilizing trophic relationships (Wagner and Van Driesche, 2010). Through nest building and foraging activities, ants alter the physical properties and nutritional status of soils through structural modifications and nutrient accumulation, and one such example is the red imported fire ant (*Solenopsis invicta*) (Lafleur *et al.*, 2005). *S. invicta* is one of the most dangerous invasive species in the world. Fortunately, even extremely low concentrations of the food additives ethyl anthranilate and butyl anthranilate effectively prevent nesting by fire ant workers, making these compounds promising for fire ant control (Chen *et al.*, 2019).

The Argentine ant (*Linepithema humile*) is another significant pest species that has been established on almost every continent, particularly in areas with a Mediterranean climate. *L. humile* has been associated with various negative effects on native communities worldwide, including impacts on belowground processes such as decomposition (Stanley and Ward, 2012). The invasion of Argentine ant is predicted to lead to the extinction of many rare, endemic plants (Bond and Slingsby, 1984). For example, *L. humile* can modify the seed dispersal system and population dynamics of the rare endemic plant *Anchusa crispa* in Corsica and Sardinia (Quilichini and Debussche, 2000). One big question in ecology concerns the reason why communities differ in their susceptibility to invasion. Based on the invasion of Argentine ant, it was found that a specific environmental factor, i.e. elevated levels of soil moisture, increased the abundance of invaders and their ability to invade native ant communities. The indirect effects of added water may increase invasibility (Menke and Holway, 2006). Few studies have investigated the impact of environmental contamination on competitive behaviour in ants. For example, selenium negatively impacted queen survival and brood production of *L. humile* (De La Riva and Trumble, 2016). Temperature extremes may also limit invasion. At 47°C, the mortality rate of all ant species tested increased, with Argentine ants shoving 100% mortality after 3 hours. At 50°C, Argentine ants displayed no survival after 1 hour, whereas *Iridomyrmex* ants exhibited ≈50% survival after 3 hours. These results may have significant implications in areas where the ground surface temperature becomes very high during the summer, as this may potentially limit the enormous spread of Argentine ants (Walters and Mackay, 2004).

Research has shown that several other invasive ant species significantly change ecosystems. For example, tramp ants are invading disturbed ecosystems worldwide. By 1993, populations of the longhorn crazy ant (*Paratrechina longicornis*), a tramp species absent only a few years earlier, had increased extremely. There were observed *P. longicornis* foragers feeding almost exclusively on the sugary excretions (honeydew) produced by vast numbers of Homoptera, primarily scale insects and mealybugs, found on many plants. High densities of ants were associated with high densities of homopterans. Such a convergent pattern occurs in disturbed tropical and subtropical ecosystems dominated by tramp ants (Wetterer *et al.*, 1999). The other species, European fire ant (*Myrmica rubra*), has invaded northern latitude coastal areas in North America. *M. rubra* exhibited relatively low thermal tolerances consistent with high-latitude and high-elevation ants. Where local *M. rubra* populations increased the most, native ant species decreased, and where local *M. rubra* populations declined, native ant species increased (Warren II *et al.*, 2019). Moreover, an incursion of the African big-headed ant (*Pheidole megacephala*) has been recorded in Australia. This species had a significant negative impact on ant species richness and diversity. Approximately 53% of native ant species present in uninvaded bushland were not sampled in areas occupied by *P. megacephala*, while the remaining species suffered declines (Callan and Majer, 2009).

Numerous mechanisms have been proposed to explain why exotic plants successfully invade natural communities with ants. The positive effects of ants as ecosystem engineers need further consideration. It was, for example, tested whether the nutrient-rich soil patches created by a native ecological engineer, such as refuse dumps from the leaf-cutting ant *Acromyrmex lobicornis*, increase the performance of exotic plants more than native plants. Since ant nests and exotic plants were more common in disturbed than in more natural environments (in line with the fluctuating resource hypothesis), the role of ant nests in promoting biological invasions should be of particular interest (Farji-Brener *et al.*, 2010).

There is also no shortage of studies comparing ant communities between uninvaded and invaded sites. Ant communities differed between the uninvaded and invaded meadows with tall goldenrod (*Solidago altissima*) and invasive Canadian goldenrod (*S. canadensis*), with a reduction of ant abundance and species richness in the latter. *Myrmica* spp. were abundant in the uninvaded meadows, whereas the black garden ant (*Lasius niger*) was the dominant species in the invaded ones. The invasion of *Solidago* plants caused changes in the abundance and species composition of ant communities through modification in microhabitat conditions (Trigos-Peral *et al.*, 2018). Uninvaded areas had significantly greater ground beetle (Coleoptera: Carabidae) and ant (Hymenoptera: Formicidae) abundance and taxonomic richness than areas invaded by common buckthorn (*Rhamnus cathartica*) (Schuh and Larsen, 2015). The influence of the Argentine ant's (*L. humile*) occurrence on seed dispersal of Mediterranean buckthorn (*Rhamnus alaternus*) was assessed. *R. alaternus* is a plant with fleshy fruits that are dispersed primarily by birds, although the seeds have an elaiosome that is attractive to ants. In invaded sites, the presence of *L. humile* was associated with the following effects: reduction of seed transport to ant nests, a shorter seed dispersal distance, an increase in seed retention time on the soil surface and increased vulnerability to predation (Gómez *et al.*, 2003). *Artemisia tridentata* (sagebrush) ecosystems are experiencing changes because of fire and exotic grass invasion. Occupancy of harvester ants (i.e. all ants that collect seeds) was more probable in areas with abundant native forbs and low shrub cover. Results showed that changes from native shrublands to grasslands dominated by non-native grasses have a positive effect on the density of harvester ants, while postfire restoration does not affect harvester ants (Holbrook *et al.*, 2016). Urban riparian zones are environments highly exposed to invasions. Even though wider corridors were likely to encompass a greater variety of ant and plant assemblages, landscape- and site-scale environmental characteristics had greater ecological significance (Ives *et al.*, 2011).

It was investigated whether the appearance of ants is in line with the directed dispersal hypothesis. For example, it turned out that the native ant community interacts with invasive leafy spurge (*E. esula*). Natural density and biomass of *E. esula* were greater on *Formica* ants' mound edges compared to random locations. However, seedling recruitment and establishment from *E. esula* seeds were not greater on mound edges than at random locations 3 m from the mound. This suggests that *Formica* ant mounds are favourable microhabitats for *E. esula* growth following seedling establishment, a likely consequence of nutrient limitation during plant growth. The results indicate positive species interactions may play an important role in biological invasions (Berg-Binder and Suarez, 2012).

Other important social insects in invasion process are the Isoptera (termites). For example, data collected from four habitats differing in their vegetation cover and fire history – annually burned savannah, savannah woodland, forest island and gallery forest – showed that soil pH appeared to be the most influential factor for fungus-growers, while tree invasion in the savannah strongly reduced the abundance of grass-feeding termite species, for example *Trinervitermes geminatus* (Dosso *et al.*, 2010). The African gamba grass (*Andropogon gayanus*), which is invading and altering Australian savannahs, also interacts with termite activity. Only 30% of *A. gayanus* litter decomposed *in situ*, compared to 61% of the native grass litter. It turned out that lower initial lignin concentrations and higher consumption by termites accounted for the greater surface decomposition rate of *A. gayanus*, so in this case termites may weaken the invasiveness of the plant (Rossiter-Rachor *et al.*, 2017). Moreover, the accumulation and spatial distribution of nutrients in termite mounds were not only related to termite activities but also depended on environmental factors, such as erosion and leaching, and invasion by plants (Chen *et al.*, 2021).

Small Arthropods: Isopoda and Non-Acari Arachnida

Soil animals belonging to the Isopoda and Arachnida have not been very common subjects in invasion studies to date. However, unexpectedly, some invasive plants were found to support high numbers of Isopoda compared to uninvaded habitats. For example, Japanese barberry (*Berberis thunbergii*) is an invasive shrub, beneath which soils harboured an average of 40% more Isopoda and more than twice as many Diplopoda (a separate class) per unit of dry soil compared with uninvaded samples (Utz *et al.*, 2023). In areas invaded by *Reynoutria* species, the abundance of Isopoda decreased (Kappes *et al.*, 2007).

Despite the importance of non-native plants in nutrient cycling and plant–soil interactions in secondary succession following agriculture abandonment, meso- and macrofauna have received much less attention than bacteria or fungi. One study compared the composition of the soil arthropod community in native remnants and plant-exotic-dominated old-field grasslands. When compared to non-invaded remnant grasslands, soil activity and soil food-web characteristics of the old-field sites included higher total arthropod abundance, particularly of Isopoda, Pseudoscorpionida and Blattaria, although a lower abundance of Hymenoptera and Entomobryomorpha (Collembola) (Vilardo *et al.*, 2018). Increased mowing intensity decreased the abundance of Isopoda, while the introduction of fresh hay used in grassland restoration positively affected nematodes (Perera *et al.*, 2022). The limited investigation of the dependence of animals such as Isopoda on invasions seems to be a serious oversight, since they show responses opposite to those of other animals.

Recent studies have indicated that the litter of exotic plants decomposes faster and releases more nutrients. The generalist consumer used in research on the rate of decomposition is, for example, rough woodlouse (*Porcellio scaber*) (Isopoda), which is a primary generalist decomposer in forest leaf litter decomposition. It was found that the consumption rate of woody species was not necessarily associated with woody plants' invasion status, but nitrogen content and composition might play an important role and might affect the biological processes of invasion to some extent. Thus, the increasing presence of alien woody plants may not influence the decomposition of plant material negatively (Liu and Liao, 2013).

Regarding spiders, studies on the impact of plant invasions on their communities are sparse. However, native forests in the Argentine Espinal region were, for example, more diverse in ground spiders assemblages than pajonal and exotic forests (Almada and Sarquis, 2016).

Molluscs: Gastropoda

The molluscs are another poorly studied group of animals regarding the impact of plant invasions on soil fauna. It is known, however, that the gastropod assemblages from *Reynoutria* stands were severely impoverished when compared with those from *Urtica* stands, with reduced densities, species richness and diversity (Kappes *et al.*, 2007). Surprisingly, seedling survivorship of common ragweed (*A. artemisiifolia*) was only slightly increased by the exclusion of molluscs and other herbivores. This suggests, contrary to appearances, a marginal relationship between plant invaders and soil molluscs. Even though ragweed lost most of its insect folivores while invading Europe, escape from herbivorous enemies like molluscs is unlikely to have provided a significant demographic advantage (MacDonald and Kotanen, 2010).

It is worth adding, however, that the introduction of exotic plants, while not necessarily invasive, is often associated with the introduction of invasive species of soil fauna, including molluscs. An interesting study examined the use of hot water treatment to destroy pests that has recently been used during the quarantine of ornamental plants. The species used were the invasive alien terrestrial New Guinea flatworm (*Platydemus manokwari*) (Platyhelminthes)' an unidentified alien species of earthworm (Annelida)' the alien snail *Acusta despecta sieboldiana* (Mollusca)' and the alien white-footed ant (*Technomyrmex albipes*) (Arthropoda). The use of hot water for protection from alien soil animal invasions may mitigate their environmental impacts, helping to protect valuable biota (Sugiura, 2008).

Soil Vertebrates as Important Herbivores

Previous studies on the relationship between plant invasions and soil vertebrates have focused primarily on vertebrate herbivores, the absence of which suggests the release of invasive plants from their consumers, consistent with the enemy release hypothesis. For example, exotic grassland plant species were less damaged than natives, experiencing less foliar damage (from both insect herbivory and fungal disease) than native species. Fertilization increased exotic cover only in communities exposed to vertebrate herbivores (Heckman *et al.*, 2016). Invasive exotic and non-invasive exotic plant species had significantly lower levels of leaf herbivory than native species (Lake and Leishman, 2004). Interestingly, soil-associated vertebrate herbivores sometimes do not appear to play a significant role in promoting plant invasiveness in this context. The long-term impact of a largely native vertebrate consumer community on native and exotic understorey plants and recruitment of native oaks in a California oak savannah showed that the presence of oaks in canopies had a strong impact on the plant community independent of consumer effects (Seabloom *et al.*, 2009). Common ragweed (*A. artemisiifolia*) is an abundant weed in its native North America, insensitive to attack by many of its natural enemies. Even though ragweed lost most of its insect folivores while invading Europe, escape from these enemies is unlikely to have a significant demographic advantage; instead, disturbance appears to have been a much more important factor in its invasion (MacDonald and Kotanen, 2010).

There is no shortage of research indicating the negative effects of plant invasions on vertebrates. One study associated capture rates of a fossorial desert anuran (amphibian), Couch's spadefoot toad (*Scaphiopus couchii*), to non-native plant invasion via soil and habitat physiognomy in native and *Tamarix*-invaded riparian habitats. Spadefoots were associated with native riparian habitat and were nearly absent from *Tamarix* sites. The *Tamarix* invasion can be deleterious to many anurans (Riddle and Bateman, 2020). In California, an invasive plant genus of great concern is *Eucalyptus* spp. Amphibians serve as models to evaluate the effect of this invasion on ground-dwelling species as they predate soil arthropods and incorporate soil microbes into their microbiotas. California slender salamander (*Batrachoseps attenuatus*) skin microbiomes were compared in southern blue gum (*Eucalyptus globulus*) sites and native California live oak (*Quercus agrifolia*) forests. *B. attenuatus* skin microbiota diversity was greater in *Quercus*-dominated habitats (Hernández-Gómez *et al.*, 2020). However, based on a review article, the effects of Japanese knotweeds (including *Reynoutria japonica*, *R. sachalinensis* and *R.* × *bohemica*) on groups of species actually varied, with a combination of losers (e.g. most arthropods and gastropods, some frogs and birds) and winners (e.g. detritivorous arthropods, certain birds) (Lavoie, 2017).

Other studies have shown how the relationships between invasive plants and vertebrates can influence population dynamics in ecosystems. The invasion of plant species grassleaf orache

(*Atriplex littoralis*) caused the late successional perennial grass, sea couch (*Elytrigia atherica*), to be outcompeted at the high marsh, and slowed down the increase of sea purslane (*Atriplex portulacoides*) at the low marsh. At the same time, herbivory by small vertebrates alternated in shaping long-term vegetation dynamics, which could be a common feature of dynamic ecosystems (Kuijper and Bakker, 2012). Plant species with vertebrate-dispersed seeds were relatively more frequent in invaded vegetation (Holmes and Cowling, 1997). Native and exotic species differ in their response to vertebrate consumer exclusion. Exotic plant species are about six times more likely to dominate communities than native species (Seabloom *et al.*, 2015).

Conclusion

The review in this chapter has revealed large disparities between the negative and positive relationships concerning invasions and the soil fauna, with a significant predominance of negative impacts of invasions on all described groups. Interestingly, the groups most negatively affected by invasions were also those characterized by relationships that potentially facilitated invasions (e.g. Oligochaeta, Nematoda, Formicidae). It is noteworthy that Collembola and Acari responded differently: they were negatively affected by invasions, whereas others facilitated their development, remained neutral or inhibited them. Isopoda, generally associated with promoting invasions, rarely showed negative effects from invasions themselves (Fig. 1.4). Tardigrades, despite being a diverse and numerous group, have not been studied.

Studies are not representative of soil fauna as a whole, because the most frequently studied groups do not constitute the most species-rich groups of soil fauna. To demonstrate this, a pie chart was prepared presenting the percentages of estimated numbers of soil organism species from the studied groups relative to the total species richness calculated based on them (Fig. 1.5). The richest taxonomic groups include microbes, protists, fungi, Collembola and Acari, as well as non-Acari Arachnida, while only Collembola and Acari seem to be relatively frequently studied. In the case of other Arachnida, one may suspect that many ground-dwelling species are not typically classified as soil organisms. Interestingly, the most frequently studied animals include those from the groups Oligochaeta, Formicidae and Isoptera, even though they are not characterized by the highest species richness (Fig. 1.5). In the case of earthworms, the proportional importance of their role in soil would increase dramatically if their biomass were taken into account. As mentioned, earthworms serve as crucial ecosystem engineers, making up the majority of soil faunal biomass in various habitats (Zaller *et al.*, 2013). In the case of Formicidae, the relatively well-studied status can be explained by their wide variety of relationships with invasions (Table 1.1).

Based on the review of articles, it can be concluded that the relationships between the external environment and invasions and soil fauna groups dominate this field of research (Table 1.1). This may be because plant and other invasions are most often treated as applied ecological problems requiring the study of simple relationships to propose rapid measures to prevent invasions. Soil organisms are rarely considered as model systems for testing deep ecological relationships. This situation can be interpreted in two ways: on the one hand, the applied nature and social responsibility of research prohibits conducting activities that prolong the existence of the problem of plant invasions; on the other hand, the key relationships between invasions and soil fauna remain poorly understood.

Table 1.1 summarizes the detected ecosystem interactions:

- biotic interactions: predation, parasitism, competition, allelopathy, amensalism (one organism harms the other without benefiting), commensalism (one organism benefits while the other does not suffer), protocooperation (both parties benefit but can live separately), mutualism, symbiosis;
- interactions with the external environment: weather factors, habitat conditions, etc.; and
- trophic interactions: relations within the food chain.

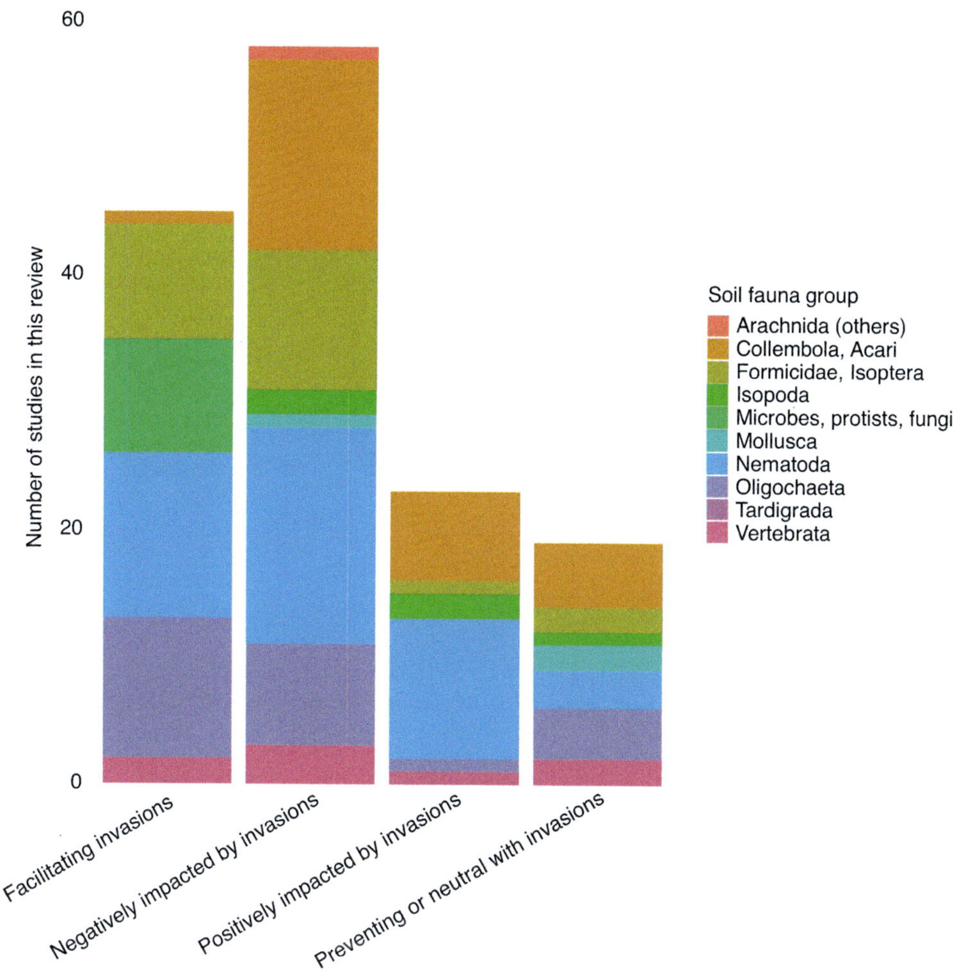

Fig. 1.4. Number of articles with detected relationships between invasions and soil fauna groups.

For example, a single study detected a mutualistic relationship between two soil microorganisms that facilitated invasion (Lebreton *et al.*, 2019). The example facultative mutualism includes cases of invasive plant dispersal by vertebrates that feed on their various parts. Interestingly, no true symbiosis between invasive plants and soil fauna or among soil fauna related to the invasion process has been detected, and no cases of amensalism related to the invasion process have been described or defined. Regarding amensalism, some relationships may not have been sufficiently understood to be considered amensalism. It seems, for example, that some soil organisms suffer from the allelopathic properties of invasive plants, but data are typically lacking as to whether allelopathy and these responding organisms remain neutral toward the invasive plant.

The importance of invasions in soil–ecosystem interactions is poorly understood. It is worth emphasizing the lack of interdisciplinary research on the importance of plant invasions for interconnected elements of soil fauna. Most studies conducted to date have focused on relatively limited numbers of taxa or functional groups of dependent organisms. Surprisingly, there are no comparisons of the relationships between invasive plants and soil fauna across their native ranges and invaded areas. Studies

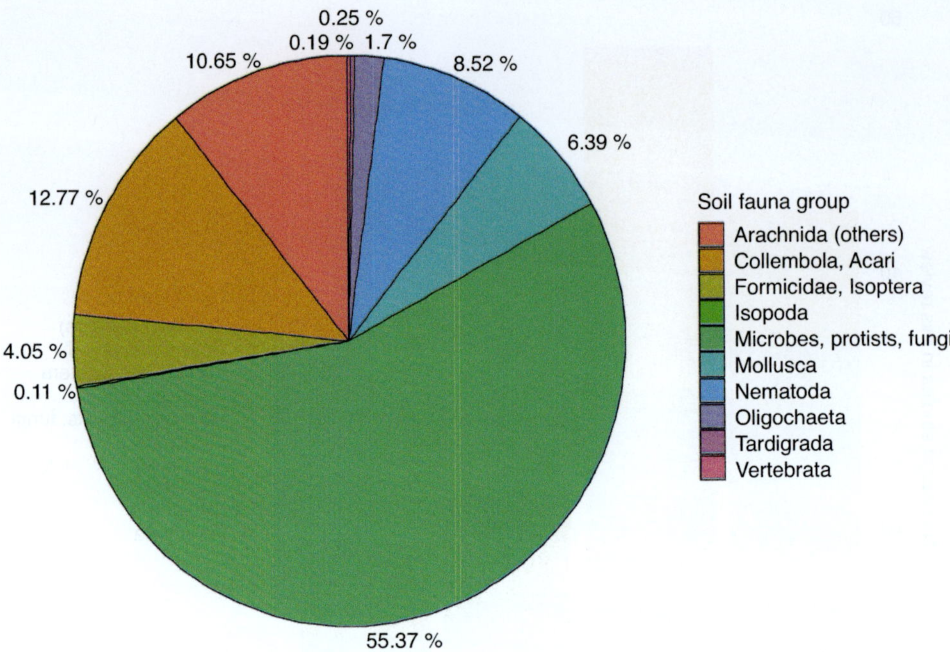

Fig. 1.5. Pie chart presenting the percentage shares of estimated numbers of soil species from individual soil fauna groups studied.

Table 1.1. Detected ecosystem interactions involving the studied groups of soil fauna and invasions.

Soil fauna	Pred	Para	Comp	Alle	Ame	Com	Prot	Mut	Sym	Env	Tro
Arachnida (others)										x	
Collembola, Acari		x	x			x				x	x
Formicidae, Isoptera	x		x	x						x	x
Isopoda										x	x
Microbes, protists, fungi		x				x		x		x	
Mollusca			x							x	
Nematoda	x	x	x	x				x		x	x
Oligochaeta	x	x	x	x		x	x			x	x
Tardigrada	NA	NA	NA	NA	NA	NA	NA	NA	NA	NA	NA
Vertebrata	x						x	x		x	x

Pred, predation; Para, parasitism (including pathogens); Comp, competition; Alle, allelopathy; Ame, amensalism; Com, commensalism; Prot, protocooperation; Mut, mutualism; Sym, symbiosis; Env, interactions with environmental factors; Tro, trophic interactions (food chain); x, relationship detected; NA, no data available.

explaining the impact of global climate change on the relationships between soil fauna and plant invasions, as well as studies based on molecular techniques, are lacking. It seems to be a promising topic in this context, deserving of global-scale modelling and analysis, as well as exploring the excellent possibility of collecting easily accessible material for genetic studies.

References

Abgrall, C., Forey, E. and Chauvat, M. (2019) Soil fauna responses to invasive alien plants are determined by trophic groups and habitat structure: A global meta-analysis. *Oikos* 128(10), 1390–1401. DOI: 10.1111/oik.06493.

Abgrall, C., Forey, E., Mignot, L. and Chauvat, M. (2018) Invasion by *Fallopia japonica* alters soil food webs through secondary metabolites. *Soil Biology and Biochemistry* 127, 100–109. DOI: 10.1016/j.soilbio.2018.09.016.

Alerding, A.B. and Hunter, R.M. (2013) Increased springtail abundance in a garlic mustard-invaded forest. *Northeastern Naturalist* 20(2), 275–288. DOI: 10.1656/045.020.0205.

Almada, M.S. and Sarquis, J.A. (2016) Ground aracnofauna (Araneae: Araneae) in native and exotic forest and grasslands of San Martín Park, Entre Ríos, Argentina [Araneofauna (Arachnida: Araneae) de suelo en bosques nativos, exóticos y pajonales del parque general San Martín, Entre Ríos Argentina]. *Ecologia Austral* 26(3), 286–292. DOI: 10.25260/EA.16.26.3.0.204.

Beckstead, J. and Parker, I.M. (2003) Invasiveness of *Ammophila arenaria*: Release from soil-borne pathogens? *Ecology* 84(11), 2824–2831. DOI: 10.1890/02-0517.

Belnap, J., Phillips, S.L., Sherrod, S.K. and Moldenke, A. (2005) Soil biota can change after exotic plant invasion: Does this affect ecosystem processes? *Ecology* 86(11), 3007–3017. DOI: 10.1890/05-0333.

Berg-Binder, M.C. and Suarez, A.V. (2012) Testing the directed dispersal hypothesis: Are native ant mounds (*Formica* sp.) favorable microhabitats for an invasive plant? *Oecologia* 169(3), 763–772. DOI: 10.1007/s00442-011-2243-2.

Berry, S.D., Leslie, G.W., Spaull, V.W. and Cadet, P. (2010) Within-field damage and distribution patterns of the stalk borer, *Eldana saccharina* (Lepidoptera: Pyralidae), in sugarcane and a comparison with nematode damage. *Bulletin of Entomological Research* 100(4), 373–385. DOI: 10.1017/S0007485309990319.

Bezemer, T.M., Graça, O., Rousseau, P. and van der Putten, W.H. (2004) Above- and below ground trophic interactions on creeping thistle (*Cirsium arvense*) in high- and low-diversity plant communities: Potential for biotic resistance? *Plant Biology* 6(2), 231–238. DOI: 10.1055/s-2004-817846.

Boag, B. and Yeates, G.W. (2001) The potential impact of the New Zealand flatworm, a predator of earthworms, in western Europe. *Ecological Applications* 11(5), 1276–1286. DOI: 10.1890/1051-0761(2001)011[1276:TPIOTN]2.0.CO;2.

Bond, W. and Slingsby, P. (1984) Collapse of an ant-plant mutalism: The Argentine ant (*Iridomyrmex humilis*) and myrmecochorous Proteaceae. *Ecology* 65(4), 1031–1037. DOI: 10.2307/1938311.

Callan, S.K. and Majer, J.K. (2009) Impacts of an incursion of African Big-headed Ants, *Pheidole megacephala* (Fabricius), in urban bushland in Perth, Western Australia. *Pacific Conservation Biology* 15(2), 102–115. DOI: 10.1071/PC090102.

Cameron, E.K., Cahill, J.F. and Bayne, E.M. (2014) Root foraging influences plant growth responses to earthworm foraging. *PLoS ONE* 9(9), e108873. DOI: 10.1371/journal.pone.0108873.

Castro-Díez, P. and Alonso, A. (2017) Effects of non-native riparian plants in riparian and fluvial ecosystems: A review for the Iberian Peninsula. *Limnetica* 36(36), 525–541. DOI: 10.23818/limn.36.19.

Čerevková, A., Miklisová, D., Bobuľská, L. and Renčo, M. (2020) Impact of the invasive plant *Solidago gigantea* on soil nematodes in a semi-natural grassland and a temperate broadleaved mixed forest. *Journal of Helminthology* 94, e51. DOI: 10.1017/S0022149X19000324.

Chen, C., Zou, X., Wu, J., Zhu, X., Jiang, X. *et al.* (2021) Accumulation and spatial homogeneity of nutrients within termite (*Odontotermes yunnanensis*) mounds in the Xishuangbanna region, SW China. *Catena* 198, 105057. DOI: 10.1016/j.catena.2020.105057.

Chen, H., Li, B., Fang, C., Chen, J. and Wu, J. (2007) Exotic plant influences soil nematode communities through litter input. *Soil Biology and Biochemistry* 39(7), 1782–1793. DOI: 10.1016/j.soilbio.2007.02.011.

Chen, S., Chen, H. and Xu, Y. (2019) Safe chemical repellents to prevent the spread of invasive ants. *Pest Management Science* 75(3), 821–827. DOI: 10.1002/ps.5184.

Chen, W., Tao, L., Zheng, R., Chen, P., Li, T. *et al.* (2012) Effects of the invasion by *Solidago canadensis* L. on the community structure of soil animals. *Acta Ecologica Sinica* 32(22), 7072–7081. DOI: 10.5846/stxb201204090493.

Chown, S.L., Janion-Scheepers, C., Marshall, A., Aitkenhead, I.J., Hallas, R. *et al.* (2023) Indigenous and introduced Collembola differ in desiccation resistance but not its plasticity in response to temperature. *Current Research in Insect Science* 3, 100051. DOI: 10.1016/j.cris.2022.100051.

Christensen, B. and Glenner, H. (2010) Molecular phylogeny of Enchytraeidae (Oligochaeta) indicates separate invasions of the terrestrial environment. *Journal of Zoological Systematics and Evolutionary Research* 48(3), 208–212. DOI: 10.1111/j.1439-0469.2009.00558.x.

Clark, R.E., Seewagen, C.L. and Redak, R. (2019) Invasive Japanese barberry, *Berberis thunbergii* (Ranunculales: Berberidaceae) is associated with simplified branch-dwelling and leaf-litter arthropod communities in a New York Forest. *Environmental Entomology* 48(5), 1071–1078. DOI: 10.1093/ee/nvz095.

Coffey, V. and Otfinowski, R. (2019) Legacies of afforestation on soil nematode community composition, structure, and diversity in a northern Canadian prairie. *Plant and Soil* 435(1–2), 437–447. DOI: 10.1007/s11104-018-03906-w.

Coleman, D.C., Geisen, S. and Wall, D.H. (2024) Soil fauna: Occurrence, biodiversity, and roles in ecosystem function. In: Paul, E.A. and Frey, S.D. (eds) *Soil Microbiology, Ecology and Biochemistry*, 5th edn. Elsevier, pp. 131–159. DOI: 10.1016/B978-0-12-822941-5.00005-3.

De Almeida, T., Forey, E. and Chauvat, M. (2022) Alien invasive plant effect on soil fauna is habitat dependent. *Diversity* 14(2), 61. DOI: 10.3390/d14020061.

De La Riva, D.G. and Trumble, J.T. (2016) Selenium exposure results in reduced reproduction in an invasive ant species and altered competitive behavior for a native ant species. *Environmental Pollution* 213, 888–894. DOI: 10.1016/j.envpol.2016.03.045.

Dickie, I.A., Yeates, G.W., St. John, M.G., Stevenson, B.A., Scott, J.T. *et al.* (2011) Ecosystem service and biodiversity trade-offs in two woody successions. *Journal of Applied Ecology* 48(4), 926–934. DOI: 10.1111/j.1365-2664.2011.01980.x.

Dehlin, H., Peltzer, D.A., Allison, V.J., Yeates, G.W., Nilsson, M.-C. *et al.* (2008) Tree seedling performance and below-ground properties in stands of invasive and native tree species. *New Zealand Journal of Ecology* 32(1), 67–79.

Dobson, A., Richardson, J. and Blossey, B. (2020) Effects of earthworms and white-tailed deer on roots, arbuscular mycorrhizae, and forest seedling performance. *Ecology* 101(1), e02903. DOI: 10.1002/ecy.2903.

Dosso, K., Konaté, S., Aidara, D. and Linsenmair, K.E. (2010) Termite diversity and abundance across fire-induced habitat variability in a tropical moist savanna (Lamto, Central Côte d'Ivoire). *Journal of Tropical Ecology* 26(3), 323–331. DOI: 10.1017/S0266467410000015.

Dumack, K., Ferlian, O., Morselli Gysi, D., Degrune, F., Jauss, R.-T. *et al.* (2022) Contrasting protist communities (Cercozoa: Rhizaria) in pristine and earthworm-invaded North American deciduous forests. *Biological Invasions* 24(5), 1345–1357. DOI: 10.1007/s10530-021-02726-x.

Eisenhauer, N., Partsch, S., Parkinson, D. and Scheu, S. (2007) Invasion of a deciduous forest by earthworms: Changes in soil chemistry, microflora, microarthropods and vegetation. *Soil Biology and Biochemistry* 39(5), 1099–1110. DOI: 10.1016/j.soilbio.2006.12.019.

Eisenhauer, N., Milcu, A., Sabais, A.C.W. and Scheu, S. (2008) Animal ecosystem engineers modulate the diversity-invasibility relationship. *PLoS ONE* 3(10), e3489. DOI: 10.1371/journal.pone.0003489.

Escudero Martinez, C.M., Guarneri, N., Overmars, H., Schaik, C., Bouwmeester, H. *et al.* (2019) Distinct roles for strigolactones in cyst nematode parasitism of *Arabidopsis* roots. *European Journal of Plant Pathology* 154(1), 129–140. DOI: 10.1007/s10658-019-01691-5.

Farji-Brener, A.G., Lescano, N. and Ghermandi, L. (2010) Ecological engineering by a native leaf-cutting ant increases the performance of exotic plant species. *Oecologia* 163(1), 163–169. DOI: 10.1007/s00442-010-1589-1.

Flaherty, L., Hills, M., Giacobbo, V., Kuczmarski, P., Momborquette, M. *et al.* (2024) Impacts of garlic mustard (*Alliaria petiolata*, Brassicaceae) invasion on oribatid mites in urban forest soils vary with the size of the invaded patch. *Pedobiologia* 103, 150933. DOI: 10.1016/j.pedobi.2024.150933.

Forey, E., Lodhar, S., Gopaul, S., Boehmer, H.J. and Chauvat, M. (2021) A functional trait-based approach to assess the impact of an alien palm invasion on plant and soil communities on a South Pacific Island. *Austral Ecology* 46(3), 398–410. DOI: 10.1111/aec.12995.

Gao, M., Xiong, C., Tsui, C.K.M. and Cai, L. (2024) Pathogen invasion increases the abundance of predatory protists and their prey associations in the plant microbiome. *Molecular Ecology* 33(3), e17228. DOI: 10.1111/mec.17228.

Germani, G. and Plenchette, C. (2005) Potential of Crotalaria species as green manure crops for the management of pathogenic nematodes and beneficial mycorrhizal fungi. *Plant and Soil* 266(1–2), 333–342. DOI: 10.1007/s11104-005-2281-9.

Gómez, C., Pons, P. and Bas, J.M. (2003) Effects of the Argentine ant *Linepithema humile* on seed dispersal and seedling emergence of *Rhamnus alaternus*. *Ecography* 26(4), 532–538. DOI: 10.1034/j.1600-0587.2003.03484.x.

Green, J., Wang, D., Lilley, C.J., Urwin, P.E. and Atkinson, H.J. (2012) Transgenic potatoes for potato cyst nematode control can replace pesticide use without impact on soil quality. *PLoS ONE* 7(2), e30973. DOI: 10.1371/journal.pone.0030973.

Gremmen, N.J.M., Chown, S.L. and Marshall, D.J. (1998) Impact of the introduced grass *Agrostis stolonifera* on vegetation and soil fauna communities at Marion Island, sub-Antarctic. *Biological Conservation* 85(3), 223–231. DOI: 10.1016/S0006-3207(97)00178-X.

Gruzdeva, L.I. and Sushchuk, A.A. (2010) Tendencies of nematodes communities to recover after soil cover degradation. *Izvestiia Akademii Nauk. Seriia Biologicheskaia* 6, 750–755.

Hale, C.M., Frelich, L.E., Reich, P.B. and Pastor, J. (2008) Exotic earthworm effects on hardwood forest floor, nutrient availability and native plants: A mesocosm study. *Oecologia* 155(3), 509–518. DOI: 10.1007/s00442-007-0925-6.

Heckman, R.W., Wright, J.P. and Mitchell, C.E. (2016) Joint effects of nutrient addition and enemy exclusion on exotic plant success. *Ecology* 97(12), 3337–3345. DOI: 10.1002/ecy.1585.

Heděnec, P., Jiménez, J.J., Moradi, J., Domene, X., Hackenberger, D. *et al.* (2022) Global distribution of soil fauna functional groups and their estimated litter consumption across biomes. *Scientific Reports* 12(1), 17362. DOI: 10.1038/s41598-022-21563-z.

Hernández-Gómez, O., Byrne, A.Q., Gunderson, A.R., Jenkinson, T.S., Noss, C.F. *et al.* (2020) Invasive vegetation affects amphibian skin microbiota and body condition. *PeerJ* 8, e8549. DOI: 10.7717/peerj.8549.

Hofstetter, R.W., Mahfouz, J.B., Klepzig, K.D. and Ayres, M.P. (2005) Effects of tree phytochemistry on the interactions among endophloedic fungi associated with the southern pine beetle. *Journal of Chemical Ecology* 31(3), 539–560. DOI: 10.1007/s10886-005-2035-4.

Holbrook, J.D., Pilliod, D.S., Arkle, R.S., Rachlow, J.L., Vierling, K.T. *et al.* (2016) Transition of vegetation states positively affects harvester ants in the Great Basin, United States. *Rangeland Ecology & Management* 69(6), 449–456. DOI: 10.1016/j.rama.2016.06.009.

Holdsworth, A.R., Frelich, L.E. and Reich, P.B. (2007a) Regional extent of an ecosystem engineer: Earthworm invasion in northern hardwood forests. *Ecological Applications* 17(6), 1666–1677. DOI: 10.1890/05-2003.1.

Holdsworth, A.R., Frelich, L.E. and Reich, P.B. (2007b) Effects of earthworm invasion on plant species richness in northern hardwood forests. *Conservation Biology* 21(4), 997–1008. DOI: 10.1111/j.1523-1739.2007.00740.x.

Holmes, P.M. and Cowling, R.M. (1997) The effects of invasion by *Acacia saligna* on the guild structure and regeneration capabilities of South African fynbos shrublands. *The Journal of Applied Ecology* 34(2), 317. DOI: 10.2307/2404879.

Hussain, N., Abbasi, T. and Abbasi, S.A. (2016) Vermicomposting transforms allelopathic parthenium into a benign organic fertilizer. *Journal of Environmental Management* 180, 180–189. DOI: 10.1016/j.jenvman.2016.05.013.

Ives, C.D., Hose, G.C., Nipperess, D.A. and Taylor, M.P. (2011) Environmental and landscape factors influencing ant and plant diversity in suburban riparian corridors. *Landscape and Urban Planning* 103(3–4), 372–382. DOI: 10.1016/j.landurbplan.2011.08.009.

Janion-Scheepers, C., Phillips, L., Sgrò, C.M., Duffy, G.A., Hallas, R. *et al.* (2018) Basal resistance enhances warming tolerance of alien over indigenous species across latitude. *Proceedings of the National Academy of Sciences USA* 115(1), 145–150. DOI: 10.1073/pnas.1715598115.

Jin, H., Chang, L., van Kleunen, M. and Liu, Y. (2022) Soil mesofauna may buffer the negative effects of drought on alien plant invasion. *Journal of Ecology* 110(10), 2332–2342. DOI: 10.1111/1365-2745.13950.

Jonsson, H., Olofsson, J., Blume-Werry, G. and Klaminder, J. (2024) Cascading effects of earthworm invasion increase graminoid density and rodent grazing intensities. *Ecology* 105(2), e4212. DOI: 10.1002/ecy.4212.

Jurová, J., Renčo, M., Gömöryová, E. and Čerevková, A. (2020) Effects of the invasive common milkweed (*Asclepias syriaca*) on nematode communities in natural grasslands. *Nematology* 22(4), 423–438. DOI: 10.1163/15685411-00003314.

Kappes, H., Lay, R. and Topp, W. (2007) Changes in different trophic levels of litter-dwelling macrofauna associated with giant knotweed invasion. *Ecosystems* 10(5), 734–744. DOI: 10.1007/s10021-007-9052-9.

Keen, S.C., Wackett, A.A., Willenbring, J.K., Yoo, K., Jonsson, H. *et al.* (2022) Non-native species change the tune of tundra soils: Novel access to soundscapes of the Arctic earthworm invasion. *Science of the Total Environment* 838, 155976. DOI: 10.1016/j.scitotenv.2022.155976.

Keith, A.M., van der Wal, R., Brooker, R.W., Osler, G.H.R., Chapman, S.J. *et al.* (2006) Birch invasion of heather moorland increases nematode diversity and trophic complexity. *Soil Biology and Biochemistry* 38(12), 3421–3430. DOI: 10.1016/j.soilbio.2006.05.013.

Khan, M.R., Khan, M.W. and Singh, K. (1997) Management of root-knot disease of tomato by the application of fly ash in soil. *Plant Pathology* 46(1), 33–43. DOI: 10.1046/j.1365-3059.1997.d01-199.x.

Knevel, I.C., Lans, T., Menting, F.B.J., Hertling, U.M. and Van Der Putten, W.H. (2004) Release from native root herbivores and biotic resistance by soil pathogens in a new habitat both affect the alien *Ammophila arenaria* in South Africa. *Oecologia* 141(3), 502–510. DOI: 10.1007/s00442-004-1662-8.

Kohyt, J. and Skubała, P. (2020) Oribatid mite (Acari: Oribatida) communities reveal the negative impact of the red oak (*Quercus rubra* L.) on soil fauna in Polish commercial forests. *Pedobiologia* 79, 150594. DOI: 10.1016/j.pedobi.2019.150594.

Kuijper, D.P.J. and Bakker, J.P. (2012) Below- and above-ground vertebrate herbivory and abiotic factors alternate in shaping salt-marsh plant communities. *Journal of Experimental Marine Biology and Ecology* 432–433, 17–28. DOI: 10.1016/j.jembe.2012.07.003.

Lafleur, B., Hooper-Bùi, L.M., Mumma, E.P. and Geaghan, J.P. (2005) Soil fertility and plant growth in soils from pine forests and plantations: Effect of invasive red imported fire ants *Solenopsis invicta* (Buren). *Pedobiologia* 49(5), 415–423. DOI: 10.1016/j.pedobi.2005.05.002.

Lake, J.C. and Leishman, M.R. (2004) Invasion success of exotic plants in natural ecosystems: The role of disturbance, plant attributes and freedom from herbivores. *Biological Conservation* 117(2), 215–226. DOI: 10.1016/S0006-3207(03)00294-5.

Lavoie, C. (2017) The impact of invasive knotweed species (*Reynoutria* spp.) on the environment: Review and research perspectives. *Biological Invasions* 19(8), 2319–2337. DOI: 10.1007/s10530-017-1444-y.

Lazarus, B.E. and Germino, M.J. (2021) Post-fire management targeting invasive annual grasses may have inadvertently released the exotic perennial forb *Chondrilla juncea* and suppressed its biocontrol agent. *Biological Invasions* 23(6), 1915–1932. DOI: 10.1007/s10530-021-02481-z.

Lazzaro, L., Mazza, G., d'Errico, G., Fabiani, A., Giuliani, C. *et al.* (2018) How ecosystems change following invasion by *Robinia pseudoacacia*: Insights from soil chemical properties and soil microbial, nematode, microarthropod and plant communities. *The Science of the Total Environment* 622–623, 1509–1518. DOI: 10.1016/j.scitotenv.2017.10.017.

Lebreton, L., Guillerm-Erckelboudt, A.-Y., Gazengel, K., Linglin, J., Ourry, M. *et al.* (2019) Temporal dynamics of bacterial and fungal communities during the infection of *Brassica rapa* roots by the protist *Plasmodiophora brassicae*. *PLoS ONE* 14(2), e0204195. DOI: 10.1371/journal.pone.0204195.

Li, P., Gu, S., Zhu, Y., Xu, T., Yang, Y. *et al.* (2023) Soil microbiota plays a key regulatory role in the outbreak of tobacco root rot. *Frontiers in Microbiology* 14, 1214167. DOI: 10.3389/fmicb.2023.1214167.

Liu, Y. and Liao, Y. (2013) Woody plant leaf litter consumption by the woodlouse *Porcellio scaber* with a choice test. *Acta Ecologica Sinica* 33(2), 475–482. DOI: 10.5846/stxb201111221783.

Liu, J., Xu, G., Yin, L., Xu, X., Armitage, D.W. *et al.* (2020) Invasive plants exert disproportionately negative allelopathic effects on the growth and physiology of the earthworm *Eisenia fetida*. *Science of The Total Environment* 747, 141534. DOI: 10.1016/j.scitotenv.2020.141534.

Long, K., Yin, R., Kardol, P., Wei, Q., Li, Y. *et al.* (2023) Bamboo invasion alters Collembola community composition varying with life-forms. *Pest Management Science* 79(7), 2517–2526. DOI: 10.1002/ps.7434.

Loss, S.R. and Blair, R.B. (2011) Reduced density and nest survival of ground-nesting songbirds relative to earthworm invasions in northern hardwood forests. *Conservation Biology* 25(5), 983–992. DOI: 10.1111/j.1523-1739.2011.01719.x.

Lu, X., He, M., Ding, J. and Siemann, E. (2018) Latitudinal variation in soil biota: Testing the biotic interaction hypothesis with an invasive plant and a native congener. *The ISME Journal* 12(12), 2811–2822. DOI: 10.1038/s41396-018-0219-5.

Lu, X.-R., Liu, T., Wang, R., Wang, H. and Duan, Y.-X. (2016) Impacts of leaf litter decomposition of invasive plant Ambrosia trifida on soil nematode community structure. *Chinese Journal of Ecology* 35(9), 2369–2378. DOI: 10.13292/j.1000-4890.201609.012.

MacDonald, A.A.M. and Kotanen, P.M. (2010) The effects of disturbance and enemy exclusion on performance of an invasive species, common ragweed, in its native range. *Oecologia* 162(4), 977–986. DOI: 10.1007/s00442-009-1557-9.

Maceda-Veiga, A., Basas, H., Lanzaco, G., Sala, M., Sostoa, A. *et al.* (2016) Impacts of the invader giant reed (*Arundo donax*) on riparian habitats and ground arthropod communities. *Biological Invasions* 18(3), 731–749. DOI: 10.1007/s10530-015-1044-7.

Mahon, M.B. and Crist, T.O. (2019) Invasive earthworm and soil litter response to the experimental removal of white-tailed deer and an invasive shrub. *Ecology* 100(5), e02688. DOI: 10.1002/ecy.2688.

Malloch, B., Tatsumi, S., Seibold, S., Cadotte, M.W. and MacIvor, J.S. (2020) Urbanization and plant invasion alter the structure of litter microarthropod communities. *Journal of Animal Ecology* 89(11), 2496–2507. DOI: 10.1111/1365-2656.13311.

McGrath, D.A. and Binkley, M.A. (2009) *Microstegium vimineum* invasion changes soil chemistry and microarthropod communities in *Cumberland plateau* forests. *Southeastern Naturalist* 8(1), 141–156. DOI: 10.1656/058.008.0113.

Melnichuk, R.D.S., Tecimen, H.B. and Görres, J.H. (2022) Do the invasive earthworms *Amynthas agrestis* (Oligochaeta: Megascolecidae) and *Lumbricus rubellus* (Oligochaeta: Lumbricidae) stimulate oxalate-based browser defenses in jack-in-the-pulpit (*Arisaema triphyllum*) by their presence or their soil biogeochemical activity? *Soil Systems* 6(1), 11. DOI: 10.3390/soilsystems6010011.

Menke, S.B. and Holway, D.A. (2006) Abiotic factors control invasion by Argentine ants at the community scale. *Journal of Animal Ecology* 75(2), 368–376. DOI: 10.1111/j.1365-2656.2006.01056.x.

Migge-Kleian, S., McLean, M.A., Maerz, J.C. and Heneghan, L. (2006) The influence of invasive earthworms on indigenous fauna in ecosystems previously uninhabited by earthworms. *Biological Invasions* 8(6), 1275–1285. DOI: 10.1007/s10530-006-9021-9.

Motard, E., Dusz, S., Geslin, B., Akpa-Vinceslas, M., Hignard, C. *et al.* (2015) How invasion by *Ailanthus altissima* transforms soil and litter communities in a temperate forest ecosystem 17(6), 1817–1832. DOI: 10.1007/s10530-014-0838-3.

Negesse, Z., Pan, K., Guadie, A., Justine, M.F., Azene, B. *et al.* (2025) Plant invasions alter soil biota and microbial activities: A global meta-analysis. *Plant and Soil* 513(1), 1031–1050. DOI: 10.1007/s11104-025-07227-7.

Nielsen, U.N., Osler, G.H.R., Campbell, C.D., Burslem, D.F.R.P. and van der Wal, R. (2010) The influence of vegetation type, soil properties and precipitation on the composition of soil mite and microbial communities at the landscape scale. *Journal of Biogeography* 37(7), 1317–1328. DOI: 10.1111/j.1365-2699.2010.02281.x.

Page, H.M., Lastra, M., Rodil, I.F., Briones, M.J.I. and Garrido, J. (2010) Effects of non-native *Spartina patens* on plant and sediment organic matter carbon incorporation into the local invertebrate community. *Biological Invasions* 12(11), 3825–3838. DOI: 10.1007/s10530-010-9775-y.

Page, M.J., McKenzie, J.E., Bossuyt, P.M., Boutron, I., Hoffmann, T.C. *et al.* (2021) The PRISMA 2020 statement: An updated guideline for reporting systematic reviews. *BMJ* 372(71), 71. DOI: 10.1136/bmj.n71.

Paudel, S., Longcore, T., MacDonald, B., McCormick, M.K., Szlavecz, K. *et al.* (2016a) Belowground interactions with aboveground consequences: Invasive earthworms and arbuscular mycorrhizal fungi. *Ecology* 97(3), 605–614. DOI: 10.1890/15-1085.

Paudel, S., Wilson, G.W.T., MacDonald, B., Longcore, T. and Loss, S.R. (2016b) Predicting spatial extent of invasive earthworms on an oceanic island. *Diversity and Distributions* 22(10), 1013–1023. DOI: 10.1111/ddi.12472.

Pearson, B.M., Minor, M.A., Robertson, A.W. and Clavijo McCormick, A.L. (2024) Plant invasion down under: Exploring the below-ground impact of invasive plant species on soil properties and invertebrate communities in the Central Plateau of New Zealand. *Biological Invasions* 26(12), 4215–4228. DOI: 10.1007/s10530-024-03441-z.

Pehle, A. and Schirmel, J. (2015) Moss invasion in a dune ecosystem influences ground-dwelling arthropod community structure and reduces soil biological activity. *Biological Invasions* 17(12), 3467–3477. DOI: 10.1007/s10530-015-0971-7.

Peralta, G., Schon, N.L., Dickie, I.A., John, M.G., Orwin, K.H. *et al.* (2019) Contrasting responses of soil nematode communities to native and non-native woody plant expansion. *Oecologia* 190(4), 891–899. DOI: 10.1007/s00442-019-04456-3.

Peralta, G., Dickie, I.A., Yeates, G.W. and Peltzer, D.A. (2020) Community- and trophic-level responses of soil nematodes to removal of a non-native tree at different stages of invasion. *PLoS ONE* 15(1), e0227130. DOI: 10.1371/journal.pone.0227130.

Perera, P.C.D., Gruss, I., Twardowski, J. and Szymura, M. (2022) Effect of *Solidago* eradication methods on soil invertebrates - preliminary studies. *Polish Journal of Environmental Studies* 31(1), 485–492. DOI: 10.15244/pjoes/138359.

Phillips, H.R.P., Cameron, E.K., Eisenhauer, N., Burton, V.J., Ferlian, O. *et al.* (2024) Global changes and their environmental stressors have a significant impact on soil biodiversity—a meta-analysis. *iScience* 27(9), 110540. DOI: 10.1016/j.isci.2024.110540.

Porazinska, D.L., Pratt, P.D. and Glblin-Davis, R.M. (2007) Consequences of *Melaleuca quinquenervia* invasion on soil nematodes in the Florida Everglades. *Journal of Nematology* 39(4), 305–312.

Porazinska, D.L., Seastedt, T.R., Gendron, E.M.S. and Schmidt, S.K. (2022) Invasive annual cheatgrass enhances the abundance of native microbial and microinvertebrate eukaryotes but reduces invasive earthworms. *Plant and Soil* 473(1–2), 591–604. DOI: 10.1007/s11104-022-05312-9.

Porco, D., Decaëns, T., Deharveng, L., James, S.W., Skarżyński, D. *et al.* (2013) Biological invasions in soil: DNA barcoding as a monitoring tool in a multiple taxa survey targeting European earthworms and springtails in North America. *Biological Invasions* 15(4), 899–910. DOI: 10.1007/s10530-012-0338-2.

Pritekel, C., Whittemore-Olson, A., Snow, N. and Moore, J.C. (2006) Impacts from invasive plant species and their control on the plant community and belowground ecosystem at Rocky Mountain National Park, USA. *Applied Soil Ecology* 32(1), 132–141. DOI: 10.1016/j.apsoil.2005.01.010.

Qin, H., Guo, W. and Li, X. (2021) Density-dependent interactions between the nematode *Meloidogyne incognita* and the biological control agent *Agasicles hygrophila* on invasive *Alternanthera philoxeroides* and its native congener *Alternantera sessilis*. *BioControl* 66(6), 837–848. DOI: 10.1007/s10526-021-10113-7.

Qin, Z., Xie, J.F., Quan, G.M., Zhang, J.E., Mao, D.J. *et al.* (2019) Changes in the soil meso- and microfauna community under the impacts of exotic *Ambrosia artemisiifolia*. *Ecological Research* 34(2), 265–276. DOI: 10.1111/1440-1703.1271.

Quan, G., Zhang, J., Xie, J., Mao, D., Xu, H. *et al.* (2011) Impact of *Mikania micrantha* invasion on soil meso- and micro-invertebrate community structure. *Ying Yong Sheng Tai Xue Bao = The Journal of Applied Ecology* 22(7), 1863–1870.

Quilichini, A. and Debussche, M. (2000) Seed dispersal and germination patterns in a rare Mediterranean island endemic (*Anchusa crispa* Viv., Boraginaceae). *Acta Oecologica* 21(6), 303–313. DOI: 10.1016/S1146-609X(00)01089-4.

Rahman, M.L. and Evans, A.A.F. (1987) Studies on host-parasite relationships of rice stem nematode, *Ditylenchus angustus* (Nematoda: Tylenchida) on rice, *Oryza sativa*, L. *Nematologica* 33(4), 451–459. DOI: 10.1163/187529287X00100.

Reid, M.L. and Emery, S.M. (2017) Native and exotic foundation grasses differ in traits and responses to belowground tri-trophic interactions. *Plant Ecology* 218(2), 173–183. DOI: 10.1007/s11258-016-0675-9.

Reid, M.L. and Emery, S.M. (2018) Scale-dependent effects of *Gypsophila paniculata* invasion and management on plant and soil nematode community diversity and heterogeneity. *Biological Conservation* 224, 153–161. DOI: 10.1016/j.biocon.2018.05.026.

Reid, M.L., Howes, A.S. and Emery, S.M. (2019) Dune soil communities primarily associated with climate factors, not exotic plant presence. *Plant and Soil* 436(1–2), 503–515. DOI: 10.1007/s11104-019-03944-y.

Ren, X., Liu, M., Sun, M., Yang, J., Wang, B. *et al.* (2023) Characteristics of protist community in banana rhizosphere soil and interactions between protists and pathogens of *Fusarium wilt* disease. *Acta Pedologica Sinica* 60(2), 558–567. DOI: 10.11766/trxb202107140361.

Renčo, M. and Baležentiené, L. (2015) An analysis of soil free-living and plant-parasitic nematode communities in three habitats invaded by *Heracleum sosnowskyi* in central Lithuania. *Biological Invasions* 17(4), 1025–1039. DOI: 10.1007/s10530-014-0773-3.

Renčo, M., Jurová, J., Gömöryová, E. and Čerevková, A. (2021) Long-term giant hogweed invasion contributes to the structural changes of soil nematofauna. *Plants* 10(10), 2103. DOI: 10.3390/plants10102103.

Renčo, M., Kornobis, F.W., Domaradzki, K., Jakubska-Busse, A., Jurová, J. *et al.* (2019) How does an invasive *Heracleum sosnowskyi* affect soil nematode communities in natural conditions? *Nematology* 21(1), 71–89. DOI: 10.1163/15685411-00003196.

Riddle, S.B. and Bateman, H.L. (2020) Habitat and soil associations of a fossorial toad in a Sonoran Desert riparian forest. *Journal of Arid Environments* 181, 104239. DOI: 10.1016/j.jaridenv.2020.104239.

Rossiter-Rachor, N.A., Setterfield, S.A., Hutley, L.B., McMaster, D., Schmidt, S. *et al.* (2017) Invasive *Andropogon gayanus* (Gamba grass) alters litter decomposition and nitrogen fluxes in an Australian tropical savannah. *Scientific Reports* 7(1), 11705. DOI: 10.1038/s41598-017-08893-z.

Roth, A.M., Whitfeld, T.J.S., Lodge, A.G., Eisenhauer, N., Frelich, L.E. *et al.* (2015) Invasive earthworms interact with abiotic conditions to influence the invasion of common buckthorn (*Rhamnus cathartica*). *Oecologia* 178(1), 219–230. DOI: 10.1007/s00442-014-3175-4.

Schittko, C. and Wurst, S. (2014) Above- and belowground effects of plant-soil feedback from exotic *Solidago canadensis* on native *Tanacetum vulgare*. *Biological Invasions* 16(7), 1465–1479. DOI: 10.1007/s10530-013-0584-y.

Schuh, M. and Larsen, K.J. (2015) *Rhamnus cathartica* (Rosales: Rhamnaceae) invasion reduces ground-dwelling insect abundance and diversity in northeast iowa forests. *Environmental Entomology* 44(3), 647–657. DOI: 10.1093/ee/nvv050.

Seabloom, E.W., Borer, E.T., Martin, B.A. and Orrock, J.L. (2009) Effects of long-term consumer manipulations on invasion in oak savanna communities. *Ecology* 90(5), 1356–1365. DOI: 10.1890/08-0671.1.

Seabloom, E.W., Borer, E.T., Buckley, Y.M., Cleland, E.E., Davies, K.F. *et al.* (2015) Plant species origin predicts dominance and response to nutrient enrichment and herbivores in global grasslands. *Nature Communications* 6, 7710. DOI: 10.1038/ncomms8710.

Shao, Y., Zhang, W., Eisenhauer, N., Liu, T., Xiong, Y. *et al.* (2017) Nitrogen deposition cancels out exotic earthworm effects on plant-feeding nematode communities. *Journal of Animal Ecology* 86(4), 708–717. DOI: 10.1111/1365-2656.12660.

Siddiqui, I.A. and Shaukat, S.S. (2003) Factors influencing the effectiveness of non-pathogenic *Fusarium solani* strain fs5 in the suppression of root-knot nematode in tomato. *Phytopathologia Mediterranea* 42(1), 17–26. DOI: 10.14601/Phytopathol_Mediterr-1687.

Skubała, P. (2012) Invasive giant knotweed (*Fallopia sachalinensis*) alters the composition of oribatid mite communities. *Biolet* 49(2), 143–155. DOI: 10.2478/v10120-012-0016-1.

Stanley, M.C. and Ward, D.F. (2012) Impacts of Argentine ants on invertebrate communities with below-ground consequences. *Biodiversity and Conservation* 21(10), 2653–2669. DOI: 10.1007/s10531-012-0324-0.

Sterzyńska, M., Shrubovych, J. and Nicia, P. (2017) Impact of plant invasion (*Solidago gigantea* L.) on soil mesofauna in a riparian wet meadows. *Pedobiologia* 64, 1–7. DOI: 10.1016/j.pedobi.2017.07.004.

Suárez, E.R., Fahey, T.J., Yavitt, J.B., Groffman, P.M. and Bohlen, P.J. (2006) Patterns of litter disappearance in a northern hardwood forest invaded by exotic earthworms. *Ecological Applications* 16(1), 154–165. DOI: 10.1890/04-0788.

Sugiura, S. (2008) Hot water tolerance of soil animals: Utility of hot water immersion in preventing invasions of alien soil animals. *Applied Entomology and Zoology* 43(2), 207–212. DOI: 10.1303/aez.2008.207.

Sun, F., Ou, Q., Yu, H., Li, N. and Peng, C. (2019) The invasive plant *Mikania micrantha* affects the soil foodweb and plant-soil nutrient contents in orchards. *Soil Biology and Biochemistry* 139, 107630. DOI: 10.1016/j.soilbio.2019.107630.

Szlavecz, K., Chang, C.-H., Bernard, M.J., Pitz, S.L., Xia, L. *et al.* (2018) Litter quality, dispersal and invasion drive earthworm community dynamics and forest soil development. *Oecologia* 188(1), 237–250. DOI: 10.1007/s00442-018-4205-4.

Terauds, A., Chown, S.L. and Bergstrom, D.M. (2011) Spatial scale and species identity influence the indigenous-alien diversity relationship in springtails. *Ecology* 92(7), 1436–1447. DOI: 10.1890/10-2216.1.

Tong, Z., Liu, W., Zhang, G., Dong, C., Zhang, Y. *et al.* (2025) The relationship between occurrence of hulless barley ear rot and population migration of grass mite (*Siteroptes* spp.). *Scientia Agricultura Sinica* 58(3), 493–506. DOI: 10.3864/j.issn.0578-1752.2025.03.007.

Tóth, Z. and Korponai, K. (2025) Diversity patterns and trait-specific variations in soil nematode communities associated with *Solidago* invasion. *Ecological Indicators* 175, 113598. DOI: 10.1016/j.ecolind.2025.113598.

Trigos-Peral, G., Casacci, L.P., Ślipiński, P., Grześ, I.M., Moroń, D. *et al.* (2018) Ant communities and *Solidago* plant invasion: Environmental properties and food sources. *Entomological Science* 21(3), 270–278. DOI: 10.1111/ens.12304.

Utz, R.M., Hart, I.S. and Rice, Q. (2023) Soil arthropod communities associated with *Berberis thunbergii* invasion in a temperate deciduous forest harbour more detritivores. *Ecological Entomology* 48(4), 512–516. DOI: 10.1111/een.13240.

Vilardo, G., Tognetti, P.M., González-Arzac, A. and Yahdjian, L. (2018) Soil arthropod composition differs between old-fields dominated by exotic plant species and remnant native grasslands. *Acta Oecologica* 91, 57–64. DOI: 10.1016/j.actao.2018.06.003.

Wagner, D.L. and Van Driesche, R.G. (2010) Threats posed to rare or endangered insects by invasions of nonnative species. *Annual Review of Entomology* 55, 547–568. DOI: 10.1146/annurev-ento-112408-085516.

Walters, A.C. and Mackay, D.A. (2004) Comparisons of upper thermal tolerances between the invasive Argentine ant (Hymenoptera: Formicidae) and two native Australian ant species. *Annals of the Entomological Society of America* 97(5), 971–975. DOI: 10.1603/0013-8746(2004)097[0971:COUTTB]2.0.CO;2.

Wang, D., Jones, L.M., Urwin, P.E. and Atkinson, H.J. (2011) A synthetic peptide shows retro- and anterograde neuronal transport before disrupting the chemosensation of plant-pathogenic nematodes. *PLoS ONE* 6(3), e17475. DOI: 10.1371/journal.pone.0017475.

Wang, Z., Luo, Y.-Q., Shi, J., Gao, R., Wang, G. *et al.* (2014) Quantitative classification and environmental interpretation of secondary forests 18 years after the invasion of pine forests by *Bursaphelenchus xylophilus* (Nematoda: Aphelenchoididae) in China. *Journal of Insect Science* 14(1), 296. DOI: 10.1093/jisesa/ieu158.

Wanner, M. and Xylander, W.E.R. (2005) Biodiversity development of terrestrial testate amoebae: Is there any succession at all? *Biology and Fertility of Soils* 41, 428–438. DOI: 10.1007/s00374-005-0850-y.

Warren II, R.J., Mathew, A., Reed, K., Bayba, S., Krupp, K. *et al.* (2019) *Myrmica rubra* microhabitat selection and putative ecological impact. *Ecological Entomology* 44(2), 239–248. DOI: 10.1111/een.12700.

Wei, C.-Y., Wang, J.-K., Shih, H.-C., Wang, H.-C. and Kuo, C.-C. (2020) Invasive plants facilitated by socioeconomic change harbor vectors of scrub typhus and spotted fever. *PLoS Neglected Tropical Diseases* 14(1), e0007519. DOI: 10.1371/journal.pntd.0007519.

Wei, Z., Shen, Z., Yang, T., Wang, X., Li, R. *et al.* (2021a) From suppressive soil to rhizosphere immunity: Towards an ecosystem thinking for soil-borne pathogen control. *Acta Pedologica Sinica* 58(4), 814–824. DOI: 10.11766/trxb202003230038.

Wei, Z., Song, Y., Xiong, W., Xu, Y. and Shen, Q. (2021b) Soil protozoa: Research methods and roles in the biocontrol of soil-borne diseases. *Acta Pedologica Sinica* 58(1), 14–22. DOI: 10.11766/trxb202002010039.

Wetterer, J.K., Miller, S.E., Wheeler, D.E., Olson, C.A., Polhemus, D.A. *et al.* (1999) Ecological dominance by *Paratrechina longicornis* (Hymenoptera: Formicidae), an invasive tramp ant, in biosphere 2. *The Florida Entomologist* 82(3), 381. DOI: 10.2307/3496865.

Wilschut, R.A. and Geisen, S. (2021) Nematodes as drivers of plant performance in natural systems. *Trends in Plant Science* 26(3), 237–247. DOI: 10.1016/j.tplants.2020.10.006.

Winsome, T., Epstein, L., Hendrix, P.F. and Horwath, W.R. (2006) Competitive interactions between native and exotic earthworm species as influenced by habitat quality in a California grassland. *Applied Soil Ecology* 32(1), 38–53. DOI: 10.1016/j.apsoil.2005.01.008.

Wolkovich, E.M., Bolger, D.T. and Holway, D.A. (2009) Complex responses to invasive grass litter by ground arthropods in a Mediterranean scrub ecosystem. *Oecologia* 161(4), 697–708. DOI: 10.1007/s00442-009-1425-7.

Wiatrowska, B., Kurek, P., Rutkowski, T., Napierała, A., Sienkiewicz, P. *et al.* (2024) Consequences of *Spiraea tomentosa* invasion in Uropodina mite (Acari: Mesostigmata) communities in wet meadows. *Experimental and Applied Acarology* 93(3), 609–626. DOI: 10.1007/s10493-024-00951-2.

Wilschut, R.A., Magnée, K.J.H., Geisen, S., van der Putten, W.H. and Kostenko, O. (2020) Plant population and soil origin effects on rhizosphere nematode community composition of a range-expanding plant species and a native congener. *Oecologia* 194(1–2), 237–250. DOI: 10.1007/s00442-020-04749-y.

Wu, S., Cheng, J., Xu, X., Zhang, Y., Zhao, Y. *et al.* (2019) Polyploidy in invasive *Solidago canadensis* increased plant nitrogen uptake, and abundance and activity of microbes and nematodes in soil. *Soil Biology and Biochemistry* 138, 107594. DOI: 10.1016/j.soilbio.2019.107594.

Wu, W., Wang, J., Yan, B., Mou, Z., Yuan, Y. *et al.* (2024) Giant African snail invasion homogenizes seasonal soil biodiversity in tropical coral islands. *Plant and Soil* 500(1–2), 571–585. DOI: 10.1007/s11104-024-06508-x.

Wu, Y., Jenkins, T., Blunden, G., Whapham, C. and Hankins, S.D. (1997) The role of betaines in alkaline extracts of *Ascophyllum nodosum* in the reduction of *Meloidogyne javanica* and *M. incognita* infestations of tomato plants. *Fundamental and Applied Nematology* 20(2), 99–102.

Xie, J., Quan, G., Zhang, J., Mao, D., Xu, H. *et al.* (2011) Impacts of *Ambrosia artemisiifolia* invasion on community structure of soil meso- and micro-fauna. *Acta Ecologica Sinica* 31(19), 5682–5690. DOI: 10.0000/j.1000-0933.2011311256825690.

Yeates, G.W. and King, K.L. (1997) Soil nematodes as indicators of the effect of management on grasslands in the New England Tablelands (NSW): Comparison of native and improved grasslands. *Pedobiologia* 41(6), 526–536. DOI: 10.1016/S0031-4056(24)00323-8.

Zaller, J.G., Parth, M., Szunyogh, I., Semmelrock, I., Sochurek, S. *et al.* (2013) Herbivory of an invasive slug is affected by earthworms and the composition of plant communities. *BMC Ecology* 13, 20. DOI: 10.1186/1472-6785-13-20.

Zhang, B., Ma, Y., Duan, W., Fan, Q. and Sun, J. (2024) Pinewood nematode induced changes in the assembly process of gallery microbiomes benefit its vector beetle's development. *Microbiology Spectrum* 12(10), 1–20. DOI: 10.1128/spectrum.01412-24.

Zhang, P., Nie, M., Li, B. and Wu, J. (2017) The transfer and allocation of newly fixed C by invasive *Spartina alterniflora* and native *Phragmites australis* to soil microbiota. *Soil Biology and Biochemistry* 113, 231–239. DOI: 10.1016/j.soilbio.2017.06.003.

Zhang, P., Neher, D.A., Li, B. and Wu, J. (2018) The impacts of above- and belowground plant input on soil microbiota: Invasive *Spartina alterniflora* versus native *Phragmites australis*. *Ecosystems* 21(3), 469–481. DOI: 10.1007/s10021-017-0162-8.

Zhang, Y., Pennings, S.C., Li, B. and Wu, J. (2019) Biotic homogenization of wetland nematode communities by exotic *Spartina alterniflora* in China. *Ecology* 100(4), e02596. DOI: 10.1002/ecy.2596.

Zhang, Y.-J., Jin, Y., Xu, J., He, H., Tao, Y. *et al.* (2022) Responses and feedback of litter properties and soil mesofauna to herbaceous plants expansion into the alpine tundra on Changbai Mountain, China. *Journal of Mountain Science* 19(2), 403–417. DOI: 10.1007/s11629-021-6751-1.

Zhang, Y., Li, B., Wu, J. and Pennings, S.C. (2020a) Contrasting latitudinal clines of nematode diversity in *Spartina alterniflora* salt marshes between native and introduced ranges. *Diversity and Distributions* 26(5), 623–631. DOI: 10.1111/ddi.13054.

Zhang, Z., Liu, J., Wang, L.-C., Wang, Y. and Deng, L.-L. (2020b) IMPACT of the invasive plant *Alternanthera philoxeroides* (Mart.) Griseb on soil meso- and microfaunal community. *Applied Ecology and Environmental Research* 18(2), 3073–3085. DOI: 10.15666/aeer/1802_30733085.

Zhao, R.J., Wu, J.H., He, W.M., Zhao, C.Y., Zhou, B. *et al.* (2024) Soil biodiversity and exotic plant invasions: Progress and perspectives. *Biodiversity Science* 32(11), 24243. DOI: 10.17520/biods.2024243.

Zhou, J., Ju, R., Li, B. and Wu, J. (2017) Responses of soil biota and nitrogen availability to an invasive plant under aboveground herbivory. *Plant and Soil* 415(1–2), 479–491. DOI: 10.1007/s11104-017-3179-z.

2 Soil Biodiversity and Its Role in Ecosystem Resilience

Mohammad Danish[1*], Mohammad Shahid[2], S. Maqbool Ahmed[1] and Andrea Čerevková[3]

[1]Section of Botany, School of Sciences, Maulana Azad National Urdu University, Hyderabad, India; [2]Marwadi University Research Centre, Department of Agriculture, Faculty of Science, Marwadi University, Rajkot, Gujrat, India; [3]Institute of Parasitology, Slovak Academy of Sciences, Košice, Slovakia

Introduction

The word biodiversity is basically a recapitulation of 'biological diversity', and the idea by itself is fairly straightforward: biodiversity is the entire extent of all living variation from genes to ecosystems (Sathvara *et al.*, 2021). It goes beyond the mere presence of species and includes their abundance, distribution and ecological roles within ecosystems. Biodiversity is one of the main factors contributing to the stability of natural systems and provides ecosystem services that include food supply, climate regulation, nutrient cycling and the aesthetics or other cultural values (Cardinale *et al.*, 2012). The term soil biodiversity refers to the living organisms below the ground that are different from the aboveground ones. Soil biodiversity represents the richness and the variety of functional organisms that live in the soil environment, stretching from the extremely small bacteria, fungi, archaea and protozoa to bigger mesofauna and macrofauna like mites, nematodes, termites and earthworms (FAO, 2020). These organisms interact with each other and with plants to regulate essential natural processes, like the breakdown of organic matter, nutrient cycling, soil aggregation and water retention (Cardinale *et al.*, 2013).

Soil is one of the most biologically diverse habitats on the planet. It is predicted that 1 g of soil may contain billions of microbial cells and thousands of microbial species (van der Putten *et al.*, 2023). In addition, most terrestrial life forms have either a direct or an indirect relationship with soil organisms, making soils the silent engine of the natural cycle. Despite the importance of soils, soil biodiversity is still poorly documented and endangered by the effects of intensive agriculture, land-use change and pollution (Key *et al.*, 2016). Hence, from the soil perspective, biodiversity is less about numbers of species and more about their functional roles and the interactions that maintain ecosystem productivity, resistance and stability.

Ecosystem Resilience and Its Importance in the Face of Disturbance

Ecosystem resilience refers to the ability of a natural system to absorb disturbances while maintaining its original structure, procedures and functions, as well as being able to reorganize itself

*Corresponding author: danish.botanica@gmail.com

DOI: 10.1079/9781836991298.0002

to continue providing the basic ecosystem services (Beck *et al.*, 2025). In ecological terms, resilience refers to both resistance (the feature of being able to resist change) and recovery (the capability to regain the initial state after disturbance). In recent times, ecosystems have become increasingly vulnerable to various sources of stress; as a result, the concept of resilience is becoming more and more important. The necessity for ecosystem resilience is illustrated by the following four major global issues (Collier, 2015).

1. Climate change: rising temperature, extended periods of drought and changed patterns of precipitation have significant effect on soil processes and species diversity. For instance, the activity of soil microbes and the carbon cycle are highly sensitive to the changes in temperature (Easton-Gomez *et al.*, 2022).
2. Land degradation: overgrazing, deforestation and unsustainable agricultural activities have led to soil erosion, a decline in soil organic matter and biodiversity loss. These are the main factors that in turn reduce an ecosystem's ability to recover (Lal, 2015).
3. Pollution: the excessive use of pesticides, herbicides and accumulation of heavy metals in the soil has altered soil microbial populations. Beneficial microorganisms have been suppressed, leading to a decrease in the functional diversity and resilience of the soil community (Gogoi *et al.*, 2024).
4. Agricultural intensification: the practice of monoculture, the overuse of fertilizers and the adoption of mechanization have led to the simplification of soil ecosystems making them more vulnerable to pests, diseases and the variability of climate (Tsiafouli *et al.*, 2015).

Resilient ecosystems are the shields that protect us against those disturbances as they maintain nutrient cycling, primary productivity and biological regulation. For instance, a diverse soil microbial community can adapt more effectively to stressful conditions, ensuring that essential functions continue even if some organisms decline (Karsai *et al.*, 2020). Therefore, the need for resilience in nature is not only a concept used for the sake of it but a real demand for the continuation of agricultural productivity, food security and climate adaptation measures.

Interrelationship between soil biodiversity and ecosystem functioning

The relationship between soil biodiversity and ecosystem functioning is context-dependent because environmental conditions vary (Eisenhauer *et al.*, 2019) or might change based on the ecosystem functions being studied. For example, Wagg *et al.* (2021) demonstrated that the efficiency of an ecosystem function such as carbon sequestration or phosphorus leaching can have differing correlations with the complexity of soil biodiversity, ranging from positive linear to negative exponential patterns. When multiple ecological functions were considered together, Wagg *et al.* (2021) found an overall positive linear association between soil biodiversity and ecosystem multifunctionality. These functions included ecosystem provisioning (food, fibre), ecosystem regulation (climate regulation, pest control), ecosystem support (nutrient cycling, soil formation) and cultural services (heritage, education) (Wagg *et al.*, 2014).

Several mechanisms illustrate how soil biodiversity contributes to ecosystem functioning, as follows.

- Nutrient cycling: microbes break down dead material, thus releasing nitrogen, phosphorus and potassium that plants can absorb. Besides, fungal mycorrhizae allow plants to get more nutrients by increasing root surface area (van der Heijden *et al.*, 2008).
- Soil structure and water regulation: the activities of earthworms and arthropods create pore spaces that improve aeration and infiltration. Organic matter, calcium compounds and fungal hyphae help stabilize soil aggregates (Six *et al.*, 2006).
- Carbon sequestration: soil biodiversity influences the balance between carbon storage and carbon emissions, an imbalance of which contributes to global heating. Diverse microbial communities play a vital role in the regulation of decomposition rates and carbon stabilization (Bardgett and van der Putten, 2014).

- Pest and disease suppression: the main contributors of microbial life (e.g. *Trichoderma* spp. or plant growth-promoting rhizobacteria; PGPR) suppress soil-borne pathogens through mechanisms like competition, antibiosis and activation of plant defence systems (Berendsen *et al.*, 2012).

Essentially, the stability and sustainability of ecosystem services depends on the interactions between the soil communities. Nevertheless, biomass may be a major factor in the provision of ecosystem functions, particularly when species are not comparable in size or traits. Soil ecologists frequently consider microbial biomass to be a dimension of biodiversity independent of species richness, because the drivers of microbial biomass differ from the drivers of specific ecosystem functions, such as carbon and nitrogen cycling (Walker *et al.*, 2018). Still, biomass can be an important factor in the delivery of ecosystem services where functional redundancy means that several species are able to act in a similar manner; therefore, the disappearance of a single species does not cause a drastic change in functions (Loreau *et al.*, 2001). At the same time, diversification of ecosystems would result in the more complete utilization of resources and therefore contribute to the stability and resilience of the biosphere (Fig. 2.1).

Soil Biodiversity: Components and Dimensions

Microorganisms: bacteria, archaea and fungi (mycorrhizae, saprophytes and pathogens)

Soil microorganisms are the main components of the largest and most diverse community of organisms on Earth and they play a fundamental role in the ecosystem functions. The major groups among them are bacteria, archaea and fungi.

Bacteria can be found everywhere in soils, and their numbers are commonly between 10^7 and 10^9 cells per 1 g of soil. They form the basis for the most necessary processes in soil, such as the decomposition of organic matter, the mineralization of nutrients, the symbiotic introduction of nitrogen in plants and the suppression of pathogens carried by the soil (Khan *et al.*, 2025). PGPR, for instance, not only facilitate nutrient absorption but also stimulate phytohormone production and activate systemic resistance in plants (Danish *et al.*, 2025).

Archaea, although less known, make a significant contribution to soil processes, mainly in the nitrogen cycle. Research has shown that ammonia-oxidizing archaea often play a dominant role in nitrification, and in some cases their activity can be even more than that of bacteria (Pester *et al.*, 2012). Besides, archaea are capable

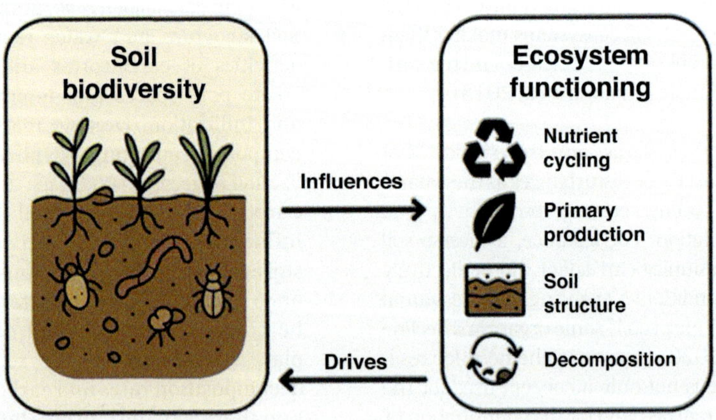

Fig. 2.1. The interrelationship between soil biodiversity and ecosystem functioning.

of surviving in extreme soil conditions and thus contribute to the stability of ecosystems under stress.

Fungi represent another major microbial group in soil. Saprophytic fungi decompose complex organic polymers such as lignin and cellulose, and thus they also recycle carbon and nutrients (Peay *et al.*, 2016). Mycorrhizal fungi, both arbuscular and ectomycorrhizal, are mutually beneficial associations with plant roots, through which they gain nutrients and water and at the same time provide protection against pathogens (Wahab *et al.*, 2023). Conversely, pathogenic fungi like *Fusarium*, *Rhizoctonia* and *Verticillium* negatively affect plant productivity; however, their populations are naturally regulated through antagonistic interactions within diverse microbial communities (Mohanan *et al.*, 2025). Therefore, microorganisms act as the 'engineers' that are responsible for soil processes, directly connecting biodiversity both above and below ground and determining the resilience and functionality of soil ecosystems.

Soil fauna: protozoa, nematodes, mites, Collembola, earthworms, termites and ants

Soils house not only single-celled organisms but also a diverse array of animals that, through feeding, grazing and the physical alteration of the soil, regulate energy flow and nutrient cycling (Fig. 2.2). Protozoa are eukaryotic organisms made up of a single cell that primary feed on bacteria, thus controlling microbial community dynamics and nutrient turnover (Cox, 2024). Along with the 'microbial loop', their activity invigorates nitrogen provision to plants. Nematodes, which are very numerous and highly diverse, live at different trophic levels depending on whether they are bacteria feeders, fungal feeders, plant parasites or predators (Ramteke *et al.*, 2024). For instance, plant-parasitic nematodes like *Meloidogyne* spp. are a major factor in agricultural yield loss, whereas the free-living nematodes play beneficial roles in nutrient cycling and maintaining soil food web stability (Ramteke *et al.*, 2024). Mites and Collembola are small soil-dwelling animals that contribute to the decomposition of organic matter, an activity that enhances microbial life and supports soil food web stability (Sanjuán *et al.*, 2022). According to Sanjuán *et al.* (2022), earthworms, referred to as ecosystem engineers, through their actions of bioturbation and casting, play a crucial role in soil aeration, aggregation and fertility. Their presence is closely associated with productivity of soils. If you consider termites and ants, you will discover that they are also key players. In tropical regions, termites are

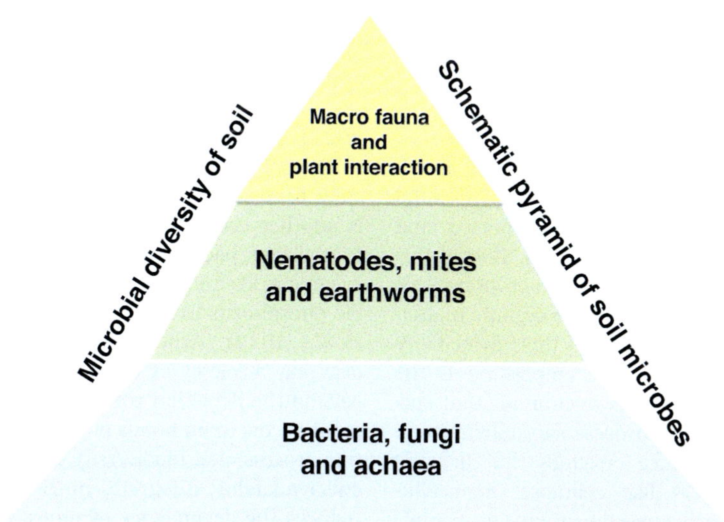

Fig. 2.2. Pyramidal structure of soil biodiversity.

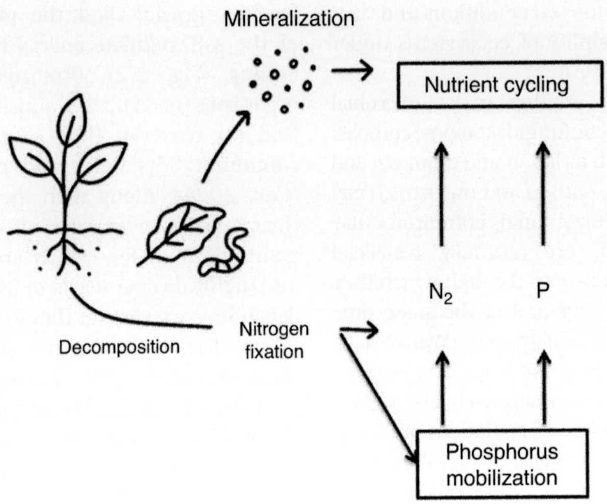

Fig. 2.3. Nutrient cycling in soil.

involved in plant decomposition and promote nutrient recycling, whereas ants impact soil porosity and seed dispersal, which assists with nutrient flow (Crotty, 2021). As a consequence, the regulation of soil fauna functions enables social interactions between microbes, plants and abiotic environment, therefore making them indispensable in ecosystem balance.

Functional Roles of Soil Biodiversity

Nutrient cycling: decomposition, mineralization, nitrogen fixation and phosphorus mobilization

Soil biodiversity is a major driver of nutrient cycling, the processes by which essential nutrients are transformed, transported and made available to plants (Fig. 2.3). Most of this cycling is carried out by the microorganisms and soil animals that consume organic matter, release nutrients and control their availability (Basiliko *et al.*, 2021). Decomposition is the decay of the dead plants or animals that surround us. During this process, saprophytic fungi and bacteria produce enzymes that degrade complex polymers like cellulose, hemicelluloses and lignin, converting them into simple compounds (McKee and Inman, 2019). This activity leads to nutrient release and humus formation, an important part of soil organic matter. Mineralization follows decomposition, during which microorganisms convert organic nutrients into inorganic forms that plants can absorb (Kociánová, 2023). For example, proteins are broken down into amino acids and, subsequently, ammonium (NH_4^+) and nitrate (NO_3^-) are produced, which plants can then take up (Zayed *et al.*, 2023). Another significant process is nitrogen fixation, carried out by nitrogen-fixing bacteria and archaea. Symbiotic nitrogen-fixers such as *Rhizobium* (in legumes), and free-living ones, like *Azotobacter*, convert atmospheric nitrogen (N_2) into ammonia (NH_3), a key contributor to soil fertility (Sepp *et al.*, 2023). Besides that, cyanobacteria in paddy soils represent an important source of biologically fixed nitrogen. Phosphorus mobilization is another crucial function, when phosphate-solubilizing bacteria and fungi that produce organic acids and phosphatases release insoluble phosphorus from mineral complexes (Song *et al.*, 2022). Arbuscular mycorrhizal fungi also play a role by extending root surface area beyond the depletion zone, thereby enhancing plant access to immobile phosphorus. Through this process, soil biodiversity maintains nutrient availability, integrates nutrient cycles and reduces the dependence of crops on chemical fertilizers, thereby supporting ecosystem sustainability and resilience (Basiliko *et al.*, 2021).

The role of microorganisms in soil formation: aggregation, porosity, aeration and water infiltration

Soil organisms play a crucial role in creating and maintaining soil structure, which is the main factor affecting water-holding capacity, air supply and root growth, in addition to influencing soil properties. Aggregation results from microbial exudates such as extracellular polysaccharides and fungal hyphae that act as 'biological glue' that binds together soil particles into stable aggregates (Al-Maliki and Ebreesum, 2020). Mycorrhizal fungi are important in stabilizing soil aggregates by binding soil particles in their hyphal networks. Porosity depends on mesofauna and macrofauna, such as mites, Collembola and earthworms, which make micropores and macropores while moving and feeding (Hartmann and Six, 2023). Aeration is also enhanced by larger burrowing animals (e.g. earthworms, termites) whose tunnels facilitate gas exchange and reduce anaerobic zones, thus promoting microbial activity. Soil biodiversity strongly influences water infiltration and retention (Smith *et al.*, 2021). With activity such as earthworm burrowing, the building of termite mounds and the interlacing of fungal networks, soil becomes more permeable to water, less susceptible to surface water runoff and more able to retain moisture (Jouquet *et al.*, 2011). Therefore, the biological inhabitants of the soil community are the main forces that bring about physical changes that make the soil more favourable for plant growth and microbial functioning.

Plant–soil interactions: mycorrhizal symbiosis, the rhizosphere microbiome and plant growth-promoting rhizobacteria

Soil biodiversity is a direct mediator of plant–soil interactions, helping crop production and natural vegetation growth. Mycorrhizal symbiosis is among the most explored of mutualisms. Arbuscular mycorrhizal fungi penetrate plant roots and spread their filaments into the surrounding soil, thus increasing the effective surface area of roots for nutrient and water absorption. In return, plants supply the fungi with carbon from photosynthesis (Ferreira *et al.*, 2021). The rhizosphere microbiome, composed of bacteria, fungi and archaea living very close to the roots, is a plant–soil interface that is dynamic and can be changed by the host (Basiliko *et al.*, 2021). Root exudates (sugars, amino acids and secondary metabolites) select and support beneficial microbes that provide the source of nutrient availability, stress tolerance and plant immunity. PGPR, for example *Pseudomonas* and *Bacillus* spp., stimulate plant growth by producing phytohormones (e.g. auxins, cytokinins) and by facilitating nitrogen fixation, phosphorus solubilization and induction of pathogen resistance (Danish *et al.*, 2025). These interactions highlight the way soil biodiversity supports agricultural sustainability, reducing dependence on chemical fertilizers and pesticides.

Pest and disease suppression: antagonistic fungi, bacteriophages and predatory nematodes

Soil biodiversity provides a natural biological control mechanism as high biodiversity means that pests and diseases are less likely to occur, thus reducing the need for synthetic pesticides. Antagonistic fungi, like *Trichoderma* and *Gliocladium*, suppress pathogens through mechanisms such as mycoparasitism, antibiosis and competition for nutrients (El Enshasy *et al.*, 2020). Non-pathogenic *Fusarium* strains may take the place of pathogenic ones, contributing to disease suppression. Bacteriophages, viruses that attack bacteria, are the regulators of soil bacterial populations and may eliminate pathogenic bacteria in the rhizosphere (Srinivasiah *et al.*, 2008). Predatory nematodes feed on plant-parasitic nematodes (e.g. *Meloidogyne* spp.), thereby reducing crop damage (Wang *et al.*, 2015) Protozoa also contribute by feeding on bacteria, helping to maintain microbial balance. The presence of diverse organisms in the soil thus ensures a variety of defence layers, making soils more resilient and less susceptible to pathogen epidemics (Eshboev *et al.*, 2024).

Carbon sequestration and climate regulation: the role of soil organisms in carbon storage and greenhouse gas emissions

Soil biodiversity is extremely important for carbon sequestration and climate regulation as it supports both the storage of organic carbon and the regulation of green gas emissions. Soil contains more carbon than the atmosphere and vegetation combined and soil microorganisms play a vital role in carbon storage (Basile-Doelsch *et al.*, 2020). Fungi and bacteria are the decomposers of organic residues, some of which are converted into humus or mineral-associated organic matter. Through their activities earthworms and termites move organic material into the deeper soil layers, enhancing long-term carbon storage. Soil organisms also help regulate greenhouse gas fluxes (Dignac *et al.*, 2017). These include:

- methanotrophic bacteria that consume methane (CH_4), reducing its emission;
- denitrifying bacteria that regulate nitrous oxide (N_2O) emissions; however, over-fertilization can alter their activity, shifting them from consuming to releasing greenhouse gases;
- fungal and microbial activities significantly affect CO_2 fluxes through decomposition processes (Bardgett and van der Putten, 2014).

Therefore, the management of soil biodiversity not only maintains agricultural productivity but also serves as a mitigation tool for climate change (Fig. 2.4).

Soil Biodiversity and Ecosystem Resilience

Functional redundancy (buffering capacity)

Functional redundancy is the concept of multiple species existing in one ecosystem and jointly fulfilling the same roles in the ecosystem. Soil ecosystems have extremely high redundancy, due to their immense microbial and faunal diversity. For example, numerous bacteria are able to fix nitrogen,

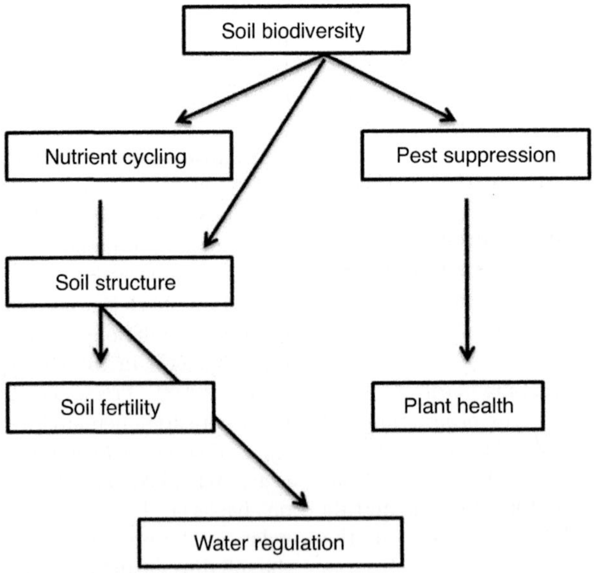

Fig. 2.4. The contribution of soils to ecosystem resilience.

whereas several fungal groups can decompose cellulose or lignin (Song *et al.*, 2022). As a result of this redundancy, a buffering effect is created: when certain taxa disappear or are affected by disturbances, such as pesticide application, drought or pollution, other organisms can continue to sustain essential ecosystem functions (Bommarco, 2024). This buffering capacity is critical for soil processes, including nutrient cycling, carbon sequestration and pathogen suppression. For instance, earthworms, termites and enchytraeids all contribute to soil aggregation, which means that the function persists even if one group declines (Bardgett and van der Putten, 2014). On the other hand, various bacterial and fungal communities can carry out nitrification or phosphorus solubilization, ensuring a continuous supply of nutrients to plants (Banerjee *et al.*, 2019). Hence, soil biodiversity acts as a safety net that reduces the risk of functional collapse within ecosystems.

Recovery after disturbances (fire, drought, floods, pollution and land-use change)

Recovery is one of the most important attributes of ecosystem resilience, and diverse soil organisms play a central role in the process of restoring ecosystem functions after disruptions. For example, after fire, fungi and bacteria in the soils rapidly inhabit burned areas, resulting in the processes of organic matter decomposition and nutrient cycling, both of which are essential for vegetation growth (Certini, 2014). Similarly, earthworms and arthropods not only provide but also facilitate the recovery of the structure of the soil as they improve porosity and aggregation as a result of the recent compaction or erosion events (Lavelle *et al.*, 2006). Drought can temporarily suppress soil respiration and nutrient turnover, which can be recovered more quickly when a diverse microbial community is present (De Silva *et al.*, 2025). Floods often lead to hypoxic soils; however, functionally diverse microbial assemblages such as denitrifiers and methanogens can re-establish biogeochemical cycling once conditions stabilize (Nguyen *et al.*, 2018). Under different anthropogenic disturbances such as pesticide contamination or heavy metal pollution, soil biodiversity provides natural bioremediation through those organisms, which are able to degrade or immobilize pollutants (Singh, 2008). Land-use change such as deforestation or conversion to intensive agriculture ranks first among all kinds of pressing issues with soil ecosystems. Nevertheless, reforestation and the application of organic fertilizers often promote the colonization by microbial and faunal communities, thereby accelerating soil fertility recovery (Morriën *et al.*, 2017). Taken together these examples demonstrate that high soil biodiversity is indispensable for the recovery of the soil ecosystem following disturbance, which is a defining feature of resilient ecosystems.

Threats to Soil Biodiversity

Intensive agriculture, monocropping and pesticide/fertilizer overuse

Intensive agriculture has greatly increased world food production; however, it often leads to the destruction of soil biodiversity. Monocropping reduces plant diversity and soil microbial and faunal communities become less diverse because there are fewer root exudates and less litter available to support diverse organisms (Gossner *et al.*, 2016). Besides that, the continuous use of chemical fertilizers alters soil pH, nutrient availability and microbial community composition, which usually leads to the growth of a few fast-growing microbial taxa at the expense of beneficial groups such as mycorrhizal fungi (Geisseler and Scow, 2014). In the same way, the overuse of pesticides also harms non-target organisms. For instance, fungicides can suppress both harmful and beneficial soil fungi, while insecticides can reduce the number of soil arthropods and nematodes that are essential for nutrient cycling (Silva *et al.*, 2019). This decrease in the number of organisms in the ecosystem causes the soil to have lower resistance and greater susceptibility to pests, diseases and climate variability. The importance of sustainable practices such as crop rotation, integrated pest management and organic amendments as ways to restore soil biodiversity in agricultural systems is the topic of many discussions nowadays (Banerjee *et al.*, 2019).

Soil erosion and degradation

Soil erosion and degradation are among the most severe threats to soil biodiversity globally. Both water and wind erosion remove topsoil, the most biologically active soil layer, where microbial biomass and soil fauna are concentrated (Yang *et al.*, 2024). The removal of topsoil decreases soil organic matter, which is the main source of energy for microbial communities. Soil degradation also occurs progressively through salinization, acidification and compaction, all of which lead to a decline in the diversity and activity of soil organisms (Yang *et al.*, 2024). For instance, compactions reduce pore space and therefore limit oxygen diffusion and water infiltration, making it very difficult for most soil fauna to survive (Bardgett and van der Putten, 2014). When soils are degraded, the functional capacity of microbial communities, such as decomposition and nutrient cycling, become weakened, leading to productivity loss and greater ecosystem instability. Among the most effective measures for enhancing soil biodiversity in degraded landscapes are conservation tillage, cover cropping and erosion control structures (Bai *et al.*, 2018).

Urbanization and land-use change

Urban expansion and land-use change have been recognized as two of the primary causes of declines in soil biodiversity. The replacement of natural ecosystems with urban landscapes or agricultural fields causes not only habitat fragmentation but also the creation of impervious surfaces and a significant change in soil microclimate (Setälä *et al.*, 2014). Soils around cities are in most cases compacted, polluted and deprived of organic inputs, and as a result microbial diversity is low and populations of soil fauna are reduced. The cutting down of forests and changing of grasslands into croplands also have negative impacts on soil communities, as these activities reduce plant–soil interactions, which in turn affects belowground diversity (Tsiafouli *et al.*, 2015). The loss of soil biodiversity due to land-use change is, on the one hand, a problem for nature conservation and, on the other hand, a cause of the breakdown of ecosystem services such as carbon sequestration, water regulation and nutrient cycling. Recovery of soil biodiversity in these human-disturbed areas after restoration, urban greening and amendments to organic matter has been demonstrated to be in its initial stages (Morriën *et al.*, 2017).

Climate change (warming, drought and altered precipitation)

Climate change is increasingly recognized as a major threat to soil biodiversity. Increasing global temperatures alter soil microbial activity and community composition, often favouring heat-loving and drought-tolerant taxa (Kaviraj *et al.*, 2024). Long droughts reduce microbial biomass and disrupt soil faunal populations, while changed precipitation patterns result in alternating flooding and drying, which stresses soil organisms and reduces their functional contributions (de Vries *et al.*, 2018). For instance, research indicates that bacterial networks are more sensitive to drought compared with fungal networks (de Vries *et al.*, 2018). Consequently, functional imbalances in decomposition and nutrient cycling may occur. In addition, changes in climate will eventually shift vegetation, and affect the quality and quantity of root exudates and litter inputs, which indirectly alter soil biodiversity (Classen *et al.*, 2015). The combined effects of warming, drought and precipitation variability threaten soil-based ecosystem services, thus making climate adaptation strategies such as organic carbon enrichment and agroecological practices for conserving soil biodiversity all the more important.

Pollution (heavy metals, plastics and nanomaterials)

Pollution is a growing threat to soil biodiversity. Heavy metals such as cadmium, lead and arsenic enter soil through industrial emissions, mining and the application of sewage

Table 2.1. Comparison of healthy and degraded soil systems.

Healthy soil system	Degraded soil system
Healthy plants (better growth and productivity)	Stressed plant (low yield, susceptible)
Abundant soil fauna (earthworms, nematodes)	Reduced fauna (poor soil structure)
Rich microbial diversity	Low microbial diversity

sludge. These metals are highly toxic to soil microorganisms and small animals, often reducing microbial biomass and enzymatic activity (Singh, 2008). Pesticide residues and hydrocarbons also change the structures of the microbial community, decreasing soil enzymatic diversity and hindering nutrient recycling. Over time, additional substances have become increasingly soil polluting. For example, microplastics disrupt soil structure, alter microbial habitats and are ingested by soil animals such as earthworms, reducing their growth and reproduction (Rillig *et al.*, 2019). On the other hand, nanoparticles of silver and titanium dioxide are extremely small, high reactive and antibacterial, killing soil bacteria, both beneficial and otherwise, and affecting soil biogeochemical processes (Kah *et al.*, 2019). Where each of these pollutants may have individual effects, their combined impact often leads to long-lasting losses in both soil biodiversity and its functions (Table 2.1). Consequently, the use of pollutants should be radically reduced and methods such as bioremediation and phytoremediation should be applied to restore polluted soils (Singh, 2008).

Emerging Frontiers in Soil Biodiversity Research

Soil biodiversity, encompassing microorganisms, fauna and related genetic diversity, is essential to the health of ecosystems, agricultural productivity and environmental sustainability. New methods for comprehending, tracking and utilizing subsurface life for sustainable ecosystem management are being shaped by recent developments and new frontiers in soil biodiversity research.

Soil microbiome sequencing and metagenomics

By enabling thorough characterization of soil microbial communities without the need for culturing, the development of high-throughput sequencing technologies has completely changed the study of soil microbial communities (Lahlali *et al.*, 2021). The taxonomic variety and functional potential of bacteria, archaea, fungi and viruses in intricate soil ecosystems are revealed by metagenomics, which sequences all of the DNA found in soil samples (Rajguru *et al.*, 2024). It enables the identification of important functional genes associated with pathogen suppression, stress resilience and nutrient cycling (Liu *et al.*, 2024). By using metagenomic techniques, scientists can monitor changes in the metabolic pathways and composition of microbial communities in response to land management techniques or environmental changes (Shah *et al.*, 2022; Kaviraj *et al.*, 2024). Integrating metagenomics with meta-transcriptomics and metabolomics further clarifies functional activity under specific environmental conditions (Prayogo *et al.*, 2020; Panicker, 2024).

Soil biodiversity indicators for ecosystem health monitoring

Accurate indicators of soil biodiversity are crucial for monitoring ecosystem sustainability. Biological indicators, such as microbial biomass, enzyme activity, microbial respiration and microbial community structure, allow rapid responses to environmental disturbances and changes in land or soil management and are often more sensitive than results from monitoring physical and chemical soil parameters (Muscolo *et al.*, 2015). Key biological indicators include respiration rates, soil biodiversity indices based on species diversity or functional groups, microbial carbon and nitrogen biomass, and enzyme activities related to nutrient cycling (e.g. dehydrogenase, phosphatase) (Nikitin *et al.*, 2022). Together, these markers reflect soil functions such as the decomposition of organic matter, nutrient cycling, soil aggregation and disease suppression (Usharani *et al.*, 2019). To simplify soil health assessment, composite indices have been developed that combine multiple microbial and biochemical indicators in a single measure. These indices help evaluate soil contamination or degradation, the impact of agricultural practices and the progress of ecosystem restoration, supporting sustainable land management and policy decisions (Pandit *et al.*, 2020; Fierer *et al.*, 2021).

Linking soil biodiversity with ecosystem services valuation

Food production, water purification, nutrient cycling, disease control and climate regulation through carbon storage are just a few of the ecosystem services that soil biodiversity supports and are essential to human well-being (Telo da Gama, 2023). Valuation techniques that emphasize the economic and ecological significance of soil biodiversity are made possible by quantifying its contribution to these services. New studies enable ecosystem service models that include biodiversity metrics by connecting particular soil organisms and microbial activities to ecosystem services (Barrios, 2007). For example, decomposer communities regulate soil organic matter dynamics and carbon sequestration, while microbial processes maintain soil fertility, resulting in healthier crops and enhancing crop yields (Condron *et al.*, 2010). By assigning a monetary value to soil biological functions, integrating biodiversity data into ecosystem service valuation aids in setting priorities for land-use planning and soil conservation initiatives (Di Minin *et al.*, 2017). Sustainable intensification strategies are guided by the trade-offs and synergies between agricultural productivity and biodiversity conservation.

The role of artificial intelligence and big data in soil biodiversity studies

Artificial intelligence (AI) and big data analytics are revolutionizing soil biodiversity research by making it possible to integrate large datasets from sequencing, soil sensors, remote sensing and environmental monitoring (Alotaibi and Nassif, 2024). Machine learning algorithms are involved to identified patterns, predict soil health and uncover new drivers of biodiversity (Chang, 2023). Barbosa *et al.* (2025) report that AI models not only classify soil types but also predict microbial community responses with respect to land-use and climate change and identify new bioindicators through multidimensional data analyses. The use of AI-enabled robotic soil sampling and automated image analysis makes large-scale biodiversity monitoring more accurate and efficient (Neupane *et al.*, 2025). Furthermore, big data integration supports meta-analyses in international soil biodiversity programmes (Russell, 2025) and enables decision-support tools for real-time soil health diagnostics and precision interventions in agriculture (Ahmad *et al.*, 2025).

Biotechnological applications: engineered microbes, biofertilizers and biocontrol agents

Utilizing soil microbes for biotechnology provides sustainable solution for crop growth, soil fertility and disease control while reducing chemical inputs (Singh *et al.*, 2024). Engineered microbes and naturally occurring beneficial rhizobacteria serve as biofertilizers by fixing atmospheric nitrogen, solubilizing phosphates, producing growth-promoting hormones or enhancing nutrient uptake (Ali *et al.*, 2020). Microbial biocontrol agents suppress soil-borne pathogens through competition, antibiosis or induced systemic resistance in plants, providing eco-friendly alternatives to chemical pesticides (Alqahtani, 2025). Advances in microbial genomics and synthetic biology enable the design of enhanced strains tailor-made for specific soil or crop needs (Nazir *et al.*, 2024). Emerging bio-inoculants stimulate plant growth and resilience under a biotic stresses such as drought, salinity and heavy metal contamination, promoting sustainable agriculture in challenging environments (Kumar *et al.*, 2025). The development and deployment of microbial consortia formulations capitalize on synergistic interactions between diverse microorganisms for improved efficacy (Shahid *et al.*, 2023). Such biotechnological applications contribute to circular agriculture by recycling nutrients and maintaining soil biodiversity, thus supporting long-term soil ecosystem sustainability. These emerging frontiers collectively advance the understanding and application of soil biodiversity knowledge in ecosystem management. Integrating molecular tools, ecological indicators, AI and biotechnological innovations paves the way towards the sustainable soil stewardship and resilient agricultural systems that are vital for global food security and environmental health.

Conclusion

Soil biodiversity, although often overlooked, is fundamental to ecosystem function. This chapter emphasizes the diversity of soil organisms from microbes and mesofauna to macrofauna that together drive nutrient cycling, the formation of soil structure, plant–soil interactions, pest suppression and carbon sequestration. These processes are the foundation of ecosystem services that are the direct source of food, agriculture, forestry and climate regulation. Soil biodiversity provides ecosystem resilience through functional redundancy and biological insurance against environmental shocks. Abundant soil communities allow ecosystems to resist, adapt and recover from shocks such as climate extremes, land degradation or pollution. It is the soil that provides stability to food production in the future and at the same time ecological integrity is being maintained. However, soil biodiversity is still suffering from different threats, such as the case with intensive agriculture, urbanization, climate change and pollution. The initiation of monoculture, overuse of chemicals and unsustainable land management have caused these degradations and also the loss of biological diversity in soil, and the disease-causing organisms in these soils have compromised the natural resilience of ecosystems. Emerging stressors such as microplastics and engineered metals exacerbate these declines.

Integrated research, conservation and policy interventions are essential, emphasizing soil biodiversity as one of the main keys to sustainability. Agroecosystem practices like crop diversification, organic amendments, reduced chemical inputs and conservation tillage can restore ecosystem services and at the same time mitigate human-induced global changes. By prioritizing soil health in land-use planning, fostering ecological awareness and aligning rural sectors in concert with nature and total ecological sustainability, agricultural policies of the future will be edging towards UN Food and Agriculture Organization-recommended practices. Protecting soil biodiversity is a matter of socio-economics and is not just an ecosystem issue. The invisible workforce beneath our soil, the survival and sustainability of ecosystems and, consequently, the standard of human life cannot be assured without protecting the soil.

References

Ahmad, S., Kaushik, R., Ghatuary, R., Kotiyal, A., Jarial, S. *et al.* (2025) Utilizing IoT and AI for soil health monitoring and enhancement in sustainable agriculture. In: Azrour, M., Mabrouki, J. and Ahmad, S. (eds) *IoT and Advanced Intelligence Computation for Smart Agriculture*. CRC Press, Boca Raton, FL, pp. 110–125. DOI: 10.1201/9781003527664-7.

Ali, R., Zulaykha, K.D. and Sajjad, N. (2020) Genetically modified microbes as biofertilizers. In: Bhat, R.A. and Hakeem, K.R. (eds) *Bioremediation and Biotechnology, Vol 4: Techniques for Noxious Substances Remediation*. Springer International Publishing, Cham, pp. 275–293. DOI: 10.1007/978-3-030-48690-7_13.

Al-Maliki, S. and Ebreesum, H. (2020) Changes in soil carbon mineralization, soil microbes, roots density and soil structure following the application of the arbuscular mycorrhizal fungi and green algae in the arid saline soil. *Rhizosphere* 14, 100203. DOI: 10.1016/j.rhisph.2020.100203.

Alotaibi, E. and Nassif, N. (2024) Artificial intelligence in environmental monitoring: In-depth analysis. *Discover Artificial Intelligence* 4(1), 84. DOI: 10.1007/s44163-024-00198-1.

Alqahtani, F.S. (2025) The utilization of microorganisms for biological control of soil-borne plant pathogens: A sustainable strategy for managing plant diseases - A comprehensive review. *Journal of Plant Pathology* 107, 1–25. DOI: 10.1007/s42161-025-01984-1.

Bai, Z., Caspari, T., Gonzalez, M.R., Batjes, N.H., Mäder, P. *et al.* (2018) Effects of agricultural management practices on soil quality: A review of long-term experiments for Europe and China. *Agriculture, Ecosystems & Environment* 265, 1–7. DOI: 10.1016/j.agee.2018.05.028.

Banerjee, S., Walder, F., Büchi, L., Meyer, M., Held, A.Y. *et al.* (2019) Agricultural intensification reduces microbial network complexity and the abundance of keystone taxa in roots. *The ISME Journal* 13(7), 1722–1736. DOI: 10.1038/s41396-019-0383-2.

Barbosa, M.I., Silva, G., Ribeiro, P., Vieira, E., Perrotta, A. *et al.* (2025) Unraveling the microbiome–environmental change nexus to contribute to a more sustainable world: A comprehensive review of artificial intelligence approaches. *Sustainability* 17(16), 7209. DOI: 10.3390/su17167209.

Bardgett, R.D. and van der Putten, W.H. (2014) Belowground biodiversity and ecosystem functioning. *Nature* 515(7528), 505–511. DOI: 10.1038/nature13855.

Barrios, E. (2007) Soil biota, ecosystem services and land productivity. *Ecological Economics* 64(2), 269–285. DOI: 10.1016/j.ecolecon.2007.03.004.

Basile-Doelsch, I., Balesdent, J. and Pellerin, S. (2020) Reviews and syntheses: The mechanisms underlying carbon storage in soil. *Biogeosciences* 17(21), 5223–5242. DOI: 10.5194/bg-17-5223-2020.

Basiliko, N., Dunfield, K. and Tenuta, M. (2021) Soil biodiversity and ecology. In: Krzic, M., Walley, F., Diochon, A., Paré, M.C. and Farrel, E.R. (eds) *Digging into Canadian Soils: An Introduction to Soil Science*. Canadian Society of Soil Science, pp. 194–233.

Beck, D., Pal, A., Zaveri, P., Zheng, A., Tiwari, A.K. *et al.* (2025) Ecosystem resilience for climate resilience, strategies and innovations: A case study from developing countries. In: Choudhury, M., Dixit, G. and Majumdar, S. (eds) *Sustainable Synergy: Harnessing Ecosystems for Climate Resilience*. Springer Nature, Cham, pp. 49–64. DOI: 10.1007/978-3-031-77957-2_4.

Berendsen, R.L., Pieterse, C.M.J. and Bakker, P.A.H.M. (2012) The rhizosphere microbiome and plant health. *Trends in Plant Science* 17(8), 478–486. DOI: 10.1016/j.tplants.2012.04.001.

Bommarco, R. (2024) Ecological redesign of crop ecosystems for reliable crop protection. A review. *Agronomy for Sustainable Development* 44(6), 51. DOI: 10.1007/s13593-024-00987-z.

Cardinale, B.J., Duffy, J.E., Gonzalez, A., Hooper, D.U., Perrings, C. *et al.* (2012) Biodiversity loss and its impact on humanity. *Nature* 486(7401), 59–67. DOI: 10.1038/nature11148.

Cardinale, B.J., Gross, K., Fritschie, K., Flombaum, P., Fox, J.W. *et al.* (2013) Biodiversity simultaneously enhances the production and stability of community biomass, but the effects are independent. *Ecology* 94(8), 1697–1707. DOI: 10.1890/12-1334.1.

Certini, G. (2014) Fire as a soil-forming factor. *Ambio* 43(2), 191–195. DOI: 10.1007/s13280-013-0418-2.

Chang, G.J. (2023) Biodiversity estimation by environment drivers using machine/deep learning for ecological management. *Ecological Informatics* 78, 102319. DOI: 10.1016/j.ecoinf.2023.102319.

Classen, A.T., Sundqvist, M.K., Henning, J.A., Newman, G.S., Moore, J.A.M. *et al.* (2015) Direct and indirect effects of climate change on soil microbial and soil microbial-plant interactions: What lies ahead. *Ecosphere* 6(8), 1–21. DOI: 10.1890/ES15-00217.1.

Collier, M.J. (2015) Novel ecosystems and social-ecological resilience. *Landscape Ecology* 30(8), 1363–1369. DOI: 10.1007/s10980-015-0243-z.

Condron, L., Stark, C., O'Callaghan, M., Clinton, P. and Huang, Z. (2010) The role of microbial communities in the formation and decomposition of soil organic matter. In: *Soil Microbiology and Sustainable Crop Production*. Springer Netherlands, Dordrecht, pp. 81–118. DOI: 10.1007/978-90-481-9479-7_4.

Cox, G.R. (2024) Response of micro-eukaryote and slime mould communities to land use change and pine invasion in Aotearoa. Thesis, The University of Canterbury.

Crotty, F. (2021) Assessing soil health by measuring fauna. In: Ottem, W. (ed.) *Advances in Measuring Soil Health*. Burleigh Dodds Science Publishing, pp. 3–24. DOI: 10.19103/AS.2020.0079.03.

Danish, M., Shahid, M., Ibrahim, S.M. and Ahamad, L. (2025) Enhancing pea plant growth, nutrient acquisition, and symbiosis in cobalt-stressed soil using metal-tolerant *Klebsiella* sp. *Journal of Basic Microbiology* 65(6), e70025. DOI: 10.1002/jobm.70025.

De Silva, S., Kariyawasam Hetti Gamage, L. and Thapa, V.R. (2025) Impact of drought on soil microbial communities. *Microorganisms* 13(7), 1625. DOI: 10.3390/microorganisms13071625.

de Vries, F.T., Griffiths, R.I., Bailey, M., Craig, H., Girlanda, M. *et al.* (2018) Soil bacterial networks are less stable under drought than fungal networks. *Nature Communications* 9(1), 3033. DOI: 10.1038/s41467-018-05516-7.

Di Minin, E., Soutullo, A., Bartesaghi, L., Rios, M., Szephegyi, M.N. *et al.* (2017) Integrating biodiversity, ecosystem services and socio-economic data to identify priority areas and landowners for conservation actions at the national scale. *Biological Conservation* 206, 56–64. DOI: 10.1016/j.biocon.2016.11.037.

Dignac, M.-F., Derrien, D., Barré, P., Barot, S., Cécillon, L. *et al.* (2017) Increasing soil carbon storage: Mechanisms, effects of agricultural practices and proxies. A review. *Agronomy for Sustainable Development* 37(2), 14. DOI: 10.1007/s13593-017-0421-2.

Easton-Gomez, S.C., Mouritz, M. and Breadsell, J.K. (2022) Enhancing emotional resilience in the face of climate change adversity: A systematic literature review. *Sustainability* 14(21), 13966. DOI: 10.3390/su142113966.

Eisenhauer, N., Schielzeth, H., Barnes, A.D., Barry, K., Bonn, A. *et al.* (2019) A multitrophic perspective on biodiversity–ecosystem functioning research. In: Eisenhauer, N., Bohan, D.A. and Dumbrell, A.J. (eds) *Advances in Ecological Research*, Vol. 61. Academic Press, pp. 1–54. DOI: 10.1016/bs.aecr.2019.06.001.

El Enshasy, H.A., Ambehabati, K.K., El Baz, A.F., Ramchuran, S., Sayyed, R.Z. *et al.* (2020) Trichoderma: Biocontrol agents for promoting plant growth and soil health. In: Yadav, A.N., Mishra, S., Kour, D., Yadav, N. and A, K. (eds) *Agriculturally Important Fungi for Sustainable Agriculture, Vol. 2: Functional Annotation for Crop Protection*. Springer International Publishing, Cham, pp. 239–259. DOI: 10.1007/978-3-030-48474-3_8.

Eshboev, F., Abdurakhmanov, J., Egamberdieva, D. and Azimova, S. (2024) Soil microbiome bioengineering for improvement of soil health and fertility. In: Parray, J.A. (ed.) *Progress in Soil Microbiome Research*. Springer Nature, Cham, pp. 443–465. DOI: 10.1007/978-3-031-71487-0_18.

FAO (2020) *State of Knowledge of Soil Biodiversity*. Rome: Food and Agriculture Organization of the United Nations.

Ferreira, D.A., da Silva, T.F., Pylro, V.S., Salles, J.F., Andreote, F.D. *et al.* (2021) Soil microbial diversity affects the plant-root colonization by arbuscular mycorrhizal fungi. *Microbial Ecology* 82(1), 100–103. DOI: 10.1007/s00248-020-01502-z.

Fierer, N., Wood, S.A. and Bueno de Mesquita, C.P. (2021) How microbes can, and cannot, be used to assess soil health. *Soil Biology and Biochemistry* 153, 108111. DOI: 10.1016/j.soilbio.2020.108111.

Geisseler, D. and Scow, K.M. (2014) Long-term effects of mineral fertilizers on soil microorganisms – a review. *Soil Biology and Biochemistry* 75, 54–63. DOI: 10.1016/j.soilbio.2014.03.023.

Gogoi, N., Sarma, A., Choudhury, M., Samanta, P., Faizan, M. *et al.* (2024) Soil heavy metal pollution and ecological risk assessment in disturbed and undisturbed soil of Morigaon, Assam. *Discover Environment* 2(1), 13. DOI: 10.1007/s44274-024-00039-z.

Gossner, M.M., Lewinsohn, T.M., Kahl, T., Grassein, F., Boch, S. *et al.* (2016) Land-use intensification causes multitrophic homogenization of grassland communities. *Nature* 540(7632), 266–269. DOI: 10.1038/nature20575.

Hartmann, M. and Six, J. (2023) Soil structure and microbiome functions in agroecosystems. *Nature Reviews Earth & Environment* 4(1), 4–18. DOI: 10.1038/s43017-022-00366-w.

Jouquet, P., Traoré, S., Choosai, C., Hartmann, C. and Bignell, D. (2011) Influence of termites on ecosystem functioning. Ecosystem services provided by termites. *European Journal of Soil Biology* 47(4), 215–222. DOI: 10.1016/j.ejsobi.2011.05.005.

Kah, M., Tufenkji, N. and White, J.C. (2019) Nano-enabled strategies to enhance crop nutrition and protection. *Nature Nanotechnology* 14(6), 532–540. DOI: 10.1038/s41565-019-0439-5.

Karsai, I., Schmickl, T. and Kampis, G. (2020) Understanding ecosystem stability and resilience through mathematical modelling. In: Karsai, I., Schmickl, T. and Kampis, G. (eds) *Resilience and Stability of Ecological and Social Systems*. Springer International Publishing, Cham, pp. 1–17. DOI: 10.1007/978-3-030-54560-4_1.

Kaviraj, M., Singh, M., Chatterjee, S. and Kumar, U. (2024) Metagenomics unveiled: Deciphering microbial responses to climate change. In: Thatoi, H., Mohapatra, S., Das, S.K. and Pradhan, S.K. (eds) *Applied Biotechnology and Bioinformatics: Agriculture, Pharmaceutical Research and Environment*. John Wiley & Sons, pp. 369–394. DOI: 10.1002/9781119896869.

Key, G., Whitfield, M.G., Cooper, J., De Vries, F.T., Collison, M. *et al.* (2016) Knowledge needs, available practices, and future challenges in agricultural soils. *SOIL* 2(4), 511–521. DOI: 10.5194/soil-2-511-2016.

Khan, A., Farid, A., Haq, F., Ullah, A. and Ahmad, I. (2025) Diversity and benefits of microbiomes in soil and environment management. In: Mohd, S., Akhtar, K. and Rehman, H. (eds) *Rhizospheric Microbiomes: The Interconnection Between Plant and Soil Health*. CRC Press, Boca Raton. DOI: 10.1201/9781003637134-1.

Kociánová, P. (2023) Role of microbes in the decomposition in headwater streams. Thesis, Department of Ecology, Univerzita Karlova, Přírodovědecká Fakulta.

Kumar, R., Farda, B., Mignini, A., Djebaili, R., Koolman, L. *et al.* (2025) Microbial solutions in agriculture: Enhancing soil health and resilience through bio-inoculants and bioremediation. *Bacteria* 4(3), 28. DOI: 10.3390/bacteria4030028.

Lahlali, R., Ibrahim, D.S.S., Belabess, Z., Kadir Roni, M.Z., Radouane, N. *et al.* (2021) High-throughput molecular technologies for unraveling the mystery of soil microbial community: Challenges and future prospects. *Heliyon* 7(10), e08142. DOI: 10.1016/j.heliyon.2021.e08142.

Lal, R. (2015) Restoring soil quality to mitigate soil degradation. *Sustainability* 7(5), 5875–5895. DOI: 10.3390/su7055875.

Lavelle, P., Decaëns, T., Aubert, M., Barot, S., Blouin, M. *et al.* (2006) Soil invertebrates and ecosystem services. *European Journal of Soil Biology* 42, S3–S15. DOI: 10.1016/j.ejsobi.2006.10.002.

Liu, L., Gao, Y., Yang, W., Liu, J. and Wang, Z. (2024) Community metagenomics reveals the processes of nutrient cycling regulated by microbial functions in soils with P fertilizer input. *Plant and Soil* 499(1–2), 139–154. DOI: 10.1007/s11104-023-05875-1.

Loreau, M., Naeem, S., Inchausti, P., Bengtsson, J., Grime, J.P. *et al.* (2001) Biodiversity and ecosystem functioning: Current knowledge and future challenges. *Science* 294(5543), 804–808. DOI: 10.1126/science.1064088.

McKee, L.S. and Inman, A.R. (2019) Secreted microbial enzymes for organic compound degradation. In: *Microbes and Enzymes in Soil Health and Bioremediation*. Springer, Singapore, pp. 225–254. DOI: 10.1007/978-981-13-9117-0_10.

Mohanan, A., Nigam, R., Yeliya, P., Daniel, G.R., Devi, N.O. *et al.* (2025) The role of plant microbiomes in suppressing soilborne pathogens: A review. *Journal of Advances in Microbiology* 25(5), 160–178. DOI: 10.9734/jamb/2025/v25i5942.

Morriën, E., Hannula, S.E., Snoek, L.B., Helmsing, N.R., Zweers, H. *et al.* (2017) Soil networks become more connected and take up more carbon as nature restoration progresses. *Nature Communications* 8, 14349. DOI: 10.1038/ncomms14349.

Muscolo, A., Settineri, G. and Attinà, E. (2015) Early warning indicators of changes in soil ecosystem functioning. *Ecological Indicators* 48, 542–549. DOI: 10.1016/j.ecolind.2014.09.017.

Nazir, A., Hussain, F.H.N. and Raza, A. (2024) Advancing microbiota therapeutics: The role of synthetic biology in engineering microbial communities for precision medicine. *Frontiers in Bioengineering and Biotechnology* 12, 1511149. DOI: 10.3389/fbioe.2024.1511149.

Neupane, A., Shahi, T.B., Koech, R., Walsh, K. and Langat, P.K. (2025) Nematode detection and classification using machine learning techniques: A review. *Agronomy* 15(11), 2481.

Nguyen, L.T.T., Osanai, Y., Lai, K., Anderson, I.C., Bange, M.P. *et al.* (2018) Responses of the soil microbial community to nitrogen fertilizer regimes and historical exposure to extreme weather events: Flooding or prolonged-drought. *Soil Biology and Biochemistry* 118, 227–236. DOI: 10.1016/j.soilbio.2017.12.016.

Nikitin, D.A., Semenov, M.V., Chernov, T.I., Ksenofontova, N.A., Zhelezova, A.D. *et al.* (2022) Microbiological indicators of soil ecological functions: A review. *Eurasian Soil Science* 55(2), 221–234. DOI: 10.1134/S1064229322020090.

Pandit, R., Parrotta, J.A., Chaudhary, A.K., Karlen, D.L., Vieira, D.L.M. *et al.* (2020) A framework to evaluate land degradation and restoration responses for improved planning and decision-making. *Ecosystems and People* 16(1), 1–18. DOI: 10.1080/26395916.2019.1697756.

Panicker, S. (2024) Meta-omics to understand the behavior of multispecies microbial communities. In: *Progress in Soil Microbiome Research*. Springer Nature, Cham, pp. 303–317. DOI: 10.1007/978-3-031-71487-0_13.

Peay, K.G., Kennedy, P.G. and Talbot, J.M. (2016) Dimensions of biodiversity in the earth mycobiome. *Nature Reviews Microbiology* 14(7), 434–447. DOI: 10.1038/nrmicro.2016.59.

Pester, M., Rattei, T., Flechl, S., Gröngröft, A., Richter, A. *et al.* (2012) AmoA-based consensus phylogeny of ammonia-oxidizing archaea and deep sequencing of amoA genes from soils of four different geographic regions. *Environmental Microbiology* 14(2), 525–539. DOI: 10.1111/j.1462-2920.2011.02666.x.

Prayogo, F.A., Budiharjo, A., Kusumaningrum, H.P., Wijanarka, W., Suprihadi, A. *et al.* (2020) Metagenomic applications in exploration and development of novel enzymes from nature: A review. *Journal, Genetic Engineering & Biotechnology* 18(1), 39. DOI: 10.1186/s43141-020-00043-9.

Rajguru, B., Shri, M. and Bhatt, V.D. (2024) Exploring microbial diversity in the rhizosphere: A comprehensive review of metagenomic approaches and their applications. *3 Biotech* (10), 224. DOI: 10.1007/s13205-024-04065-9.

Ramteke, S., Dewangan, M., Ali, M. and Thiske, S. (2024) A brief review on biodiversity of soil nematode. *World Journal of Advanced Research and Reviews* 24(1), 359–372. DOI: 10.30574/wjarr.2024.24.1.3008.

Rillig, M.C., Lehmann, A., de Souza Machado, A.A. and Yang, G. (2019) Microplastic effects on plants. *The New Phytologist* 223(3), 1066–1070. DOI: 10.1111/nph.15794.

Russell, D.J. (2025) Quality review enhances the benefits of data publication for soil biodiversity conservation. *Applied Soil Ecology* 206, 105893. DOI: 10.1016/j.apsoil.2025.105893.

Sanjuán, F.A.B., González, L.C. and de Mello Prado, R. (2022) Mesofauna edáfica, algunos estudios realizados: Revisión. [Edaphic mesofauna, some studies done: A review]. *INGE CUC* 18(2), 197–208. DOI: 10.17981/ingecuc.18.2.2022.16.

Sathvara, P., Anuradha, J., Tripathi, S. and Sanjeevi, R. (2024) Current scenario of biodiversity loss due to developmental activities. In: *Impact of Societal Development and Infrastructure on Biodiversity Decline*. IGI Global, pp. 50–69. DOI: 10.4018/979-8-3693-6950-0.ch005.

Sepp, S.-K., Vasar, M., Davison, J., Oja, J., Anslan, S. *et al.* (2023) Global diversity and distribution of nitrogen-fixing bacteria in the soil. *Frontiers in Plant Science* 14, 1100235. DOI: 10.3389/fpls.2023.1100235.

Setälä, H., Bardgett, R.D., Birkhofer, K., Brady, M., Byrne, L. *et al.* (2014) Urban and agricultural soils: Conflicts and trade-offs in the optimization of ecosystem services. *Urban Ecosystems* 17(1), 239–253. DOI: 10.1007/s11252-013-0311-6.

Shah, A.M., Khan, I.M., Shah, T.I., Bangroo, S.A., Kirmani, N.A. *et al.* (2022) Soil microbiome: A treasure trove for soil health sustainability under changing climate. *Land* 11(11), 1887. DOI: 10.3390/land11111887.

Shahid, M., Khan, M.S. and Singh, U.B. (2023) Pesticide-tolerant microbial consortia: Potential candidates for remediation/clean-up of pesticide-contaminated agricultural soil. *Environmental Research* 236, 116724. DOI: 10.1016/j.envres.2023.116724.

Silva, V., Mol, H.G.J., Zomer, P., Tienstra, M., Ritsema, C.J. *et al.* (2019) Pesticide residues in European agricultural soils – a hidden reality unfolded. *Science of The Total Environment* 653, 1532–1545. DOI: 10.1016/j.scitotenv.2018.10.441.

Singh, A. (2008) Bioremediation of soils contaminated with pesticides: Concepts, method and recent developments. *Indian Journal of Microbiology* 48, 35–40. DOI: 10.1007/s12088-008-0004-7.

Singh, S., Ashoka, P., Ahlawat, U., Changdeo, W.B., Rehsawla, R. *et al.* (2024) Mechanisms and applications of microbial biotechnology in soil health and agricultural productivity: A review. *Journal of Advances in Biology & Biotechnology* 27(7), 1420–1438. DOI: 10.9734/jabb/2024/v27i71104.

Six, J., Frey, S.D., Thiet, R.K. and Batten, K.M. (2006) Bacterial and fungal contributions to carbon sequestration in agroecosystems. *Soil Science Society of America Journal* 70(2), 555–569. DOI: 10.2136/sssaj2004.0347.

Smith, P., Keesstra, S.D., Silver, W.L. and Adhya, T.K. (2021) The role of soils in delivering nature's contributions to people. *Philosophical Transactions of the Royal Society B* 376(1834), 20200169. DOI: 10.1098/rstb.2020.0169.

Song, X., Zhang, J., Li, D. and Peng, C. (2022) Nitrogen-fixing cyanobacteria have the potential to improve nitrogen use efficiency through the reduction of ammonia volatilization in red soil paddy fields. *Soil and Tillage Research* 217, 105274. DOI: 10.1016/j.still.2021.105274.

Srinivasiah, S., Bhavsar, J., Thapar, K., Liles, M., Schoenfeld, T. *et al.* (2008) Phages across the biosphere: Contrasts of viruses in soil and aquatic environments. *Research in Microbiology* 159(5), 349–357. DOI: 10.1016/j.resmic.2008.04.010.

Telo da Gama, J. (2023) The role of soils in sustainability, climate change, and ecosystem services: Challenges and opportunities. *Ecologies* 4(3), 552–567. DOI: 10.3390/ecologies4030036.

Tsiafouli, M.A., Thébault, E., Sgardelis, S.P., de Ruiter, P.C., van der Putten, W.H. *et al.* (2015) Intensive agriculture reduces soil biodiversity across Europe. *Global Change Biology* 21(2), 973–985. DOI: 10.1111/gcb.12752.

Usharani, K.V., Roopashree, K.M. and Naik, D. (2019) Role of soil physical, chemical and biological properties for soil health improvement and sustainable agriculture. *Journal of Pharmacognosy and Phytochemistry* 8, 1256–1267.

van der Heijden, M.G.A., Bardgett, R.D. and van Straalen, N.M. (2008) The unseen majority: Soil microbes as drivers of plant diversity and productivity in terrestrial ecosystems. *Ecology Letters* 11(3), 296–310. DOI: 10.1111/j.1461-0248.2007.01139.x.

van der Putten, W.H., Bardgett, R.D., Farfan, M., Montanarella, L., Six, J. *et al.* (2023) Soil biodiversity needs policy without borders. *Science* 379(6627), 32–34. DOI: 10.1126/science.abn7248.

Wagg, C., Bender, S.F., Widmer, F. and van der Heijden, M.G.A. (2014) Soil biodiversity and soil community composition determine ecosystem multifunctionality. *Proceedings of the National Academy of Sciences USA* 111(14), 5266–5270. DOI: 10.1073/pnas.1320054111.

Wagg, C., Hautier, Y., Pellkofer, S., Banerjee, S., Schmid, B. *et al.* (2021) Diversity and asynchrony in soil microbial communities stabilizes ecosystem functioning. *eLife* 10, e62813. DOI: 10.7554/eLife.62813.

Wahab, A., Muhammad, M., Munir, A., Abdi, G., Zaman, W. *et al.* (2023) Role of arbuscular mycorrhizal fungi in regulating growth, enhancing productivity, and potentially influencing ecosystems under abiotic and biotic stresses. *Plants (Basel, Switzerland)* 12(17), 3102. DOI: 10.3390/plants12173102.

Walker, T.W.N., Kaiser, C., Strasser, F., Herbold, C.W., Leblans, N.I.W. *et al.* (2018) Microbial temperature sensitivity and biomass change explain soil carbon loss with warming. *Nature Climate Change* 8(10), 885–889. DOI: 10.1038/s41558-018-0259-x.

Wang, K.H., Myers, R.Y., Srivastava, A. and Sipes, B.S. (2015) Evaluating the predatory potential of carnivorous nematodes against *Rotylenchulus reniformis* and *Meloidogyne incognita*. *Biological Control* 88, 54–60. DOI: 10.1016/j.biocontrol.2015.04.023.

Yang, Q., Peng, J., Ni, S., Zhang, C., Wang, J. *et al.* (2024) Soil erosion-induced decline in aggregate stability and soil organic carbon reduces aggregate-associated microbial diversity and multifunctionality of agricultural slope in the Mollisol region. *Land Degradation & Development* 35(11), 3714–3726. DOI: 10.1002/ldr.5163.

Zayed, O., Hewedy, O.A., Abdelmoteleb, A., Ali, M., Youssef, M.S. *et al.* (2023) Nitrogen journey in plants: From uptake to metabolism, stress response, and microbe interaction. *Biomolecules* 13(10), 1443. DOI: 10.3390/biom13101443.

3 Invasive Plants and Their Effects on Soil Fauna and Ecosystem Health

Qaiser Javed[1], Babar Iqbal[2*], Muhammad Bouhadi[1], Monika Jakubus[3], Amir Abdullah Khan[2] and Marko Černe[1*]

[1]Institute of Agriculture and Tourism, Poreč, Croatia; [2]School of Environment and Safety Engineering, Jiangsu University, Zhenjiang, China; [3]Department of Soil Science and Microbiology, Poznan University of Life Sciences, Poland

General Soil Fauna and Ecosystem Interactions

Introduction to soil fauna and their ecological roles

Soil fauna refers to the heterotrophic organisms that inhabit the soil ecosystem, ranging in size from microscopic nematodes to larger invertebrates, for example, beetles and earthworms. They are typically grouped by size: microfauna (e.g. protozoa and nematodes), mesofauna (e.g. collembolans and mites) and macrofauna (e.g. earthworms, millipedes and termites) (Görres and Amador, 2021). These organisms represent key functional components of terrestrial ecosystems and play a crucial role in essential ecological processes (Potapov *et al.*, 2022); for example, through their feeding, burrowing and interaction with other biota, soil fauna influence the physical structure and chemical composition of soil (Görres and Amador, 2021). They also create microhabitats, aid in aeration and regulate the flow and retention of water, all of which are critical for plant growth and microbial function (Ertiban, 2019).

One of the most vital roles of soil fauna is in the decomposition of organic matter and the mineralization of nutrients (Frouz, 2018). Macrofauna, such as earthworms and termites, fragment organic material and transport it deeper into the soil profile (Gongalsky, 2021), while meso- and microfauna continue the breakdown process by feeding on organic matter and associated microorganisms (Bonfanti *et al.*, 2025). This stepwise decomposition process releases essential nutrients such as N, P and S in forms that plants can absorb (Medina-Sauza *et al.*, 2019). For example, earthworm activity increases soil mineral nitrogen concentrations, possibly due to the consumption of soil microbial biomass (Araujo *et al.*, 2004). Lang *et al.* (2023) investigated the impact of earthworms on soil ammonium (NH_4^+), nitrate (NO_3^-) and total mineral nitrogen (ammonium + nitrate) using meta-analytic techniques. Earthworms were found to increase NO_3^- (+88%) and total mineral nitrogen (+63%) but not affect NH_4^+. It was assumed that earthworms affect total mineral nitrogen mainly by their impact on NO_3^-. Endogeic and epigeic earthworms significantly increased NO_3^- and total mineral nitrogen, whereas no clear effect of anecic

*Corresponding authors: babar@ujs.edu.cn and marko@iptpo.hr

DOI: 10.1079/9781836991298.0003

earthworms was found. However, nematodes, as part of the microfauna, also regulate microbial populations and have both direct and indirect roles in nitrogen mineralization.

Bacterivorous and fungivorous nematodes feed on microbial biomass and excrete excess nitrogen (primarily as NH_4^+) into the soil, thus increasing plant-available nitrogen levels (Gebremikael *et al.*, 2014). In another study, Gebremikael *et al.* (2016) found that the presence of nematodes significantly increased plant biomass production (+9%), net N (+25%) and net P (+23%) availability compared to their absence, demonstrating that nematodes link below- and aboveground processes, primarily through increasing nutrient availability.

Soil fauna also exhibit complex interactions with plant communities, both above and below ground (Wubs *et al.*, 2019; Forey *et al.*, 2021). These interactions can be mutualistic or antagonistic, influencing plant health. Plants and soil organisms interact in complex feedback loops, which influence key ecosystem processes such as primary productivity, organic matter decomposition and nutrient cycling (Winck *et al.*, 2020; Brousseau *et al.*, 2025). Traditionally, leaf litter has been recognized as the primary driver of the soil food web; however, recent studies have also demonstrated that plant roots play a significant role (Brousseau *et al.*, 2025). Many soil invertebrates stimulate root growth through their feeding activity or indirectly by enhancing nutrient availability (Wolters, 1991). Furthermore, arthropod activity modifies soil structure, enhancing porosity and aggregation, which improves root penetration and soil gas exchange. This structural enhancement supports the microbial community, fostering a dynamic and resilient soil food web (Bagyaraj *et al.*, 2016).

The introduction and subsequent proliferation of invasive plant species can markedly disrupt the intricate trophic and functional interactions among soil microfauna, mesofauna and macrofauna (Abgrall *et al.*, 2019; De Almeida *et al.*, 2022). Such species often alter both the quantity and biochemical composition of organic inputs, particularly leaf litter and root exudates, thereby modifying habitat conditions and restructuring the resource base that sustains soil faunal communities. In terrestrial ecosystems, litter decomposition rates are on average 117% higher for invasive plants than for native species. Higher litter masses imply higher rates of element input into the soil (Ehrenfeld, 2010). These shifts in soil biotic networks can cascade to affect key ecosystem processes, including nutrient mineralization, organic matter decomposition and soil carbon sequestration, ultimately influencing ecosystem stability and resilience (Forey *et al.*, 2021).

Alterations to the soil biotic community ultimately influence ecosystem-level processes, including primary productivity, carbon storage and food web stability. Reduced activity of soil fauna can lead to compaction, lower infiltration rates and reduced nutrient availability, making the soil less hospitable not only to native plants but also to the microbial and faunal communities it once supported. Such feedback loops can favour invasive species, hinder restoration efforts and result in long-term ecosystem degradation (Fig. 3.1). Therefore, understanding the role of soil fauna in ecosystem functioning is crucial not only for appreciating biodiversity below ground but also for managing biological invasions and restoring degraded landscapes.

Symbiotic relationships between soil fauna and invasive plants

Symbiotic relationships between soil fauna and plant species play a vital role in ecosystem functioning, particularly in nutrient cycling, organic matter decomposition, soil formation and the maintenance of plant productivity. While these interactions are often finely tuned between native flora and native soil fauna, invasive plant species frequently establish novel and sometimes disruptive associations in their introduced ranges. While invasive plants negatively affect the richness and composition of native plant communities, they have a lesser impact on mammals, birds and reptiles (Schirmel *et al.*, 2016). Invasive plants affect soils and native species not only through resource uptake but also via the release of litter and root exudates, which reshape nutrient cycling, microbial communities and soil properties. These changes can generate positive feedback that favours invasion or negative feedback that degrades soil functioning and affects surrounding native plant communities (Fig. 3.2).

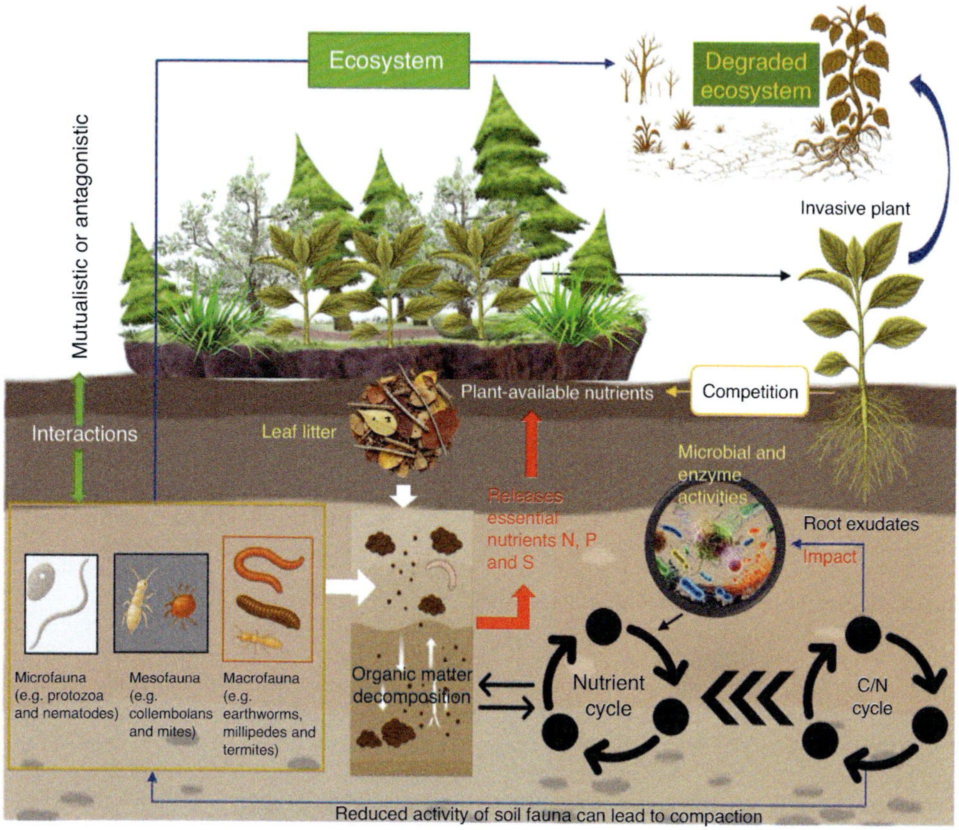

Fig. 3.1. Ecosystem functions mediated by soil fauna and the impacts of invasive plants on these interactions.

One of the key ways invasive plants succeed in new habitats is by forming mutualistic relationships with resident soil fauna. Earthworms, mycorrhizal fungi, termites and microarthropods are often co-opted to facilitate the growth of invasive plants. For example, mycorrhizal fungi can enhance the uptake of essential nutrients such as phosphorus (P), and this relationship has been observed to be particularly strong in the invasive *Acacia* species in South Africa (Wood, 2017; Jacobs *et al.*, 2020). These species associate with both arbuscular mycorrhizal fungi (AMF) and nitrogen-fixing rhizobia, and their litter, enriched with symbiotically fixed nitrogen, has been shown to improve soil fertility over time (Jacobs *et al.*, 2020). Such symbiotic benefits can provide invasive plants with a competitive edge over native flora, particularly in nutrient-poor ecosystems. Additionally, detritivorous fauna such as ants and collembolans may contribute to seed dispersal and microsite preparation for invasive seedlings, thereby increasing their establishment success (Christian, 2001).

Beyond mutualism, invasive plants may engage in exploitative interactions, whereby they benefit from the activity of soil fauna without providing any reciprocal advantages (Gioria *et al.*, 2023). Invasive plant litter can alter soil food web dynamics by introducing chemically distinct substrates that differ significantly from native inputs (He *et al.*, 2022). Invasive species alter the physical and chemical properties of the soil, including pH, nutrient content, moisture levels and organic matter composition, in a manner that facilitates their easy adaptation to new environments (Javed *et al.*, 2023). These changes can make the soil inhospitable for native species of soil fauna while encouraging the proliferation of generalist or

Fig. 3.2. Soil feedback mechanisms of invasive plants.

opportunistic species. For instance, Stinson *et al.* (2006) demonstrated that soils conditioned by garlic mustard (*Alliaria petiolata* (M. Bieb.) Cavara & Grande) severely reduced colonization of AMF in seedlings of native tree species (such as sugar maple, red maple and white ash). The study further isolated glucosinolate hydrolysis products (like allyl isothiocyanate, benzyl isothiocyanate and glucotropaeolin) from garlic mustard root extracts and showed they suppress both AMF colonization and spore germination, confirming the allelochemical mechanism behind the suppression. This suppression not only disrupts mutualistic relationships between plants and fungi, but also indirectly lowers abundances of fungivorous soil fauna, such as springtails (Collembola) and oribatid mites (McCary and Wise, 2019). Similarly, *Spartina alterniflora* Loisel invasion in Chinese coastal wetlands alters sediment redox conditions and organic matter content changes that are associated with reduced density, biomass and diversity of macrobenthic invertebrate communities (e.g. polychaetes, amphipods) in invaded habitats vs native or uninvaded areas (Caiyun *et al.*, 2014; Chen *et al.*, 2023; Lyu *et al.*, 2023; Zheng *et al.*, 2023).

In some ecosystems, soil fauna may actively facilitate the invasion process through a phenomenon known as soil engineering. Invasive earthworms, such as *Lumbricus terrestris*, act as soil engineers through burrowing, casting and mucus secretion. These activities reshape soil structure, improving porosity, aggregation, aeration and nutrient distribution (forming the so-called drilosphere) (Medina-Sauza *et al.*, 2019; Kim *et al.*, 2022). At the same time, earthworms facilitate the establishment of invasive plants by burying and dispersing seeds and creating nutrient-rich patches through their casts (Eisenhauer *et al.*, 2008; Clause *et al.*, 2015). However, in previously earthworm-free ecosystems their invasion often leads to declines in soil micro- and invertebrate biodiversity (Ferlian *et al.*, 2018).

According to the 2018 report by the US National Park Service, invasive earthworms (notably *L. terrestris*) in North American forests act as soil engineers by rapidly removing forest leaf litter and exposing bare soil, which is highly conducive to the germination and establishment of garlic mustard seedlings. Meta-analytic and field studies show that increased earthworm biomass is linked to reduced leaf litter, diminished native plant cover and enhanced spread of invasive species, including garlic mustard. Moreover, plots invaded by garlic mustard support higher earthworm biomass, while eradication of garlic mustard reduces earthworm populations, which suggests a reinforcing invasion feedback loop (Craven *et al.*, 2017; Stinson *et al.*, 2018). Similarly, collembolans and mites enhance fungal biomass regulation, micro aggregation and litter decomposition, stimulating microbial activity and nutrient mineralization (e.g. a single springtail species can increase microbial biomass by ≈56% and increase decay rates by ≈30%) (Liu *et al.*, 2019). In invaded soils, mesofauna activity has been shown to enhance soil nutrient mineralization, potentially supporting the nutrient demands of invasive plants (Jin *et al.*, 2022).

Invasive plant species can significantly disrupt soil fauna communities, leading to declines in biodiversity and impairments in essential ecological functions such as nutrient cycling, soil structure formation and disease regulation. For instance, *Eucalyptus globulus* subsp. *bicostata* (Maiden, Blakely & Simmonds) plantations in northwestern Spain have been associated with decreased macroinvertebrate richness due to the accumulation of eucalypt leaf litter, which alters the composition and abundance of soil invertebrate communities (Rivas-Torres *et al.*, 2025). In tropical ecosystems, the spread of *Chromolaena odorata* (L.) has led to significant shifts in soil carbon and nitrogen contents, negatively affecting soil fauna diversity and ecosystem processes (Wei *et al.*, 2017). These examples underscore the complex and often detrimental effects of invasive plants on soil ecosystems.

The intricate interactions between invasive plants and soil fauna form self-reinforcing feedbacks that amplify the impacts of invasion while destabilizing native soil microbes and plant communities. This disruption reduces soil biodiversity and functional resilience, heightening vulnerability to environmental stresses. Because soil fauna are vital to nutrient cycling and ecosystem functioning, their impairment risks cascading ecological consequences. Effective management must therefore integrate restoration of belowground biotic networks alongside invasive species control to rebuild ecosystem stability and resilience. Recognizing and addressing these belowground dynamics is key to sustaining long-term ecosystem health.

Changes in Soil Chemistry and Physical Structure

Impact of invasive plants on soil properties

The impact of invasive plants on the soil environment needs to be considered in relation to soil functionality and soil health. Generally, soil health plays a crucial role as a vital living ecosystem that sustains plants, animals and humans, ensuring the production of biomass for food, feed, fibre and biofuels (Fausak *et al.*, 2024; Jakubus *et al.*, 2025). Human activity degrades soil health, resulting in unfavourable changes that reduce soil productivity. However, plants, including invasive ones, can also degrade soil health and fertility, and their impact should be assessed accordingly. However, the literature on this topic is inconsistent, and the impact of invasive plants is highly dependent on both biotic and abiotic factors. In the context of soil, the effects of invasive plants can result from climate and land-use changes (Afzal *et al.*, 2023). Above all, the intensity of growth of invasive plants, dictated by their biology or the availability of water and nutrients in the soil, should be taken into account. The importance of the root system's structure and its potential for biomass production should not be overlooked.

To properly assess such an impact, a detailed and in-depth analysis of all soil parameters is necessary. Unfortunately, this is not a widespread practice, which leads to limitations, incomplete conclusions and inaccurate assessments of the actual state. As Wohlgemuth *et al.* (2022) claim, this approach is dictated by analytical difficulties and the fact that most researchers focus on easily measurable soil parameters. The justification for this approach may be that the selected soil properties are directly related to soil microbiological activity and to ensuring basic plant well-being. Such an approach is not correct, but unfortunately it prevails in studies devoted to this issue. The soil parameters most frequently analysed in this context are consistent with those commonly used in soil health indices. The authors Jakubus *et al.* (2025) and Lehmann *et al.* (2020), for instance, propose a holistic approach to soil health, indicating a large group of soil indicators that allow, to a varying extent, the evaluation of ecosystem services. Among them, total organic carbon (TOC), total nitrogen, soil pH, bioavailable nutrients and nitrogen mineralization are of particular importance, but rarely is the impact of invasive plants on the soil determined through the analysis of quantitative changes on soil cation exchange capacity or the availability of K, Ca, Mg and P (Wohlgemuth *et al.*, 2022).

In a 2022 review of European literature, Wohlgemuth *et al.* found that invasive trees had no impact on soil properties in 16% of studies, a positive impact in 45% and an adverse impact in 39%. While the authors generally found the impact of invasive plants on soil properties to be slight, they noted that these plants could, in certain specific situations, actually enhance soil fertility. A study by Unger *et al.* (2022) also highlights the diverse and ambiguous effects of invasive plants, suggesting this ambiguity may arise from their impact on carbon cycling and the increased microbial decomposition of organic matter. According to Vujanović *et al.* (2022), invasive plants usually have negative or neutral environmental impacts, with beneficial effects seldom observed, although potential benefits may arise from interactions between woody plants and nitrogen-fixing species in specific communities.

The invasion of a soil environment by multiple species frequently alters soil biogeochemistry, often resulting in increased amounts of C, N, organic matter and a higher abundance of nitrifying bacteria. These observations were confirmed by Vilà *et al.* (2024), who reported that invasive nitrogen-fixing plants can increase N availability in soil. Duda *et al.* (2003) observed elevated levels of NO_3, P and K in soils invaded by *Halogeton glomeratus* (M. Bieb.) Ledeb., although the magnitude of these increases varied across sites. Rodgers *et al.* (2008) also document increases in available N, P, Ca and Mg, as well as an increase in soil pH, under the influence of the invasive plant *A. petiolata* (M. Bieb.) Cavara & Grande. Another invasive species widespread across North America is *Bromus tectorum* L., which has also been associated with increased availability of Mg, N, P, Cu, Fe, Ca and K, highlighting species-specific effects on soil biogeochemistry. However, this pattern is not universal. For example, *Robinia pseudoacacia* L., a nitrogen-fixing pioneer tree, has been shown to reduce soil quality by altering nutrient cycling and chemistry (Li *et al.*, 2022).

A key mechanism for this is the release of allelochemicals, toxic compounds that can further alter the pH, hinder nutrient cycling and inhibit beneficial soil microbes, such as mycorrhizal fungi and symbiotic bacteria (Afzal *et al.*, 2023). The allelopathic effects of invasive plants make it challenging and costly to restore temporarily overgrown areas to agricultural use (Afzal *et al.*, 2023).

Invasive plant root systems also play a crucial role in shaping soil properties, with significant effects on structure, water retention and aeration (Weidenhamer and Callaway, 2010). These authors explain how the roots of invasive species can have both constructive and adverse effects. A fast-growing, extensive root system promotes the formation of a nodular structure with proper porosity. Root secretions promote the formation of aggregates and, consequently, optimal water and air balance. In addition, plants reduce soil density through root channeling and litter deposition.

In contrast, poorly developed root systems, such as shallow or weak taproots, can worsen soil conditions, leading to soil compaction, reduced porosity and unfavourable water and air conditions. Moreover, rapid wetting and drying cycles cause shrinkage and strengthening of the soil. Some invasive species even establish monocultures with shallow root systems, which reduces soil stability and increases the risk of erosion. This risk is further increased in soils with high proportions of fine mineral fractions such as clay and silt. For example, Li *et al.* (2022) demonstrate how the invasion by *R. pseudoacacia* L. leads to a reduction in sand, accompanied by a simultaneous increase in the silt and clay fractions in the soil.

Compounds secreted by the roots of invasive plants are characterized by the production of significant biomass such as leaf litter (Weidenhamer and Callaway, 2010; Vujanović *et al.*, 2022; Afzal *et al.*, 2023). An excess of generated biomass in the soil can be interpreted in two ways: either as a substantial source of nutrients released during the decomposition of organic matter or as a potential pathway for the soil to undergo intensively depletion of nutrients, leading to habitat impoverishment.

Afzal *et al.* (2023) observed that invasive plants often generate abundant biomass with high nutrient content, resulting in rapid litter decomposition and accelerated nutrient cycling. However, this also increases nutrient loss through leaching or volatilization, which can destabilize long-term soil fertility. The rapid decomposition of accumulated biomass favours a positive N or C balance in the soil, enhancing their cycling and the net primary production. However, it can also release excessive amounts of N and intensify nitrification, increasing the activity of nitrifying bacteria. These conditions disadvantage native plants while facilitating the growth and spread of invasive species, ultimately reducing plant diversity (Vujanović *et al.*, 2022).

Larger amounts of organic acids also lower soil pH, causing acidification, which in turn disrupts the biogeochemical cycles and reduces microbiological activity. Some species, however, produce slowly decomposing biomass litter, which negatively affects nutrient transformations, particularly nitrogen (Vujanović *et al.*, 2022). The invasive species *Phragmites australis* subsp. *Australis* and *Aegilops triuncialis* also produce litter that decomposes slowly, potentially reducing mineralization rates and increasing nitrate leaching. Similarly, the invasive species *Fallopia japonica* (Houtt.) Ronse Decr., also alters the dynamics of organic matter decomposition, disrupting carbon and nitrogen cycling (Koutika *et al.*, 2007). Although these authors emphasize that the impact of invasive plants cannot be generalized, as it depends on local site characteristics (soil properties, parental material and plant cover), which determine the magnitude and direction of the invasion effects on ecosystems. Therefore, in addition to excessive mineralization of organic compounds, the opposite phenomena can be observed when invasive plants form dense covers that limit humus formation and reduce nutrient availability.

Climate change, invasive plants and soil fauna

Biological invasion is reportedly a prominent component of global environmental change while also acting synergistically with its other elements (e.g. global warming, increases in N deposition and atmospheric carbon dioxide, alters precipitation as well as bringing about land-use change) (Dukes and Mooney, 1999; Sun *et al.*, 2023). As recently reported, numerous invasive species share traits that enable them to capitalize on a wide variety of elements under global change. Thus, invasive plants often benefit more from global environmental change than native plants; environmental change promotes plant invasion by disturbing soil fauna (Liu *et al.*, 2017; Javed *et al.*, 2023).

Temperature and nitrogen are key limiting factors for plant performance and are significantly impacted by anthropogenic environmental changes (Ren *et al.*, 2021). Fang *et al.* (2021) have argued that climate change is the most critical variable affecting the spread of alien plant invasions. Climate warming can generally affect plant performance by changing source–sink relationships and thermal requirements of species. Therefore, global warming may facilitate the invasion of alien species from warmer ranges into a region where it is otherwise too cold for an invader to compete with native species (Peng *et al.*, 2019). Atmospheric nitrogen deposition is estimated to increase by two to three times during this century (Yang *et al.*, 2021), which can affect plant performance by changing both their nitrogen supply and relative content of phytohormones. Increasing evidence now supports the view of Yang *et al.* (2021) that nutrient enrichment is occurring, which may favour the fitness of invasive species over that of native species (Uddin and Robinson, 2018). Moreover, climate warming is known to increase nutrient availability by promoting soil organic mineralization and increasing plants' photosynthetic activity (Ma *et al.*, 2011). In several cases, a warmer climate and increased nitrogen deposition may be responsible for many successful plant invasions (Peng *et al.*, 2019; Song *et al.*, 2021). Climate change is projected to intensify hydrological extremes, with heavier precipitation events expected in many regions, including North America, Central and Northern Europe, Asia and Eastern Africa (Pachauri *et al.*, 2014), while longer and more frequent droughts have already been observed in areas such as the Mediterranean, Central America/Mexico, the Amazon, north-east Brazil and South Africa, with further intensification projected under future scenarios (Orlowsky and Seneviratne, 2013; Pachauri *et al.*, 2014). Changes in precipitation and temperature, which affect evaporation, will inevitably alter the soil moisture dynamics of many terrestrial ecosystems, increasing the risk of both prolonged waterlogging after extreme rainfall and more frequent arid conditions during drought. These changes will alter how invasive plants interact with soil fauna.

Invasive species often possess drought-resistant traits, such as deeper root systems or water-conserving physiology, allowing them to outcompete native species under water-limited conditions (Funk, 2013). For example, yellow starthistle (*Centaurea solstitialis*) significantly depletes soil moisture reserves in the grasslands of California and Oregon, USA. Its deep root system allows it to exploit deep soil moisture late in the growing season, which reduces soil moisture recharge for the following season. According to the US Department of Agriculture's Fire Effects Information System, yellow starthistle can extract soil moisture from depths greater than ≈2 m and, in dense infestations, can rapidly deplete moisture from 30 cm to 1.8 m, preventing seasonal recharge (Innes and Zouhar, 2021). Also, dense infestations displace native plants, degrade habitat and reduce forage availability (Enloe *et al.*, 2004).

Soil microbial communities and soil enzyme activities are involved in the production of primary greenhouse gases, such as carbon dioxide (CO_2) and nitrous oxide (N_2O). Farmland is considered to be a significant source of greenhouse gas emissions, and CO_2, N_2O and methane are among the major gases that contribute to global heating. Increased microbial activity underneath exotic *Sphagneticola trilobata* (L.) may be advantageous because of its superior ability to uptake nutrients, thereby directly reducing nutrient losses from surface runoff. Increases in soil enzymatic activities involved in carbon and nitrogen cycling in invaded soils suggest that *S. trilobata* may more effectively utilize soil organic matter and increase N_2O and CO_2 emission rates compared to non-invaded soils (Javed *et al.*, 2023). Furthermore, Javed *et al.* (2025) noticed that microbial biomass carbon, nitrogen and CO_2 emissions were higher in the soils of native plant communities invaded by invasive *S. trilobata*. However, this change does not necessarily benefit the soil food web. It has also been observed that total soil carbon stocks declined by approximately 12% in invaded areas (invaded 2140 $g{\cdot}C{\cdot}m^{-2}$ vs uninvaded 2429 $g{\cdot}C{\cdot}m^{-2}$), although this varied with environmental context. Microbial biomass decreased by about 26% in invaded plots (invaded 7.40 $g{\cdot}C{\cdot}m^{-2}$ vs uninvaded 10.05 $g{\cdot}C{\cdot}m^{-2}$), with greater reductions where the invader was dense (Litton *et al.*, 2008).

Floods, wildfires, heatwaves and even soil pollution often create open niches or weakened ecosystems and soil fauna that invasive plants readily colonize. In the Brazilian Cerrado, molasses grass (*Melinis minutiflora*) accumulates biomass more rapidly than native plants, increasing both fire frequency and intensity. Such accumulation can lead to severe fires spreading from the grass layer to the canopy. In a dry-season inferno, flames can reach up to 12 m, with heat felt from 200 m away (Rossi *et al.*, 2014). In another example, D'Antonio *et al.* (2011) discovered that in Hawai'i, sites burned in 1970 remained dominated by molasses grass 37 years later, with minimal recovery of native species, despite decades of fire suppression. Invasive plants also survive and colonize areas polluted by heavy metals. Javed *et al.* (2023, 2025) found that invasive *S. trilobata* also possesses superior traits, such as rapid colonization, strong allelopathy and high phenotypic plasticity, and its ability to produce more relative biomass compared to surrounding native species, which enables it to adapt to polluted soils and sustain its invasion in a harsh environment. Such dynamics risk accelerating soil carbon losses, amplifying climate warming in a reinforcing cycle.

Responses of Soil Microorganisms to Biological Invasion

Soil microbial communities play a dual role in plant invasions, both responding to and shaping invasion outcomes. For example, invasive plants influence soil communities in three principal mechanisms: (1) exotic plants often experience neutral or positive plant–soil feedback, while native species face negative ones; (2) exotics modify soil biota by increasing pathogen abundance or by disrupting mutualistic symbionts, but are themselves less affected; and (3) through the release of allelochemicals that local soils cannot detoxify, or that soil microbes convert into even more phytotoxic compounds (Van der Putten, 2010).

In contrast, soil microbial communities play a significant role in the establishment of exotic plants, often contributing to their invasive nature by giving exotic plants a competitive advantage over native plant species, since many native plants are subject to negative feedback effects (Callaway *et al.*, 2004; Petermann *et al.*, 2008; Gibbons *et al.*, 2017). A specific example is when exotic plants gain an advantage by accumulating local pathogens in their rhizosphere, which suppresses native species). This phenomenon is known as the accumulation of local pathogens hypothesis (Eppinga *et al.*, 2006). The hypothesis is supported by evidence of soil pathogen build-up in regions invaded by *C. odorata* in India (Mangla *et al.*, 2008). However, South African studies show no such effect, suggesting that pathogen accumulation depends on the specific invader-native combination and habitat (Wood, 2017).

Until recently, it had been proposed that invasive exotic plants could efficiently utilize AMF, as interactions with these fungi had been assumed to be relatively non-specific (Florianová *et al.*, 2023). However, the weed-like nature of many invasive exotic plants has been shown to correlate with lower mycorrhizal dependence (Vogelsang and Bever, 2009), although there are relatively few examples of studies that actually test mycorrhizal effects on invasive plants. In a comparative study, the mycorrhizal status of invasive exotic plants was found to depend on the plant species with which they co-occur and the habitat they invade (Majewska *et al.*, 2018).

Woody plant invaders can also restructure microbial networks by importing or amplifying ectomycorrhizal (EM) fungal partners. For example, invasive *Pinus* species often co-invade grasslands and shrublands with EM fungi, and in some cases a single EM species can be sufficient to enable pine establishment and expansion. This co-invasion not only benefits the pines by enhancing nutrient acquisition in novel soils but also rewires local mutualist networks, thereby reinforcing the invader's dominance over native vegetation (Dickie *et al.*, 2010, 2017; Hayward *et al.*, 2015). Callaway *et al.* (2008) found that garlic mustard (*A. petiolata*) suppressed root colonization by AMF and EM fungi, and particular native mycorrhizal species in the non-native range. No similar impacts were observed when garlic mustard was grown in native soils. Moreover, EM fungi enhance pine nutrient acquisition in novel soils, and pines subsidize EM networks with carbon. The reinforced symbiosis facilitates further spread (Hayward *et al.*, 2015).

Many invaders leave long-lasting ('legacy') shifts in nitrification rates and nitrifier community composition that persist even after the plants are removed (Afzal *et al.*, 2023). Additionally, nitrogen-fixing species, such as Australian acacias (*Acacia dealbata* and *Acacia longifolia* (Andrews) Willd.), enhance soil nitrogen through symbiotic fixation and alter bacterial communities, extracellular enzyme activities and C–N–P cycling. Invader-derived litter chemistry and root exudates also influence decomposition pathways, often altering the balance between fungal and bacterial activity, changing soil respiration and affecting the stabilization of organic matter (Le Maitre *et al.*, 2011; Rai and Singh, 2020). Abiotic–biotic legacy interactions, such as elevated atmospheric nitrogen deposition, can amplify invasion legacies in soils, prolonging altered microbial processes (e.g. nitrification) and making restoration more challenging even after the invaders have been removed (Cui *et al.*, 2023).

Soil fauna invasions can also influence microbial responses; the specific impacts of different invasive species on microbial communities, along with their underlying mechanisms, are summarized in Table 3.1. Meta-analyses and field studies document consistent changes to soil chemistry and microbiomes following an invasion (Ferlian *et al.*, 2020; Lang *et al.*, 2023; Thouvenot *et al.*, 2024). These fauna-driven changes can interact with plant invasions by accelerating nutrient turnover and changing microsite conditions.

Impact on Soil Mesofauna and Macrofauna

Biological invasions not only reshape plant and microbial communities but also alter the soil mesofauna and macrofauna that underpin key ecosystem processes, such as decomposition, nutrient cycling and soil structure formation (Lazzaro

Table 3.1. Examples of invasive plants and soil fauna-altering soil microbial communities and ecosystem functions (author' own table).

Invader (type)	System/region	Microbial mechanism	Key microbial change	Context	Source(s)
Solidago canadensis l. and *Erigeron canadensis* L. (plant)	Old fields, China	Co-invasion reshapes soil microbiomes	Bacterial/fungal community reassembly under *Solidago* and *Erigeron* co-invasion	Outcomes vary with soil origin and lineage	He *et al.* (2024)
Phragmites australis (Cav.) Trin. ex Steud. (plant)	Wetlands, North America	Differential plant–soil feedbacks vs native lineage	Distinct rhizosphere microbial communities and stronger plant–soil feedbacks	Invader gains an advantage via apparent competition	Lee *et al.* (2024)
Reynoutria japonica Houtt. (plant)	Riparian sites, Europe	Disruption of mutualists (AMF) + allelopathic rhizosphere	↓ AMF abundance/ structure irrespective of season/soil	Effects vary by host and habitat	Stefanowicz *et al.* (2021), Shalmiyev *et al.* (2024)
Impatiens glandulifera Royle (plant)	Riparian forests, Europe	Mycorrhiza-mediated competitive release	↓ AMF benefits to neighbours; ↓ fungal hyphal biomass	Salinity/ tidal context dependent	Gaggini *et al.* (2019), Ab Razak et al. (2023)
Spartina alterniflora (plant)	Coastal marshes, China	Network and function rewiring in sediments	↓ Bacterial/fungal diversity; altered co-occurrence networks and metabolism	Alters nutrient cycles in arid soils	Li *et al.* (2020), Zhang *et al.* (2024)
Prosopis juliflora (Sw.) DC. (plant)	Arid lands, Middle East/ Africa	N-cycle acceleration via microbiome shifts	↑ N transformations; ↑ trace-gas emissions	Patterns vary with invasion age	Yagle *et al.* (2024)
Lantana camara L. (plant)	Semi-arid woodlands, India	Soil property and microbial biomass shifts	↑ Microbial biomass C; ↑ N; ↓ Acidobacteria	Strongest near active mounds	Hansda *et al.* (2024)
Microstegium vimineum (Trin.) A.Camus (plant)	Forest, central New Jersey, USA	Pathogen build-up over time	Invader-associated foliar/↑ soil fungi, triggering disease	Strongest in long-invaded sites	Schramm and Ehrenfeld (2010)
Red imported fire ant (*Solenopsis invicta*) (soil fauna)	Multiple habitats, China	Nesting reorganizes soil bacteria	Convergent bacterial communities in nests; ↑ Actinobacteria	Modified soils can cascade to plant–microbe interactions	Shi *et al.* (2023), Song *et al.* (2023)
Invasive earthworms (e.g. *Lumbricus* spp.) (soil fauna)	Temperate forests, North America	Gut–soil microbiome exchange; litter mixing	Earthworm microbiomes adapt and influence soil microbial assembly	Restructured microbial pathways that alter nutrient cycling and plant performance	Ferlian *et al.* (2024)
Wild boar (*Sus scrofa*) (soil fauna)	Temperate forests, Changbai Mountain area, Jilin Province	Bioturbation shifts soil food webs	Grubbing changes nematode community structure (microbial grazers)	Indirect effects on microbe–plant interactions and nutrient turnover	Sun *et al.* (2024)

Arrows indicate the direction of change: ↑, increase; ↓ decrease.

et al., 2018). Mesofauna and macrofauna are highly responsive to changes in litter quality, soil chemistry (Hickerson, 2010) and microclimate induced by invasive plants (De Almeida *et al.*, 2022). These faunal groups also interact closely with soil microorganisms, meaning that invasion-driven shifts can cascade across trophic levels (DeVore, 2011).

Nematodes, which occupy multiple trophic levels (bacterivores, fungivores, herbivores, omnivores and predators), are highly sensitive to changes in plant inputs and soil nutrient availability. Consequently, invasive plants can significantly alter the abundance, diversity and functional composition of nematodes. For instance, invasion by *S. alterniflora* in Chinese coastal marshes increases bacterivorous nematodes but reduces herbivorous taxa, reflecting shifts toward bacterial decomposition pathways in invaded sediments (Chen *et al.*, 2007; Zhang *et al.*, 2018). Similarly, *A. longifolia* in Portuguese coastal dunes has been shown to increase enrichment and structure indices of nematode assemblages due to elevated nitrogen and labile carbon inputs from nitrogen-fixing litter (Ulm, 2019). Such changes can reconfigure the soil food web and accelerate nutrient turnover, often reinforcing the invader's dominance.

Soil mites (Acari) are crucial decomposers and micro-predators, and their diversity often declines under plant invasions that homogenize litter and soil microhabitats (Prabhakar, 2023). In Mediterranean grasslands invaded by *Oxalis pes-caprae*, oribatid mite richness and abundance dropped significantly, likely due to altered litter chemistry and reduced structural complexity in agricultural localities (Petsikos *et al.*, 2007). In some forest systems, however, certain predatory mites may increase if their prey (e.g. nematodes or collembolans) proliferate in the altered environment, indicating that mite responses can be taxon-specific and linked to broader shifts in the soil community (Walter and Proctor, 2013).

Collembola (springtails) are among the most abundant mesofaunal decomposers, highly sensitive to moisture and litter quality (Čuchta *et al.*, 2019). For example, the invasion of *Impatiens glandulifera* (Himalayan balsam) along European riverbanks has led to a reduction in collembolan abundance and diversity, attributed to the plant's nutrient-rich but rapidly decomposing litter, which reduces habitat stability for these detritivores (Hejda *et al.*, 2009). Similarly, North American forests invaded by garlic mustard have shown declines in microarthropod abundance, potentially linked to reduced fungal biomass following mycorrhizal suppression, which deprives fungivorous collembola of their key food sources (Anthony, 2019).

Earthworms can be both victims and drivers of invasion impacts. Some plant invasions reduce native earthworm diversity by altering the litter quality or the soil pH. For example, the invasion of *Carpobrotus edulis* in Mediterranean coastal systems has been linked to a decrease in the abundance of native earthworms due to high salt and altered organic matter content in the soil (Novoa *et al.*, 2014). Conversely, invasive earthworms can act as ecosystem engineers, transforming soil structure, accelerating the turnover of organic matter and promoting the growth of nutrient-demanding invasive plants. In northern North American forests, European *Lumbricus* species mix organic horizons into the mineral soil, reducing forest floor thickness, altering microbial communities and disadvantaging native flora that are adapted to thick litter layers (Hale *et al.*, 2008).

Above- and belowground arthropods respond to changes in vegetation structure and microclimate induced by invasion. On tropical islands invaded by the yellow crazy ant (*Anoplolepis gracilipes*), native soil arthropod diversity has declined significantly, partly due to direct predation and partly due to altered litter decomposition dynamics (Wetterer, 2005). Even in temperate systems, plant invasions can restructure ground-dwelling arthropod communities; for example, *F. japonica* stands in Europe reduce beetle and spider diversity, likely via shading and changes in litter microhabitat (Topp *et al.*, 2008).

Gastropods (snails and slugs) are particularly vulnerable to habitat and microclimate changes induced by invasive plants. Dense monocultures can alter leaf litter quality and humidity, directly affecting detritivorous snails. Invasive *Tradescantia fluminensis* in New Zealand forests reduces native gastropod abundance by creating a thick groundcover that alters humidity regimes and changes leaf litter composition (Standish *et al.*, 2004). In some cases, invasive plants may indirectly increase slug abundance by creating moist, shaded conditions that favour these pests, as observed in European grasslands invaded by *Solidago canadensis* (Schittko, 2015).

Across taxa, the effects of plant invasions on soil mesofauna and macrofauna are context-dependent but often involve a common set of drivers: altered litter quality and chemistry, changes in soil moisture and temperature, shifts in microbial food resources and direct physical habitat modification. These changes can weaken native soil food webs, facilitate further invasion and alter ecosystem processes from decomposition to nutrient cycling. Recent studies have documented diverse and taxon-specific responses of soil fauna to invasions across different ecosystems and climatic regions. To provide a comprehensive overview, Table 3.2 summarizes these findings, highlighting the taxa affected, the invasive species involved, observed impacts on soil fauna and the proposed ecological mechanisms. This synthesis illustrates both the direct

Table 3.2. Summary of recent studies on the effects of invasive plant and animal species on soil mesofauna and macrofauna. The table highlights taxon-specific responses across different ecosystems and geographic regions, along with the proposed mechanisms driving these effects.

Taxon affected	Invasive species	Ecosystem/ location	Reported effect on fauna	Proposed driver/ mechanism	Source (s)
Nematodes	*Solidago gigantea* subsp. *serotina* (Kuntze) McNeill (goldenrod)	Riparian meadows, Slovakia	Increased abundance of bacterivorous and fungivorous nematodes, accompanied by a decrease in functional diversity	Altered litter quality and nutrient inputs from invasive plants	Jakubcsiková *et al.* (2024)
Nematodes	*Acer negundo* f. *angustifolium* Schwer. (boxelder)	Floodplain forests, Slovakia	No significant change in nematode abundance or diversity compared to native species	Similar soil conditions across invasive and native stands	Renčo *et al.* (2025)
Collembola (springtails)	*Phyllostachys edulis* f. *abbreviatus* G.H. Lai (moso bamboo)	Subtropical forests, China	Reduced abundance and diversity of Collembola communities	Changes in soil pH and microhabitat due to bamboo litter	Long *et al.* (2023)
Collembola (springtails)	*Robinia pseudoacacia* L. (black locust)	Urban forests, Italy	Decreased abundance and diversity of Collembola communities	Altered soil structure and litter quality from the invasive tree	Zizolfi *et al.* (2025)
Oribatid mites	*Quercus rubra* L. (red oak)	Temperate forests, Poland	Reduced diversity and abundance of oribatid mites	Differences in litter chemistry and microbial communities	Kohyt and Skubała (2020)
Soil fauna	*Ligustrum lucidum* W.T. Aiton (Privet)	Subtropical montane forests, Argentina	Decreased abundance and contribution of soil fauna to litter decomposition; increased decomposition rate	Changes in litter quality and microenvironment	Fernandez *et al.* (2022)
Ground-active invertebrates	*Phragmites australis* (Cav.) Trin. ex Steud. (common reed)	Great Lakes marshes, USA	Reduced diversity and altered assemblages of ground-active invertebrates	Changes in vegetation structure and microhabitat	Lower *et al.* (2024)
Soil microhabitat	*Amynthas spp.* (jumping worms)	Temperate forests, USA	Reduced leaf litter biomass; minimal effects on soil conditions and native herpetofauna	Consumption of leaf litter by invasive earthworms	Bennett *et al.* (2024)

and indirect consequences of invasions on soil biodiversity, emphasizing the context-dependent nature of these interactions. Recognizing these cross-taxa patterns is essential for designing restoration strategies that address not only plant removal but also the recovery of soil faunal communities and their functional roles.

Management Implications and Future Directions

Implications for ecosystem restoration and sustainable management approaches

Although the presence of invasive species in natural habitats is well documented, there is no clear understanding of their impact on soil, as their influence can range from negative to neutral to positive. However, the functionality, together with the tourism and recreational attractiveness of areas occupied by invasive species, becomes limited. Monocultures of invasive plants are often unattractive to people, reduce landscape values and limit possibilities for utilization. As a consequence, this can lead to a decrease in the economic value of land and a decrease in income from tourism or agriculture in areas where invasive species spread. It is essential to emphasize that we are obligated to implement specialized management and restoration strategies, and these approaches will likely vary depending on the specific environmental conditions and the resources and knowledge available at a given time.

The presence of invasive plants also has serious economic consequences, especially in the agricultural sector (Arun *et al.*, 2022). These species affect the functionality of agricultural land, not only by significantly reducing crop yields but also by overgrowing the banks of ditches and riverbeds and weakening the function of field margins. They also harm the populations of pollinators and predators that control pests, further exacerbating the situation in agriculture. In addition, some invasive species, such as Sosnowsky's hogweed (*Heracleum sosnowskyi* Manden.), can pose a direct threat to human health (Słowiński *et al.*, 2024).

Future research needs

Combating invasive plants requires an interdisciplinary approach that combines modern technologies, biological and mechanical methods, social education and cooperation in many aspects. Continuous monitoring of threats and scientific studies to develop more effective and environmentally friendly control methods is also of key importance. Only such comprehensive actions can effectively reduce the negative impact of invasive plants on ecosystems and the economy. The key directions for development and action are as follows.

Integrated control methods

Controlling invasive plants requires the implementation of various techniques. Weidlich *et al.* (2020) emphasized that there are non-chemical methods (such as cutting, soil preparation, mowing and fire) and chemical ones (including selective herbicides and growth regulators). However, the decision about which control method to use depends heavily on the invasive plant species' growth, the local economic situation and resources available for control. Significant successes have been reported with the use of electromagnetic radiation (microwaves). This method is classified as a thermal technique for weed control, but it is gaining popularity in controlling the populations of Sosnowski hogweed and Japanese *F. japonica*. Furthermore, compared to the other thermal methods, microwaves are more environmentally friendly as they does not require gases to produce heat (Słowiński *et al.*, 2024).

It is also possible to use biological methods (introduction of natural enemies of plants, pathogens, or competitive species). However, research shows that biological control programmes have both positive and negative consequences, differentially affecting various groups of stakeholders (Seastedt, 2015). The primary point of uncertainty in the widespread use of biological control is related to potential additional interference in the ecosystem and the unpredictability of the effects of such actions. Despite these uncertainties, biological control could play a significant role in mitigation and adaptation strategies used to maintain biological diversity, as well as contribute to human well-being by protecting food and fibre resources. Regardless, the systematic and properly planned use of the various controlling methods allows for restoring the balance of ecosystems and limiting the further expansion of invasive species.

Constant monitoring and mapping

The use of modern geographic information systems (GIS) enables precise mapping and monitoring of areas at risk of invasion, facilitating a rapid response and more effective management. In relation to this, it is proposed to use of sensory classification methods that combine GPS-based field surveys and drone-operated hyperspectral aerial photography, which can be effectively used to monitor and map invaded areas (Papp *et al.*, 2021).

Education and introduction of special programmes

In the face of the growing global problem of invasive plants, it is necessary to expand and strengthen education, as well as increase social participation in controlling the spread of invasive species. Introducing educational programmes that raise the ecological awareness of local communities and land users reduces the risk of unknowingly introducing and spreading invasive plants. According to Verbrugge *et al.* (2021), comprehensive educational programmes should be introduced that target key stakeholder groups, including policy makers and authorities (local and national), nature and water managers, professionals working outdoors, retailers and traders. An important aspect in implementing educational and monitoring programmes, as well as protecting the environment against invasive plants, is systemic financing and legislative support. Unfortunately, in many countries there is no such systematic approach to this issue. Therefore, such support should be developed not only on a local scale but also internationally, because, as the current situation indicates, invasive plants are 'citizens' of the world. The development of formal strategies and legal regulations for combating biological invasions, which will enable better coordination of activities at the national and regional levels (Verbrugge *et al.*, 2021).

Scientific research and development of restoration methods

More research is needed on the mechanisms behind how invasive plants impact ecosystems and the development of effective methods for restoring biodiversity and soil functionality after their removal. This research should focus on ecologically safe control methods that use natural enemies, as well as proper management of the soil environment to maintain an optimal pH, nutrient availability and level of organic matter. It is also important to understand the long-term effects of invasive species on the properties and functioning of the soil and, once the problem has been dealt with, methods for restoring its natural functions. A promising direction of research is the analysis of interactions between invasive species and soil microorganisms, which play a key role in maintaining soil fertility and health. A better understanding of these interactions could contribute to the development of more effective methods for the remediation of degraded ecosystems using natural biological processes.

References

Abgrall, C., Forey, E. and Chauvat, M. (2019) Soil fauna responses to invasive alien plants are determined by trophic groups and habitat structure: A global meta-analysis. *Oikos* 128, 1390–1401. DOI: 10.1111/oik.06493.

Ab Razak, N., Gange, A.C., Sutton, B.C. and Mansor, A. (2023) The invasive plant *Impatiens glandulifera* manipulates microbial associates of competing native species. *Plants* 12, 1552. DOI: 10.3390/plants12071552.

Afzal, M.R., Naz, M., Ashraf, W. and Du, D. (2023) The legacy of plant invasion: Impacts on soil nitrification and management implications. *Plants* 12, 2980. DOI: 10.3390/plants12162980.

Anthony, M.A. (2019) Fungal community and soil responses to invasive *Alliaria petiolata* (garlic mustard), soil warming, and nitrogen deposition. Dissertation, University of New Hampshire. Available at: https://scholars.unh.edu/dissertation/2435

Araujo, Y., Luizão, F.J. and Barros, E. (2004) Effect of earthworm addition on soil nitrogen availability, microbial biomass and litter decomposition in mesocosms. *Biology and Fertility of Soils* 39, 146–152. DOI: 10.1007/s00374-003-0696-0.

Arun, M.N., Kumar, R.M., Sreedevi, B., Padmavathi, G.P., Revathi, P. *et al.* (2022) The rising threat of invasive alien plant species in agriculture. In: Ondrasek, G. and Zhang, L. (eds) *Resource Management in Agroecosystems*. IntechOpen. DOI: 10.5772/intechopen.106742.

Bagyaraj, D., Nethravathi, C. and Nitin, K. (2016) Soil biodiversity and arthropods: Role in soil fertility. In: Chakravarthy, A.K. and Sridhara, S. (eds) *Economic and Ecological Significance of Arthropods in Diversified Ecosystems: Sustaining Regulatory Mechanisms*. Springer, pp. 17–51. DOI: 10.1007/978-981-10-1524-3_2.

Bennett, S., Phillips, H.R., Dalziel, A.C., Manzer, L.R. and Cameron, E.K. (2024) Testing the impacts of invasive jumping worms at their northern range limit. *European Journal of Soil Biology* 120, 103590. DOI: 10.1016/j.ejsobi.2023.103590.

Bonfanti, J., Potapov, A.M., Angst, G., Ganault, P., Briones, M.J. *et al.* (2025) Linking effect traits of soil fauna to processes of organic matter transformation. *Functional Ecology* 39, 446–461. DOI: 10.1111/1365-2435.14720.

Brousseau, P.M., Forey, E., Santonja, M. and Chauvat, M. (2025) Functional traits of plant roots and Collembola determine their tri-trophic interactions with soil microbes. *Functional Ecology* 39, 462–477. DOI: 10.1111/1365-2435.14714.

Caiyun, Z., Xiaoyan, L., Jiade, B., Fengchun, L. and Junsheng, L. (2014) Impact of *Spartina alterniflora* on benthic macro-invertebrates communities on mangrove wetland in Xicungang Estuary, Guangxi. *Biodiversity Science* 22, 630. DOI: 10.3724/SP.J.1003.2014.00109.

Callaway, R.M., Thelen, G.C., Rodriguez, A. and Holben, W.E. (2004) Soil biota and exotic plant invasion. *Nature* 427, 731–733. DOI: 10.1038/nature02322.

Callaway, R.M., Cipollini, D., Barto, K., Thelen, G.C., Hallett, S.G. *et al.* (2008) Novel weapons: Invasive plant suppresses fungal mutualists in America but not in its native Europe. *Ecology* 89, 1043–1055. DOI: 10.1890/07-0370.1.

Chen, H., Li, B., Hu, J., Chen, J. and Wu, J. (2007) Effects of *Spartina alterniflora* invasion on benthic nematode communities in the Yangtze Estuary. *Marine Ecology Progress Series* 336, 99–110.

Chen, Q., Yu, Z. and Wang, M. (2023) Effects of *Spartina alterniflora* invasion on macrobenthic faunal community in different habitats of a mangrove wetland in Zhanjiang, South China. *Regional Studies in Marine Science* 66, 103148. DOI: 10.1016/j.rsma.2023.103148.

Christian, C.E. (2001) Consequences of a biological invasion reveal the importance of mutualism for plant communities. *Nature* 413, 635–639. DOI: 10.1038/35098093.

Clause, J., Forey, E., Lortie, C.J., Lambert, A.M. and Barot, S. (2015) Non-native earthworms promote plant invasion by ingesting seeds and modifying soil properties. *Acta Oecologica* 64, 10–20. DOI: 10.1016/j.actao.2015.02.004.

Craven, D., Thakur, M.P., Cameron, E.K., Frelich, L.E., Beauséjour, R. *et al.* (2017) The unseen invaders: Introduced earthworms as drivers of change in plant communities in North American forests (a meta-analysis). *Global Change Biology* 23, 1065–1074. DOI: 10.1111/gcb.13446.

Čuchta, P., Kaňa, J. and Pouska, V. (2019) An important role of decomposing wood for soil environment with a reference to communities of springtails (Collembola). *Environmental Monitoring and Assessment* 191, 222. DOI: 10.1007/s10661-019-7363-x.

Cui, M., Yu, H., Fan, X., Nawaz, M., Lian, J. *et al.* (2023) Nitrogen deposition amplifies the legacy effects of plant invasion. *Plants* 13, 72. DOI: 10.3390/plants13010072.

D'Antonio, C.M., Hughes, R.F. and Tunison, J.T. (2011) Long-term impacts of invasive grasses and subsequent fire in seasonally dry Hawaiian woodlands. *Ecological Applications* 21, 1617–1628. DOI: 10.1890/10-0638.1.

De Almeida, T., Forey, E. and Chauvat, M. (2022) Alien invasive plant effect on soil fauna is habitat dependent. *Diversity* 14, 61. DOI: 10.3390/d14020061.

DeVore, J.L. (2011) An exercise in complexity: Indirect influences of invasion by an exotic grass (*Microstegium vimineum*) on forest floor food webs. PhD thesis, The University of Georgia.

Dickie, I.A., Bolstridge, N., Cooper, J.A. and Peltzer, D.A. (2010) Co-invasion by *Pinus* and its mycorrhizal fungi. *New Phytologist* 187, 475–484. DOI: 10.1111/j.1469-8137.2010.03277.x.

Dickie, I.A., Bufford, J.L., Cobb, R.C., Desprez-Loustau, M.L., Grelet, G.P. *et al.* (2017) The emerging science of linked plant–fungal invasions. *New Phytologist* 215, 1314–1332. DOI: 10.1111/nph.14657.

Duda, J.J., Freeman, D.C., Emlen, J.M., Belnap, J., Kitchen, S.G. *et al.* (2003) Differences in native soil ecology associated with invasion of the exotic annual chenopod, *Halogeton glomeratus*. *Biology and Fertility of Soils* 38, 72–77. DOI: 10.1007/s00374-003-0638-x.

Dukes, J.S. and Mooney, H.A. (1999) Does global change increase the success of biological invaders? *Trends in Ecology and Evolution* 14, 135–139. DOI: 10.1016/S0169-5347(98)01554-7.

Ehrenfeld, J.G. (2010) Ecosystem consequences of biological invasions. *Annual Review of Ecology, Evolution, and Systematics* 41, 59–80. DOI: 10.1146/annurev-ecolsys-102209-144650.

Eisenhauer, N., Milcu, A., Sabais, A.C. and Scheu, S. (2008) Animal ecosystem engineers modulate the diversity-invasibility relationship. *PLoS ONE* 3, e3489. DOI: 10.1371/journal.pone.0003489.

Enloe, S.F., DiTomaso, J.M., Orloff, S.B. and Drake, D.J. (2004) Soil water dynamics differ among rangeland plant communities dominated by yellow starthistle (*Centaurea solstitialis*), annual grasses, or perennial grasses. *Weed Science* 52, 929–935. DOI: 10.1614/WS-03-156R.

Eppinga, M.B., Rietkerk, M., Dekker, S.C., Ruiter, P.C., Van der Putten, W.H. *et al.* (2006) Accumulation of local pathogens: A new hypothesis to explain exotic plant invasions. *Oikos* 114, 168–176. DOI: 10.1111/j.2006.0030-1299.14625.x.

Ertiban, S.M. (2019) Soil fauna as webmasters, engineers and bioindicators in ecosystems: Implications for conservation ecology and sustainable agriculture. *American Journal of Life Sciences* 7, 17–26. DOI: 10.11648/j.ajls.20190701.14.

Fang, Y., Zhang, X., Wei, H., Wang, D., Chen, R. *et al.* (2021) Predicting the invasive trend of exotic plants in China based on the ensemble model under climate change: A case for three invasive plants of Asteraceae. *Science of the Total Environment* 756, 143841. DOI: 10.1016/j.scitotenv.2020.143841.

Fausak, L.K., Bridson, N., Diaz-Osorio, F., Jassal, R.S. and Lavkulich, L.M. (2024) Soil health–a perspective. *Frontiers in Soil Science* 4, 1462428. DOI: 10.3389/fsoil.2024.1462428.

Ferlian, O., Eisenhauer, N., Aguirrebengoa, M., Camara, M., Ramirez-Rojas, I. *et al.* (2018) Invasive earthworms erode soil biodiversity: A meta-analysis. *Journal of Animal Ecology* 87, 162–172. DOI: 10.1111/1365-2656.12746.

Ferlian, O., Thakur, M.P., Castañeda González, A., San Emeterio, L.M., Marr, S. *et al.* (2020) Soil chemistry turned upside down: A meta-analysis of invasive earthworm effects on soil chemical properties. *Ecology* 101, e02936. DOI: 10.1002/ecy.2936.

Ferlian, O., Goldmann, K., Bonkowski, M., Dumack, K., Wubet, T. *et al.* (2024) Invasive earthworms shift soil microbial community structure in northern North American forest ecosystems. *Iscience* 27. DOI: 10.1016/j.isci.2024.108889.

Fernandez, R.D., Moreno, M.L., Aragón, R. and Harguindeguy, N.P. (2022) *Ligustrum lucidum* invasion decreases abundance and relative contribution of soil fauna to litter decomposition but increases decomposition rate in a subtropical montane forest of northwestern Argentina. *Canadian Journal of Forest Research* 52, 261–268. DOI: 10.1139/cjfr-2021-0169.

Florianová, A., Hanzelková, V., Drtinová, L., Pánková, H., Cajthaml, T. *et al.* (2023) Plant–soil interactions in the native range of two congeneric species with contrasting invasive success. *Oecologia* 201, 461–477. DOI: 10.1007/s00442-023-05329-6.

Forey, E., Lodhar, S., Gopaul, S., Boehmer, H.J. and Chauvat, M. (2021) A functional trait-based approach to assess the impact of an alien palm invasion on plant and soil communities on a South Pacific island. *Austral Ecology* 46, 398–410. DOI: 10.1111/aec.12995.

Frouz, J. (2018) Effects of soil macro-and mesofauna on litter decomposition and soil organic matter stabilization. *Geoderma* 332, 161–172. DOI: 10.1016/j.geoderma.2017.08.039.

Funk, J.L. (2013) The physiology of invasive plants in low-resource environments. *Conservation Physiology* 1, cot026. DOI: 10.1093/conphys/cot026.

Gaggini, L., Rusterholz, H.P. and Baur, B. (2019) The annual invasive plant *Impatiens glandulifera* reduces hyphal biomass of soil fungi in deciduous forests. *Fungal Ecology* 39, 242–249. DOI: 10.1016/j.funeco.2018.12.004.

Gebremikael, M.T., Buchan, D. and De Neve, S. (2014) Quantifying the influences of free-living nematodes on soil nitrogen and microbial biomass dynamics in bare and planted microcosms. *Soil Biology and Biochemistry* 70, 131–141. DOI: 10.1016/j.soilbio.2013.12.006.

Gebremikael, M.T., Steel, H., Buchan, D., Bert, W. and De Neve, S. (2016) Nematodes enhance plant growth and nutrient uptake under C and N-rich conditions. *Scientific Reports* 6, 32862. DOI: 10.1038/srep32862.

Gibbons, S.M., Lekberg, Y., Mummey, D.L., Sangwan, N., Ramsey, P.W. *et al*. (2017) Invasive plants rapidly reshape soil properties in a grassland ecosystem. *MSystems* 2. DOI: 10.1128/msystems.00178-16.

Gioria, M., Hulme, P.E., Richardson, D.M. and Pyšek, P. (2023) Why are invasive plants successful? *Annual Review of Plant Biology* 74, 635–670. DOI: 10.1146/annurev-arplant-070522-071021.

Gongalsky, K.B. (2021) Soil macrofauna: Study problems and perspectives. *Soil Biology and Biochemistry* 159, 108281. DOI: 10.1016/j.soilbio.2021.108281.

Görres, J.H. and Amador, J.A. (2021) The soil fauna. In: Gentry, T.J., Fuhrmann, J.J. and Zuberer, D.A. (eds) *Principles and Applications of Soil Microbiology. Elsevier*. pp. 191–212. DOI: 10.1016/B978-0-12-820202-9.00008-3.

Hale, C.M., Frelich, L.E., Reich, P.B. and Pastor, J. (2008) Exotic earthworm effects on hardwood forest floor, nutrient availability and native plants: A mesocosm study. *Oecologia* 155, 509–518. DOI: 10.1007/s00442-007-0925-6.

Hansda, P., Kumar, S., Singh, S. and Garkoti, S.C. (2024) Assessing the influence of invasion of *Lantana camara* on vegetation attributes and soil properties across varied disturbance gradients in semi-arid forests of Aravali hills, Delhi. *Plant Ecology* 225, 893–906. DOI: 10.1007/s11258-024-01441-6.

Hayward, J., Horton, T.R., Pauchard, A. and Nuñez, M.A. (2015) A single ectomycorrhizal fungal species can enable a *Pinus* invasion. *Ecology* 96, 1438–1444. DOI: 10.1890/14-1100.1.

He, M., Liu, T., Liu, Y., Long, Y., Li, Y. *et al*. (2024) Impacts of *Solidago canadensis* and *Erigeron canadensis* co-invasion on soil microbial communities. *Conservation Science and Practice* 6, e13202. DOI: 10.1111/csp2.13202.

He, Y.H., Rutherford, S., Javed, Q., Wan, J.S., Ren, G.Q. *et al*. (2022) Mixed litter and incubation sites drive non-additive responses in seed germination and seedling growth of lettuce. *Biochemical Systematics and Ecology* 105, 104479. DOI: 10.1016/j.bse.2022.104479.

Hejda, M., Pyšek, P. and Jarošík, V. (2009) Impact of invasive plants on the species richness, diversity and composition of invaded communities. *Journal of Ecology* 97, 393–403. DOI: 10.1111/j.1365-2745.2009.01480.x.

Hickerson, C.A.M. (2010) Interactions among top-down regulators in a temperate forest floor ecosystem; effects on macrofauna, mesofauna, microbes and litter decay. Dissertation. Cleveland State University. Available at: https://engagedscholarship.csuohio.edu/etdarchive/131

Innes, R.J. and Zouhar, K. (2021) *Centaurea solstitialis*, yellow starthistle. Fire Effects Information System. US Department of Agriculture, Forest Service, Rocky Mountains Research Station, Fire Sciences Laboratory. Available at: www.fs.usda.gov/database/feis/plants/forb/censol/all.html

Jacobs, K., Conradie, T. and Jacobs, S. (2020) Microbial communities in the fynbos region of South Africa: What happens during woody alien plant invasions. *Diversity* 12, 254. DOI: 10.3390/d12060254.

Jakubcsiková, M., Demková, L., Renčo, M. and Čerevková, A. (2024) Impact of *Solidago gigantea* invasion on nematode communities in grasslands. *Nematology* 26, 779–793. DOI: 10.1017/S0022149X19000324.

Jakubus, M., Černe, M., Palčić, I., Pasković, I., Ban, S.G. *et al*. (2025) The application of sewage sludge-derived compost or biochar as a nature-based solution (NBS) for healthier soil. *Sustainability* 17, 1630. DOI: 10.3390/su17041630.

Javed, Q., Sun, J., Rutherford, S., Li, J., Iqbal, B. *et al*. (2023) Soil pollution and the invasion of congener *Sphagneticola* in crop lands. *Journal of Environmental Management* 340, 118013. DOI: 10.1016/j.jenvman.2023.118013.

Javed, Q., Sun, J., Černe, M., Iqbal, B., Bouhadi, M. *et al*. (2025) The effect of native plant diversity on the success of invasive species in polluted soils. *Ecotoxicology* 1–17. DOI: 10.1007/s10646-025-02905-9.

Jin, H., Chang, L., van Kleunen, M. and Liu, Y. (2022) Soil mesofauna may buffer the negative effects of drought on alien plant invasion. *Journal of Ecology* 110, 2332–2342. DOI: 10.1111/1365-2745.13950.

Kim, G., Jo, H., Kim, H.S., Kwon, M. and Son, Y. (2022) Earthworm effects on soil biogeochemistry in temperate forests focusing on stable isotope tracing: A review. *Applied Biological Chemistry* 65, 88. DOI: 10.1186/s13765-022-00758-y.

Kohyt, J. and Skubała, P. (2020) Oribatid mite (Acari: Oribatida) communities reveal the negative impact of the red oak (*Quercus rubra* L.) on soil fauna in Polish commercial forests. *Pedobiologia* 79, 150594. DOI: 10.1016/j.pedobi.2019.150594.

Koutika, L.S., Vanderhoeven, S., Chapuis-Lardy, L., Dassonville, N. and Meerts, P. (2007) Assessment of changes in soil organic matter after invasion by exotic plant species. *Biology and Fertility of Soils* 44, 331–341. DOI: 10.1007/s00374-007-0210-1.

Lang, B., Betancur-Corredor, B. and Russell, D.J. (2023) Earthworms increase mineral soil nitrogen content–a meta-analysis. *Soil Organisms* 95, 1–16. DOI: 10.25674/so95iss1id308.

Lazzaro, L., Mazza, G., d'Errico, G., Fabiani, A., Giuliani, C. *et al.* (2018) How ecosystems change following invasion by *Robinia pseudoacacia*: Insights from soil chemical properties and soil microbial, nematode, microarthropod and plant communities. *Science of the Total Environment* 622, 1509–1518. DOI: 10.1016/j.scitotenv.2017.10.017.

Le Maitre, D.C., Gaertner, M., Marchante, E., Ens, E.J., Holmes, P.M. *et al.* (2011) Impacts of invasive Australian acacias: Implications for management and restoration. *Diversity and Distributions* 17, 1015–1029. DOI: 10.1111/j.1472-4642.2011.00816.x.

Lee, S.F., Chapman, S.K., Mozdzer, T.J., Eller, F. and Langley, J.A. (2024) Does introduced European *Phragmites australis* experience belowground microbial enemy release in North America? *Ecosphere* 15, e4952. DOI: 10.1002/ecs2.4952.

Lehmann, J., Bossio, D.A., Kögel-Knabner, I. and Rillig, M.C. (2020) The concept and future prospects of soil health. *Nature Reviews Earth & Environment* 1, 544–553. DOI: 10.1038/s43017-020-0080-8.

Li, K., Ni, R., Lv, C., Xue, L., Zhang, C. *et al.* (2022) The effect of *Robinia pseudoacacia* expansion on the soil particle size distribution on Mount Tai, China. *Catena* 208, 105774. DOI: 10.1016/j.catena.2021.105774.

Li, Y.X., Laborda, P., Xie, X.L., Zhou, R., Chen, Y. *et al.* (2020) *Spartina alterniflora* invasion alters soil microbial metabolism in coastal wetland of China. *Estuarine, Coastal and Shelf Science* 245, 106982. DOI: 10.1016/j.ecss.2020.106982.

Litton, C.M., Sandquist, D.R. and Cordell, S. (2008) A non-native invasive grass increases soil carbon flux in a Hawaiian tropical dry forest. *Global Change Biology* 14, 726–739. DOI: 10.1111/j.1365-2486.2008.01546.x.

Liu, Y., Oduor, A.M., Zhang, Z., Manea, A., Tooth, I.M. *et al.* (2017) Do invasive alien plants benefit more from global environmental change than native plants? *Global Change Biology* 23, 3363–3370. DOI: 10.1111/gcb.13579.

Liu, Y., Wang, L., He, R., Chen, Y., Xu, Z. *et al.* (2019) Higher soil fauna abundance accelerates litter carbon release across an alpine forest-tundra ecotone. *Scientific Reports* 9, 10561. DOI: 10.1038/s41598-019-47072-0.

Long, K., Yin, R., Kardol, P., Wei, Q., Li, Y. *et al.* (2023) Bamboo invasion alters Collembola community composition varying with life-forms. *Pest Management Science* 79, 2517–2526. DOI: 10.1002/ps.7434.

Lower, E., Sturtevant, R., Iott, S., Martinez, F., Rutherford, E. *et al.* (2024) The Great Lakes' most unwanted: Characterizing the impacts of the top ten Great Lakes aquatic invasive species. *Journal of Great Lakes Research* 50, 102365. DOI: 10.1016/j.jglr.2024.102365.

Lyu, C., Zhang, S., Ren, X., Liu, M., Leung, K.S.K. *et al.* (2023) The effect of *Spartina alterniflora* eradication on waterbirds and benthic organisms. *Restoration Ecology* 31, e14023. DOI: 10.1111/rec.14023.

Majewska, M.L., Rola, K., Stefanowicz, A.M., Nobis, M., Błaszkowski, J. *et al.* (2018) Do the impacts of alien invasive plants differ from expansive native ones? An experimental study on arbuscular mycorrhizal fungi communities. *Biology and Fertility of Soils* 54, 631–643. DOI: 10.1007/s00374-018-1283-8.

Ma, L.N., Lü, X.T., Liu, Y., Guo, J.X., Zhang, N.Y. *et al.* (2011) The effects of warming and nitrogen addition on soil nitrogen cycling in a temperate grassland, northeastern China. *PLoS ONE* 6, e27645. DOI: 10.1371/journal.pone.0027645.

Mangla, S., Inderjitand Callaway, R.M. (2008) Exotic invasive plant accumulates native soil pathogens which inhibit native plants. *Journal of Ecology* 96, 58–67. DOI: 10.1111/j.1365-2745.2007.01312.x.

McCary, M.A. and Wise, D.H. (2019) Plant invader alters soil food web via changes to fungal resources. *Oecologia* 191, 587–599. DOI: 10.1007/s00442-019-04510-0.

Medina-Sauza, R.M., Álvarez-Jiménez, M., Delhal, A., Reverchon, F., Blouin, M. *et al.* (2019) Earthworms building up soil microbiota, a review. *Frontiers in Environmental Science* 7, 81. DOI: 10.3389/fenvs.2019.00081.

Novoa, A., Rodríguez, R., Richardson, D. and González, L. (2014) Soil quality: A key factor in understanding plant invasion? The case of *Carpobrotus edulis* (L.) NE Br. *Biological Invasions* 16, 429–443. DOI: 10.1007/s10530-013-0531-y.

Orlowsky, B. and Seneviratne, S.I. (2013) Elusive drought: Uncertainty in observed trends and short-and long-term CMIP5 projections. *Hydrology and Earth System Sciences* 17, 1765–1781. DOI: 10.5194/hess-17-1765-2013.

Pachauri, R.K., Allen, M.R., Barros, V.R., Broome, J., Cramer, W. *et al.* (2014) In: Pachauri, R. (ed.) *Climate Change 2014: Synthesis Report. Contribution of Working Groups I, II and III to the Fifth Assessment Report of the Intergovernmental Panel on Climate Change*. IPCC, Geneva.

Papp, L., Leeuwen, B., Szilassi, P., Tobak, Z., Szatmári, J. *et al.* (2021) Monitoring invasive plant species using hyperspectral remote sensing data. *Land* 10, 29. DOI: 10.3390/land10010029.

Peng, Y., Yang, J.X., Zhou, X.H., Peng, P.H., Li, J.J. *et al.* (2019) An invasive population of *Solidago canadensis* is less sensitive to warming and nitrogen-addition than its native population in an invaded range. *Biological Invasions* 21, 151–162. DOI: 10.1007/s10530-018-1812-2.

Petermann, J.S., Fergus, A.J., Turnbull, L.A. and Schmid, B. (2008) Janzen-Connell effects are widespread and strong enough to maintain diversity in grasslands. *Ecology* 89, 2399–2406. DOI: 10.1890/07-2056.1.

Petsikos, C., Dalias, P. and Troumbis, A.Y. (2007) Effects of *Oxalis pes-caprae* L. invasion in olive groves. *Agriculture, Ecosystems & Environment* 120, 325–329. DOI: 10.1016/j.agee.2006.10.019.

Potapov, A.M., Beaulieu, F., Birkhofer, K., Bluhm, S.L., Degtyarev, M.I. *et al.* (2022) Feeding habits and multifunctional classification of soil-associated consumers from protists to vertebrates. *Biological Reviews* 97, 1057–1117. DOI: 10.1111/brv.12832.

Prabhakar, S. (2023) Influence of soil invertebrates on soil decomposition. *Journal of Survey in Fisheries Sciences* 10, 618–629.

Rai, P.K. and Singh, J. (2020) Invasive alien plant species: Their impact on environment, ecosystem services and human health. *Ecological Indicators* 111, 106020. DOI: 10.1016/j.ecolind.2019.106020.

Ren, G.Q., Zou, C.B., Wan, L.Y., Johnson, J.H., Li, J. *et al.* (2021) Interactive effect of climate warming and nitrogen deposition may shift the dynamics of native and invasive species. *Journal of Plant Ecology* 14, 84–95. DOI: 10.1093/jpe/rtaa071.

Renčo, M., Čerevková, A. and Gömöryová, E. (2025) Floodplain forest soil nematode communities as influenced by non-native *Acer negundo* L. invasion. *Diversity* 17, 376. DOI: 10.3390/d17060376.

Rivas-Torres, A., Graça, M.A., Landeira-Dabarca, A., Álvarez, M., Juen, L. *et al.* (2025) *Eucalyptus globulus* afforestation reduces invertebrate richness and diversity in streams. *Hydrobiology* 4, 16. DOI: 10.3390/hydrobiology4020016.

Rodgers, V.L., Wolfe, B.E., Werden, L.K. and Finzi, A.C. (2008) The invasive species *Alliaria petiolata* (garlic mustard) increases soil nutrient availability in northern hardwood-conifer forests. *Oecologia* 157, 459–471. DOI: 10.1007/s00442-008-1089-8.

Rossi, R.D., Martins, C.R., Viana, P.L., Rodrigues, E.L. and Figueira, J.E.C. (2014) Impact of invasion by molasses grass (*Melinis minutiflora* P. Beauv.) on native species and on fires in areas of campo-cerrado in Brazil. *Acta Botanica Brasilica* 28, 631–637. DOI: 10.1590/0102-33062014abb3390.

Schirmel, J., Bundschuh, M., Entling, M.H., Kowarik, I. and Buchholz, S. (2016) Impacts of invasive plants on resident animals across ecosystems, taxa, and feeding types: A global assessment. *Global Change Biology* 22, 594–603. DOI: 10.1111/gcb.13093.

Schittko, C. (2015) On species invasions in urban grasslands: Interactions with soil biota, functional diversity, and global change. PhD dissertation, Freie Universität, Berlin.

Schramm, J.W. and Ehrenfeld, J.G. (2010) Leaf litter and understory canopy shade limit the establishment, growth and reproduction of *Microstegium vimineum*. *Biological Invasions* 12, 3195–3204. DOI: 10.1007/s10530-010-9711-1.

Seastedt, T.R. (2015) Biological control of invasive plant species: A reassessment for the A nthropocene. *New Phytologist* 205, 490–502. DOI: 10.1111/nph.13065.

Shalmiyev, G., Shmulenson, A. and Zhao, D. (2024) Allelopathy of *F. japonica* as an invasive species (January 16, 2024), SSRN. Available at: https://papers.ssrn.com/sol3/Delivery.cfm?abstractid=4739895

Shi, L., Liu, F. and Peng, L. (2023) Impact of red imported fire ant nest-building on soil properties and bacterial communities in different habitats. *Animals* 13. DOI: 10.3390/ani13122026.

Słowiński, K., Grygierzec, B., Wajs-Bonikowska, A., Baran, A., Tabor, S. *et al.* (2024) Biochemistry of microwave controlled *Heracleum sosnowskyi* (Manden.) roots with an ecotoxicological aspect. *Scientific Reports* 14, 14260. DOI: 10.1038/s41598-024-65164-4.

Song, J., Xia, J., Hui, D., Zheng, M., Wang, J. *et al.* (2021) Plant functional types regulate non-additive responses of soil respiration to 5-year warming and nitrogen addition in a semi-arid grassland. *Functional Ecology* 35, 2593–2603. DOI: 10.1111/1365-2435.13902.

Song, J., Tang, Z., Zhao, X., Yin, Y., Li, X. *et al.* (2023) Red imported fire ant nesting affects the structure of soil microbial community. *Frontiers in Cellular and Infection Microbiology* 13, 1221996. DOI: 10.3389/fcimb.2023.1221996.

Standish, R.J., Williams, P.A., Robertson, A.W., Scott, N.A. and Hedderley, D.I. (2004) Invasion by a perennial herb increases decomposition rate and alters nutrient availability in warm temperate lowland forest remnants. *Biological Invasions* 6, 71–81. DOI: 10.1023/B:BINV.0000010127.06695.f4.

Stefanowicz, A.M., Kapusta, P., Stanek, M., Frąc, M., Oszust, K. *et al.* (2021) Invasive plant *Reynoutria japonica* produces large amounts of phenolic compounds and reduces the biomass but not activity of soil microbial communities. *Science of the Total Environment* 767, 145439. DOI: 10.1016/j.scitotenv.2021.145439.

Stinson, K.A., Campbell, S.A., Powell, J.R., Wolfe, B.E. and Callaway, R.M. (2006) Invasive plant suppresses the growth of native tree seedlings by disrupting belowground mutualisms. *PLoS Biology* 4, e140. DOI: 10.1371/journal.pbio.0040140.

Stinson, K.A., Frey, S.D., Jackson, M., Coates-Connor, E., Anthony, M. *et al.* (2018) Responses of non-native earthworms to experimental eradication of garlic mustard and implications for native vegetation. *Ecosphere* 9, e02353. DOI: 10.1002/ecs2.2353.

Sun, J., Khattak, W.A., Abbas, A., Nawaz, M., Hameed, R. *et al.* (2023) Ecological adaptability of invasive weeds under environmental pollutants: A review. *Environmental and Experimental Botany* 215, 105492. DOI: 10.1016/j.envexpbot.2023.105492.

Sun, Z.J., Setälä, H., Cui, Y.X., Meng, C., Cui, M.M. *et al.* (2024) Effects of wild boar grubbing on the soil nematode community subject to seasonal variation in a broad-leaved Korean pine forest in Northeast China. *Forest Ecology and Management* 551, 121549. DOI: 10.1016/j.foreco.2023.121549.

Thouvenot, L., Ferlian, O., Horn, L., Jochum, M. and Eisenhauer, N. (2024) Effects of earthworm invasion on soil properties and plant diversity after two years of field experiment. *NeoBiota* 94, 31–56. DOI: 10.3897/neobiota.94.119307.

Topp, W., Kappes, H. and Rogers, F. (2008) Response of ground-dwelling beetle (Coleoptera) assemblages to giant knotweed (*Reynoutria* spp.) invasion. *Biological Invasions* 10, 381–390. DOI: 10.1007/s10530-007-9137-6.

Uddin, M.N. and Robinson, R.W. (2018) Can nutrient enrichment influence the invasion of *Phragmites australis*? *Science of the Total Environment* 613, 1449–1459. DOI: 10.1016/j.scitotenv.2017.06.131.

Ulm, F. (2019) Impacts of *Acacia longifolia* invasion on soil nutrient cycles: From invasion to solution. PhD thesis, Universidade de Lisboa.

Unger, I.M., Kremer, R.J., Veum, K.S. and Goyne, K.W. (2022) Immediate and long-term effects of invasive plant species on soil characteristics. *Soil Ecology Letters* 4, 276–288. DOI: 10.1007/s42832-021-0104-4.

Van der Putten, W.H. (2010) Impacts of soil microbial communities on exotic plant invasions. *Trends in Ecology and Evolution* 25, 512–519. DOI: 10.1016/j.tree.2010.06.006.

Verbrugge, L.N., Dawson, M., Gettys, L.A., Leuven, R.S., Marchante, H. *et al.* (2021) Novel tools and best practices for education about invasive alien species. *Management of Biological Invasions* 12(1), 8–24. DOI: 10.3391/mbi.2021.12.1.02.

Vilà, M., Trillo, A., Castro-Díez, P., Gallardo, B. and Bacher, S. (2024) Field studies of the ecological impacts of invasive plants in Europe. *NeoBiota* 90, 139–159. DOI: 10.3897/neobiota.90.112368.

Vogelsang, K.M. and Bever, J.D. (2009) Mycorrhizal densities decline in association with nonnative plants and contribute to plant invasion. *Ecology* 90, 399–407. DOI: 10.1890/07-2144.1.

Vujanović, D., Losapio, G., Milić, S. and Milić, D. (2022) The impact of multiple species invasion on soil and plant communities increases with invasive species co-occurrence. *Frontiers in Plant Science* 13, 875824. DOI: 10.3389/fpls.2022.875824.

Walter, D.E. and Proctor, H.C. (2013) Mites in soil and litter systems. In: Walter, D.E. and Proctor, H.C. (eds) *Mites: Ecology, Evolution & Behaviour: Life at a Microscale*. Springer, pp. 161–228. DOI: 10.1007/978-94-007-7164-2_6.

Wei, H., Xu, J., Quan, G., Zhang, J. and Qin, Z. (2017) Invasion effects of *Chromolaena odorata* on soil carbon and nitrogen fractions in a tropical savanna. *Ecosphere* 8, e01831. DOI: 10.1002/ecs2.1831.

Weidenhamer, J.D. and Callaway, R.M. (2010) Direct and indirect effects of invasive plants on soil chemistry and ecosystem function. *Journal of Chemical Ecology* 36, 59–69. DOI: 10.1007/s10886-009-9735-0.

Weidlich, E.W., Flórido, F.G., Sorrini, T.B. and Brancalion, P.H. (2020) Controlling invasive plant species in ecological restoration: A global review. *Journal of Applied Ecology* 57, 1806–1817. DOI: 10.1111/1365-2664.13656.

Wetterer, J.K. (2005) Worldwide distribution and potential spread of the long-legged ant, *Anoplolepis gracilipes* (Hymenoptera: Formicidae). *Sociobiology* 45, 77–97.

Winck, B.R., Chauvat, M., Coulibaly, S.F., Santonja, M., Saccol de Sá, E.L. *et al.* (2020) Functional collembolan assemblages induce different plant responses in *Lolium perenne*. *Plant and Soil* 452, 347–358. DOI: 10.1007/s11104-020-04579-0.

Wohlgemuth, T., Gossner, M.M., Campagnaro, T., Marchante, H., van Loo, M. *et al.* (2022) Impact of non-native tree species in Europe on soil properties and biodiversity: A review. *NeoBiota* 78, 45–69. DOI: 10.3897/neobiota.78.87022.

Wolters, V. (1991) Soil invertebrates - Effects on nutrient turnover and soil structure - A Review. *Zeitschrift Für Pflanzenernährung Und Bodenkunde* 154, 389–402.

Wood, A.R. (2017) Fungi and invasions in South Africa. *Bothalia-African Biodiversity & Conservation* 47, 1–16. DOI: 10.4102/abc.v47i2.2124.

Wubs, E.J., van der Putten, W.H., Mortimer, S.R., Korthals, G.W., Duyts, H. *et al.* (2019) Single introductions of soil biota and plants generate long-term legacies in soil and plant community assembly. *Ecology Letters* 22, 1145–1151. DOI: 10.1111/ele.13271.

Yagle, I., Levinzon, A., Grünzweig, J.M., Dufour-Dror, J.M., Zurgil, U. *et al.* (2024) Invasion of prosopis trees into arid ecosystem alters soil carbon and nitrogen processes and soil trace gases emissions. *Soil Biology and Biochemistry* 198, 109558. DOI: 10.1016/j.soilbio.2024.109558.

Yang, Y., Li, T., Wang, Y., Cheng, H., Chang, S.X. *et al.* (2021) Negative effects of multiple global change factors on soil microbial diversity. *Soil Biology and Biochemistry* 156, 108229. DOI: 10.1016/j.soilbio.2021.108229.

Zhang, P., Neher, D.A., Li, B. and Wu, J. (2018) The impacts of above-and belowground plant input on soil microbiota: Invasive *Spartina alterniflora* versus native *Phragmites australis*. *Ecosystems* 21, 469–481. DOI: 10.1007/s10021-017-0162-8.

Zhang, T., Song, B., Wang, L., Li, Y., Wang, Y. *et al.* (2024) *Spartina alterniflora* invasion reduces soil microbial diversity and weakens soil microbial inter-species relationships in coastal wetlands. *Frontiers in Microbiology* 15, 1422534. DOI: 10.3389/fmicb.2024.1422534.

Zheng, X., Javed, Z., Liu, B., Zhong, S., Cheng, Z. *et al.* (2023) Impact of *Spartina alterniflora* invasion in coastal wetlands of China: Boon or bane? *Biology* 12, 1057. DOI: 10.3390/biology12081057.

Zizolfi, M., Maisto, G., Santini, G., Memoli, V., Trifuoggi, M. *et al.* (2025) Invasive plants do not exert univocal responses on soil abiotic and biotic characteristics. *Forests* 16, 209. DOI: 10.3390/f16020209.

4 Impacts of Invasive Alien Plants on Soil Properties, Fauna and Ecosystem Functions under Climate Change

Renáta Petrikovszki*, Barbara Simon and Gergely Boros
Hungarian University of Agriculture and Life Sciences, Gödöllő, Hungary

Introduction

The appearance of invasive alien plants is a global phenomenon, and at the same time one of the main direct drivers of ecosystem change (IPBES, 2019). The invasion rate of these species is predicted to increase rapidly in the coming decades, due to human activities such as the expansion of global trade, agriculture and climate change (Vitousek *et al.*, 1997; Mack *et al.*, 2000).

Globally, Argentina, South Africa and France have the highest number (over 150) of invasive plant species on their own management list. Meanwhile, Thailand, Argentina, Korea, Hungary and Uruguay have the highest percentages (over 30%) of invasive plants listed in the Global Register of Introduced and Invasive Species (GRIIS) database. Generally, every continent and country has its own invasive plants, which have adapted to the environmental conditions of the given area and have spread successfully. Terrestrial invasive plants pose the greatest threat, among which herbs are the most frequent growth forms, followed by trees and shrubs. In terms of taxonomy, *Poaceae*, *Fabaceae*, *Asteraceae*, *Rosaceae* and *Cactaceae* are the most invasive plant species-rich families worldwide (Li *et al.*, 2025).

The increasing invasion of plant species poses a major threat to native biodiversity as they modify the entire ecosystem (Ehrenfeld, 2010; Pyšek *et al.*, 2012; Chen *et al.*, 2015; Stefanowicz *et al.*, 2017). Invasive plants can also modify their environment – and, through this, the soil in which they grow – in many different ways, such as the mechanical impact of roots, or indirectly through secondary metabolites exuded by roots and contained in litter (Meier and Bowman, 2008) (Fig. 4.1). Together with all these aboveground plant communities, soil fauna composition and belowground processes are strongly influenced; however, most effects of invasive plants are case-specific (Belnap and Phillips, 2001; Kourtev *et al.*, 2003). Stefanowicz *et al.* (2018) demonstrated that invasive plants significantly influence most soil properties; however, the direction and magnitude of this influence are species-related.

This is supported by several authors, who found positive impacts of invasive plants on microbiological and soil properties, soil moisture content and species richness compared to non-invasive sites (Vilà *et al.*, 2011; Pyšek *et al.*, 2012; Castro-Díez *et al.*, 2014; Lobe *et al.*, 2014; Stefanowicz *et al.*, 2017; Dar *et al.*, 2023). Nevertheless, these positive results cannot be generalized, since a lot of contradictory effects

*Corresponding author: Petrikovszki.Renata@uni-mate.hu

DOI: 10.1079/9781836991298.0004

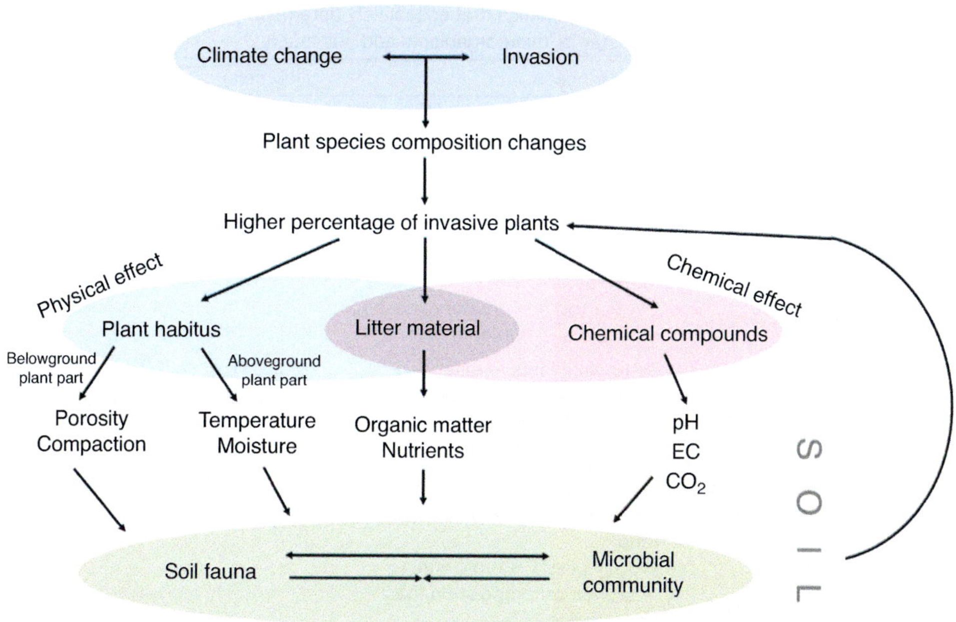

Fig. 4.1. Illustration of the effect and interaction of climate change and invasive plant species on soil's physical, chemical and biological aspects. EC, soil electrical conductivity.

have also been reported (Aguilera *et al.*, 2010; Ehrenfeld, 2010; Mincheva *et al.*, 2014; Prescott and Zukswert, 2016; Jones *et al.*, 2019) due to differences in C:N ratio, lignin content of plant matter and environmental conditions, etc. Thus, the effects of these invasive plants must be considered in a site- and species-specific way (Callaway and Aschehoug, 2000). The environmental conditions (texture, original nutrient and moisture content of soil, chemical and physical soil properties, climatic conditions, etc.) and the heterogeneous characteristics and life forms of invasive plants (annual or perennial, herbaceous or ligneous) also need to be taken into consideration (Table 4.1).

Climate Change, Invasive Plants and Soil Fauna Changes

Climate change has many aspects: it can cause extreme weather conditions, including rising temperatures and unpredictable precipitation patterns, as well as increasing atmospheric CO_2 levels, which enhance the greenhouse effect. These changes not only affect ecosystems above ground but also directly and indirectly affect soil, soil fauna and the processes within (Pritchard, 2011; Coyle *et al.*, 2017).

In a comprehensive meta-analysis by Phillips *et al.* (2024), the impact of environmental stressors of global change on soil fauna was assessed. According to their analyses, environmental pollution causes the greatest negative impact on soil fauna, while invasion (either plant or animal invaders) is the fourth in their ranking. However, it is important to note that fewer published studies are available on the topic of invasion (under 200) compared to pollution (around 800), which proves that this topic is currently under-researched and has great potential. Based on their evaluation, drought tends to be the most influential stressor to soil fauna. However, drought and other factors (like increasing CO_2 levels and extreme temperature changes) are difficult to separate and could reinforce each other, which is why a complex view is necessary (Phillips *et al.*, 2024).

Table 4.1. The most important plant characteristics that collectively determine how invasive plants can alter plant and soil properties, nutrient status, microorganisms and soil fauna.

Plant trait	Importance	Source(s)
Structure of plants (size; leaf area; canopy structure; root structure)	Greater size + leaf area → greater surface area to capture sunlight → increased photosynthesis → greater carbon assimilation → more biomass → more litter input Denser canopy → more soil moisture retained Deeper roots → greater excess to moisture and nutrients	Ehrenfeld (2003), de Graaf *et al.* (2013), Bi and Zhou (2021), Raheem *et al.* (2024)
Life span (annual; perennial)	Perennial plants → extended growing season → larger amount of carbon load into the soil compared to annual plants	McKown *et al.* (2013), Liu *et al.* (2023), Tang *et al.* (2023)
Tissue chemistry (herbaceous; semi-woody; woody)	Decomposition rate → determined by lignin, polyphenol content, C:N ratio Woody plants → more lignin → slower decomposition than herbaceous	Talbot *et al.* (2012), Tamura *et al.* (2017)
Symbiotic relationship (none; mycorrhizal; leguminous)	Symbiosis with N-fixing organisms → increases N accessibility → stimulate photosynthesis → enhance carbon load into the soil Mycorrhizal association → helps nutrient (especially P) absorption	Baldwin (2016), Mbaabu *et al.* (2019), Dar *et al.* (2023), Qiu *et al.* (2023)
Allelopathy	Allelopathic chemicals → influence the development and viability of surrounding plants → decrease biomass It can negatively affect growth and respiration of *Eisenia fetida* earthworm	Cheema *et al.* (2013), Liu *et al.* (2020)

As a result of human impacts and a changing climate, the spread of invasive plants has increased. These plants, due to their reproduction strategies, dispersion traits, allelopathy, genetic adaptation and phenotypic plasticity, successfully adapt to new climatic conditions, colonize new habitats and displace native plants, transforming the local ecosystem (Merilä and Hendry, 2014; Gioria *et al.*, 2023; Syngkli *et al.*, 2025). Faster decomposition of invasive plants emits more CO_2 in the soil, which is partly released into the atmosphere, leading to a warmer climate. This warmer and drier climate allows more resistant and more quickly adaptable invasive plants to thrive, which in turn reinforces this cycle (Raheem *et al.*, 2024).

Climate change and the effect of invasion cannot be separated; there is a reciprocal interaction between them (Syngkli *et al.*, 2025). Therefore, the effect of one on soil fauna cannot be interpreted without considering the other.

Invasive Plants Alter Soil Physical, Chemical and Biological Properties

Generally, interactions between plants and soil are complex and reciprocal. Soil characteristics can influence the appearance and survival of plants, and plants can also have impacts on soil's physical, chemical and biological properties (Ehrenfeld, 2010). For example, an increased biomass of soil fauna could enhance aboveground plant productivity across ecosystems (Sackett *et al.*, 2010). There are no general patterns regarding the effect of invasive plants on the alterations of physical and chemical properties of soil (Koutika *et al.*, 2007; Stefanowicz *et al.*, 2017), since these effects could be negative, positive or neutral (Coyle *et al.*, 2017; Renčo *et al.*, 2021). However, the influence of invasive plants on soil conditions is mainly direct and may increase as more invasive species occur (Vujanović *et al.*, 2022). These changes generated by invasive plants in the soil properties environment may trigger feedbacks that allow invasive plants to colonize more easily, while also hindering the re-establishment of native plants (Stefanowicz *et al.*, 2018).

In most cases, invasive plants contain a higher amount of nitrogen combined with a low C:N and C:P ratio (Ashton *et al.*, 2005; McTee *et al.*, 2017). Nitrogen and phosphorus are released more quickly during decomposition, which accelerates the process (Allison and Vitousek, 2004) because this provides favourable conditions for decomposers (Loranger-Merciris *et al.*, 2008; De Almeida *et al.*, 2022). Liao *et al.* (2008) reported that invasive species were correlated with much higher litter decomposition rates, and the same results were published from Hawai'i (Allison and Vitousek, 2004) where the litter decay rates for exotics were much higher as well. Therefore, the key to the impact of invasive plants on soil properties may act through decomposition processes. Decomposition processes are controlled by the quality of litter, abundance of decomposer organisms, and climate and geographical location. For example, decomposition is generally faster in warmer regions which is due to enhanced activity of microorganisms and other decomposers (Coûteaux *et al.*, 1995; Keiser and Bradford, 2017; Helsen *et al.*, 2018).

Results on the effect of invasive species on decomposition are not univocal; however, the three main drivers (climate, litter quality and decomposer communities) are considered very concordant in the literature (Gholz *et al.*, 2000; Hättenschwiler *et al.*, 2005; Cornwell *et al.*, 2008; Bradford *et al.*, 2016; Djukic *et al.*, 2018).

The effect of invasive plants on soil physical properties

There are few publications that examine the direct changes that invasive plants induce in the physical properties of soil. Based on the model of Robinson *et al.* (2022), soil compaction is related to organic matter content, but this observation is dependent on soil type. Since a larger amount of organic matter is produced under invasive plants and decomposition is also increased, it can be concluded that porosity is higher in areas of invasive fauna, as Dar *et al.* (2023) found lower bulk density values under *Anthemis cotula*-invaded plots.

One of the keys to their success is that invasive plants are undemanding with regard to soil type (Haukka *et al.*, 2013). As Sun *et al.* (2022) summarized, clay soils could be more compacted and less aerated which negatively influences microbial activity and soil structure and, thus, slows down the decomposition process. The litter produced by invasive plants can improve the properties of soils that have a clay texture.

Under invasive plants, shading of the soil surface creates a microclimate with decreased temperature and increased soil moisture (Ashton *et al.*, 2005), which could affect the decomposition rate as well (McTee *et al.*, 2017).

Soil moisture, another important soil property, is strongly affected by the rooting depth of plants and the season of active uptake (Williams and Baruch, 2000; Gerlach, 2004; Enloe *et al.*, 2004; Potts *et al.*, 2008). Invasive plants have deeper roots and faster growth which can lead to enhanced water uptake and evaporation (Wolf *et al.*, 2004; Ehrenfeld, 2010). Although invasive species with large leaf areas shade the soil surface, they exploit groundwater resources from deeper depths at the same time.

The effect of invasive plants on soil chemical properties

Invasive plant litter releases nitrogen and phosphorus faster during decomposition, and thus these nutrients become available sooner, which speeds up the decomposition process (Allison and Vitousek, 2004). However, organic matter may not accumulate in soil under invasive plants due to rapid decomposition, and nutrients released from litter may therefore leach from the soil system quickly (Allison and Vitousek, 2004). Another explanation is that certain invasive plants have a higher demand for nutrient uptake (Dassonville *et al.*, 2008), which generates a higher quality of invasive plant litter (Feng *et al.*, 2007). In addition, the phenomenon that the rapidly decomposing litter of invaders may promote the quicker decomposition of co-occurring native species has also been covered by other authors (Ashton *et al.*, 2005; Rodgers *et al.*, 2008; Chen *et al.*, 2013; Zhang *et al.*, 2014).

Stefanowicz *et al.* (2017) examined the effects of three invasive plant species (*Reynoutria japonica*, *Rudbeckia laciniata* and *Solidago gigantea*) and found that plant invasions did not affect the majority of soil physical and chemical properties, except for the total phosphorus content, which declined under the invasion areas. In another study, *S. gigantea* similarly altered soil pool sizes of nitrogen and phosphorus in some sites (Scharfy *et al.*, 2009) but not others (Vanderhoeven *et al.*, 2006), which may be because decomposer communities specialize in the waste materials they encounter during their feeding (Veen *et al.*, 2014). In contrast, Dar *et al.* (2023) found that *Anthemis cotula* invasion increased the values of soil organic carbon, total and available nitrogen, total and available potassium and total phosphorus. However, available phosphorus was lower in these areas, which might be because *A. cotula* can form strong mycorrhizal associations, which helps in the absorption of phosphorus from deeper soil horizons (Shah *et al.*, 2008).

A 5 year-long *Ligustrum sinense* invasion area had significantly greater pH values and concentrations of NH_4^+ and NO_3^- (Lobe *et al.*, 2014). *R. japonica* increased the content of NO_3^-, but it was lower under *S. gigantea* vegetation, compared to the control area (Stefanowicz *et al.*, 2017). Electrical conductivity of soil can be also increased in invaded areas (Varughese and Joseph, 2023).

However, the effects of invasive plants on soil properties are not consistently the same: Jamin *et al.* (2022) showed that invasive plant species *Impatiens glandulifera* did not have a significant impact on either total or labile soil organic matter content.

The effect of invasive plants on soil biological properties

Accelerated decomposition and the differences between invaded and uninvaded areas may be related to changes in microbial and invertebrate communities (Ashton *et al.*, 2005).

Different fungal and bacterial communities colonize shoots and roots, and this may differ depending on whether plants are forbs or grasses. At the initial stage of decomposition, *r*-strategist decomposers break down labile matter, whereas later *K*-strategists take their place and consume recalcitrant matter (Dilly *et al.*, 2004).

For example, certain invasive species (*Lonicera maackii*) have their own microbial community, thanks to their morphological characteristics, which could enhance microbial colonization and increase the decomposition process (Yadav *et al.*, 2005; Arthur *et al.*, 2012). These new, more diverse and dense microbial communities may provide alternative food resources for bacterivores and fungivores.

According to Harkes *et al.* (2021), increased abundance of Basidiomycota fungi could correlate with the lower, more acidic pH in soil, while Scharfy *et al.* (2010) explains that an observed microbial community became dominated by fungi due to antibacterial compounds (e.g. sesquiterpenes) and/or fungal growth-promoting compounds (e.g. acetic acid) released from *Solidago* roots, or an increased presence of root pathogenic fungi.

Members of soil fauna can belong to different feeding groups, and depending on habitat, it is difficult to draw general conclusions about the effects of invasive plants (McCary *et al.*, 2016). As Tanner *et al.* (2013) noticed, there could be contrary effects of invasive species (namely *Impatiens glandulifera*) on below- and aboveground communities: the abundance of foliage-feeders and ground-dwelling species decreased, whereas belowground invertebrate abundance seemed neutral or even positive under invasive plants. The difference between decomposers living on the soil surface and in the soil may be due to differences in the specialist/generalist ratio.

Generally, decomposers prefer leaf materials with high nitrogen and a low C:N ratio. For example, earthworms could be found in higher abundance under nitrogen-rich invasive species (Chen *et al.*, 2013). In some cases, invasive earthworms could be more dominant under invasive plants than native Lumbricids. These findings may be related to the higher pH caused by invasive *Ligustrum sinense* which might have provided an advantage for European Lumbricid earthworms to disperse (Lobe *et al.*, 2014). In addition, invasive plants could serve as shelter or food for macrodetritivores, and after the plant dies detritivores may remain in the area to search for new alternative nutrients like the litter from native plants (Mayer *et al.*, 2005).

The differences between feeding groups can be shown in free-living nematodes. While invasive *Asclepias syriaca* did not change the composition of nematode trophic groups or species diversity (Jurová *et al.*, 2020; Jakubcsiková *et al.*, 2021), *S. gigantea* enhanced nematode density and decreased diversity; in addition, only herbivorous nematodes appeared in higher numbers in invaded grassland (Čerevková *et al.*, 2020). Since *S. gigantea* could change the bacterial/fungal balance towards fungal dominance, certain fungivore nematode genera could be there in higher abundance (Scharfy *et al.*, 2010; Harkes *et al.*, 2021). However, this phenomenon is not always observed (Quist *et al.*, 2014), which may be related to the duration of invasion.

In addition to soil mesofauna decomposers, the higher levels of soil food webs are also sensitive to plant invasions. Garlic mustard (*Alliaria petiolata*) is a well-known and studied example: McCary and Wise (2019) showed that soil fungal composition reduced the overall density of fungal hyphae, which indirectly led to a decrease in soil fungivores, such as oribatid mites and collembolans. Collembola abundance decreased in the presence of invasive *Reynoutria*; additionally, habitat (forest or meadow) and invasion status together could have a stronger influence on Collembola life forms: invasion did not influence euedaphic springtails which could be traced back to their soil-dwelling life form, avoiding the effects of litter from invasive species (De Almeida *et al.*, 2022).

Allelopathy and Invasive Plant–Soil Fauna Interactions

Many variables, including pests, diseases, microorganisms and microclimatic conditions impact the production and the fate in the soil of invasive plant allelochemicals. All these variables could be influenced by climate change (Ehlers *et al.*, 2020).

Invasive species can not only change chemical and physical soil parameters, but also produce certain allelochemicals that can influence the function and structure of soil microbial communities (Thorpe *et al.*, 2009; Coats and Rumpho, 2014; Li *et al.*, 2017), native plants (Qin *et al.*, 2013; Zheng *et al.*, 2015) and soil fauna components (Xiao *et al.*, 2013; Peralta *et al.*, 2019; Liu *et al.*, 2020) as well. Both leaves and roots of invasive plants could introduce allelochemicals into the soil (Bais *et al.*, 2003; Xu *et al.*, 2020), but the allelopathic effects of leaf extracts could be stronger than those of root extracts (Liu *et al.*, 2020).

The above- and belowground subsystems are closely linked because the root exudates of plants support the rhizosphere community, which in turn provides nutrients to plants by decomposing organic matter (Kulmatiski *et al.*, 2008). Invasive plants have a so-called legacy effect on their environment, which is a long-term impact after the removal of the species and which can be positive, negative or neutral. For example, allelochemicals can enhance soil microbial activity, but decrease the abundance of soil microorganisms (Unger *et al.*, 2022; Afzal *et al.*, 2023) in different cases. On the other hand, soil microorganisms may enhance or prevent the spread of invasive plants, depending on their capacity to break down the allelopathic substances that such plants produce (Li *et al.*, 2015).

Certain invasive *Asteraceae* spp. can decrease body weight and respiration rate and increase the DNA damage biomarkers of earthworms (*Eisenia fetida*) under laboratory conditions (Liu *et al.*, 2020). In the field, high levels of cardenolides released from *A. syriaca* roots in the soil can attract entomopathogenic nematodes that act against its root pests (Rasmann *et al.*, 2011), and allelopathical compounds (e.g. catechin) of *Reynoutria* spp. can create unsuitable conditions for plant-parasitic nematodes (De Almeida *et al.*, 2022). Kato-Noguchi and Kato (2023) showed that pyrrolizidine alkaloids released from the below- and aboveground parts of the invasive *Chromolaena odorata* can cause mortality and suppress herbivorous insects and plant-parasitic nematodes.

In conclusion, the allelochemicals from invasive plant leaf litter could often be toxic for non-target soil invertebrates, but more examination under field conditions is needed to prove this statement (Ehlers *et al.*, 2020). The fact that allelopathic effects can be demonstrated under controlled laboratory conditions does not necessarily mean that they will occur in a natural system. It has not been proven that these putative allelopathic substances are present in concentrations sufficient to inhibit the development of other (native) plants. On the other hand, the effect of allelopathic substances released from the roots may depend on plant characteristics (e.g. density or root growth parameters) and the microhabitat (like climatic conditions, soil properties or microbial activity) (Hierro and Callaway, 2003).

Conclusion

Human activity and climate change are facilitating the spread of invasive plants. Native plants are slower or unable to adapt to extreme and changing weather, which also favours the spread of invasive plants. In addition, climate change and the invasion of plants may also stimulate each other. Therefore, it is difficult to separate and determine which factor affects the other first.

Based on our research in the available literature, we conclude that there are numerous articles summarizing the allelopathic effects of invasive plants. However, most of them focus on the interactions between plants. There are only a few articles concentrating on the interactions between invasive plants and microorganisms or the interactions between invasives and herbivores. There are a lot of experiments in the literature, in which the effects of plant extracts on soil fauna are examined. However, these are mostly native plants or some other plant species selected based on some specific reason. There are only a very small number of articles which specifically focus on the extracts of invasive plants.

Furthermore, there is not always a chemical analysis carried out to determine the active substances used by an invasive species. In many cases allelopathy does not even come up as a possible process; not even as a speculation. Finally, in many cases, these experiments are carried out under laboratory conditions and the concentrations of the active substances applied on the test animals are often much higher than those occurring in natural ecosystems. Thus, we do not really know if these would harm these animals under natural circumstances in normal concentrations, or if we are observing the phenomenon of hormesis.

Another interesting phenomenon we noticed is that, in the literature, it is not always mentioned how long an invasion has existed in a given area, or when the invasive plants first appeared. Often, a system is being examined that has been in place for several years and is likely already established. It would be crucial to know how invasive plants change the microclimate, soil factors and fauna from the beginning, tracking it step by step.

Considering the above findings, more field observations are encouraged. The setting up and tracing of controlled invasions could be undertaken, taking place adjacent to invaded areas. These experiments would allow researchers to observe real-time invasion and its mechanism on a 'clear field', which may provide new information even for the control of invasive plant species.

References

Afzal, M.R., Naz, M., Ashraf, W. and Du, D. (2023) The legacy of plant invasion: Impacts on soil nitrification and management implications. *Plants (Basel, Switzerland)* 12(16), 2980. DOI: 10.3390/plants12162980.

Aguilera, A.G., Alpert, P., Dukes, J.S. and Harrington, R. (2010) Impacts of the invasive plant *Fallopia japonica* (Houtt.) on plant communities and ecosystem processes. *Biological Invasions* 12(5), 1243–1252. DOI: 10.1007/s10530-009-9543-z.

Allison, S.D. and Vitousek, P.M. (2004) Rapid nutrient cycling in leaf litter from invasive plants in Hawai'i. *Oecologia* 141(4), 612–619. DOI: 10.1007/s00442-004-1679-z.

Arthur, M.A., Bray, S.R., Kuchle, C.R. and McEwan, R.W. (2012) The influence of the invasive shrub, *Lonicera maackii*, on leaf decomposition and microbial community dynamics. *Plant Ecology* 213(10), 1571–1582. DOI: 10.1007/s11258-012-0112-7.

Ashton, I.W., Hyatt, L.A., Howe, K.M., Gurevitch, J. and Lerdau, M.T. (2005) Invasive species accelerate decomposition and litter nitrogen loss in a mixed deciduous forest. *Ecological Applications* 15(4), 1263–1272. DOI: 10.1890/04-0741.

Bais, H.P., Vepachedu, R., Gilroy, S., Callaway, R.M. and Vivanco, J.M. (2003) Allelopathy and exotic plant invasion: From molecules and genes to species interactions. *Science (New York, N.Y.)* 301(5638), 1377–1380. DOI: 10.1126/science.1083245.

Baldwin, S.A.Y.A.H. (2016) Archaeal rhizosphere communities differ between the native and invasive lineages of the wetland plant *Phragmites australis* (common reed) in a Chesapeake Bay subestuary. *Biological Invasions* 18(9), 2717–2728. DOI: 10.1007/s10530-016-1144-z.

Belnap, J. and Phillips, S.L. (2001) Soil biota in an ungrazed grassland: Response to annual grass (*Bromus tectorum*) invasion. *Ecological Applications* 11(5), 1261–1275. DOI: 10.1890/1051-0761(2001)011[1261:SBIAUG]2.0.CO;2.

Bi, Y. and Zhou, H. (2021) Changes in peanut canopy structure and photosynthetic characteristics induced by an arbuscular mycorrhizal fungus in a nutrient-poor environment. *Scientific Reports* 11(1), 14832. DOI: 10.1038/s41598-021-94092-w.

Bradford, M.A., Berg, B., Maynard, D.S., Wieder, W.R. and Wood, S.A. (2016) Understanding the dominant controls on litter decomposition. *Journal of Ecology* 104(1), 229–238. DOI: 10.1111/1365-2745.12507.

Callaway, R.M. and Aschehoug, E.T. (2000) Invasive plants versus their new and old neighbors: A mechanism for exotic invasion. *Science (New York, N.Y.)* 290(5491), 521–523. DOI: 10.1126/science.290.5491.521.

Castro-Díez, P., Godoy, O., Alonso, A., Gallardo, A. and Saldaña, A. (2014) What explains variation in the impacts of exotic plant invasions on the nitrogen cycle? A meta-analysis. *Ecology Letters* 17(1), 1–12. DOI: 10.1111/ele.12197.

Čerevková, A., Miklisová, D., Bobuľská, L. and Renčo, M. (2020) Impact of the invasive plant *Solidago gigantea* on soil nematodes in a semi-natural grassland and a temperate broadleaved mixed forest. *Journal of Helminthology* 94, 1–14. DOI: 10.1017/S0022149X19000324.

Cheema, Z.A., Farooq, M. and Wahid, A. (2013) *Allelopathy: Current Trends and Future Applications*. Springer, Berlin, Heidelberg, Germany, pp. 978–3. DOI: 10.1007/978-3-642-30595-5.

Chen, B.-M., Peng, S.-L., D'Antonio, C.M., Li, D.-J. and Ren, W.-T. (2013) Non-additive effects on decomposition from mixing litter of the invasive *Mikania micrantha* H.B.K. with native plants. *PLoS ONE* 8(6), e66289. DOI: 10.1371/journal.pone.0066289.

Chen, X., Wang, R., Cao, Q., Zhang, H. and Ge, X. (2015) The relationship between the distribution of invasive plant *Alternanthera philoxeroides* and soil properties is scale-dependent. *Polish Journal of Environmental Studies* 24, 1931–1938. DOI: 10.15244/pjoes/43367.

Coats, V.C. and Rumpho, M.E. (2014) The rhizosphere microbiota of plant invaders: An overview of recent advances in the microbiomics of invasive plants. *Frontiers in Microbiology* 5, 368. DOI: 10.3389/fmicb.2014.00368.

Cornwell, W.K., Cornelissen, J.H.C., Amatangelo, K., Dorrepaal, E., Eviner, V.T. *et al.* (2008) Plant species traits are the predominant control on litter decomposition rates within biomes worldwide. *Ecology Letters* 11(10), 1065–1071. DOI: 10.1111/j.1461-0248.2008.01219.x.

Coûteaux, M.-M., Bottner, P. and Berg, B. (1995) Litter decomposition, climate and liter quality. *Trends in Ecology and Evolution* 10(2), 63–66. DOI: 10.1016/S0169-5347(00)88978-8.

Coyle, D.R., Nagendra, U.J., Taylor, M.K., Campbell, J.H., Cunard, C.E. *et al.* (2017) Soil fauna responses to natural disturbances, invasive species, and global climate change: Current state of the science and a call to action. *Soil Biology and Biochemistry* 110, 116–133. DOI: 10.1016/j.soilbio.2017.03.008.

Dar, M.A., Ahmad, M., Singh, R., Kumar Kohli, R., Singh, H.P. *et al.* (2023) Invasive plants alter soil properties and nutrient dynamics: A case study of *Anthemis cotula* invasion in Kashmir Himalaya. *Catena* 226, 107069. DOI: 10.1016/j.catena.2023.107069.

Dassonville, N., Vanderhoeven, S., Vanparys, V., Hayez, M., Gruber, W. *et al.* (2008) Impacts of alien invasive plants on soil nutrients are correlated with initial site conditions in NW Europe. *Oecologia* 157(1), 131–140. DOI: 10.1007/s00442-008-1054-6.

De Almeida, T., Forey, E. and Chauvat, M. (2022) Alien invasive plant effect on soil fauna is habitat dependent. *Diversity* 14(2), 61. DOI: 10.3390/d14020061.

de Graaf, M.A., Six, J., Jastrow, J.D., Schadt, C.W. and Wullschleger, S.D. (2013) Variation in root architecture among switchgrass cultivars impacts root decomposition rates. *Soil Biology and Biochemistry* 58, 198–206. DOI: 10.1016/j.soilbio.2012.11.015.

Dilly, O., Bloem, J., Vos, A. and Munch, J.C. (2004) Bacterial diversity in agricultural soils during litter decomposition. *Applied and Environmental Microbiology* 70(1), 468–474. DOI: 10.1128/AEM.70.1.468-474.2004.

Djukic, I., Kepfer-Rojas, S., Schmidt, I.K., Larsen, K.S., Beier, C. *et al.* (2018) Early stage litter decomposition across biomes. *Science of the Total Environment* 628–629, 1369–1394. DOI: 10.1016/j.scitotenv.2018.01.012.

Ehlers, B.K., Berg, M.P., Staudt, M., Holmstrup, M., Glasius, M. *et al.* (2020) Plant secondary compounds in soil and their role in belowground species interactions. *Trends in Ecology and Evolution* 35(8), 716–730. DOI: 10.1016/j.tree.2020.04.001.

Ehrenfeld, J.G. (2003) Effects of exotic plant invasions on soil nutrient cycling processes. *Ecosystems* 6(6), 503–523. DOI: 10.1007/s10021-002-0151-3.

Ehrenfeld, J.G. (2010) Ecosystem consequences of biological invasions. *Annual Review of Ecology, Evolution, and Systematics* 41(1), 59–80. DOI: 10.1146/annurev-ecolsys-102209-144650.

Enloe, S.F., DiTomaso, J.M., Orloff, S.B. and Drake, D.J. (2004) Soil water dynamics differ among rangeland plant communities dominated by yellow starthistle (*Centaurea solstitialis*), annual grasses, or perennial grasses. *Weed Science* 52(6), 929–935. DOI: 10.1614/WS-03-156R.

Feng, Y.L., Auge, H. and Ebeling, S.K. (2007) Invasive *Buddleja davidii* allocates more nitrogen to its photosynthetic machinery than five native woody species. *Oecologia* 153(3), 501–510. DOI: 10.1007/s00442-007-0759-2.

Gerlach, J.D. (2004) The impacts of serial land-use changes and biological invasions on soil water resources in California, USA. *Journal of Arid Environments* 57(3), 365–379. DOI: 10.1016/S0140-1963(03)00102-2.

Gholz, H.L., Wedin, D.A., Smitherman, S.M., Harmon, M.E. and Parton, W.J. (2000) Long-term dynamics of pine and hardwood litter in contrasting environments: Toward a global model of decomposition. *Global Change Biology* 6(7), 751–765. DOI: 10.1046/j.1365-2486.2000.00349.x.

Gioria, M., Hulme, P.E., Richardson, D.M. and Pyšek, P. (2023) Why are invasive plants successful? *Annual Review of Plant Biology* 74, 635–670. DOI: 10.1146/annurev-arplant-070522-071021.

Harkes, P., van Heumen, L.J.M., Elsen, S.J.J., Mooijman, P.J.W., Vervoort, M.T.W. *et al.* (2021) Characterization of the habitat- and season-independent increase in fungal biomass induced by the invasive giant goldenrod and its impact on the fungivorous nematode community. *Microorganisms* 9, 437. DOI: 10.3390/microorganisms9020437.

Hättenschwiler, S., Tiunov, A.V. and Scheu, S. (2005) Biodiversity and litter decomposition in terrestrial ecosystems. *Annual Review of Ecology, Evolution, and Systematics* 36(1), 191–218. DOI: 10.1146/annurev.ecolsys.36.112904.151932.

Haukka, A.K., Dreyer, L.L. and Esler, K.J. (2013) Effect of soil type and climatic conditions on the growth and flowering phenology of three *Oxalis* species in the Western Cape, South Africa. *South African Journal of Botany* 88, 152–163. DOI: 10.1016/j.sajb.2013.07.012.

Helsen, K., Smith, S.W., Brunet, J., Cousins, S.A.O., De Frenne, P. *et al.* (2018) Impact of an invasive alien plant on litter decomposition along a latitudinal gradient. *Ecosphere* 9(1), e02097. DOI: 10.1002/ecs2.2097.

Hierro, J.L. and Callaway, R.M. (2003) Allelopathy and exotic plant invasion. *Plant and Soil* 256(1), 29–39. DOI: 10.1023/A:1026208327014.

IPBES Díaz, S., Settele, J., Brondízio, E.S., Ngo, H.T. and Guèze, M. (eds). (2019) *Summary for Policymakers of the Global Assessment Report on Biodiversity and Ecosystem Services of the Intergovernmental Science-Policy Platform on Biodiversity and Ecosystem Services*. IPBES Secretariat, Bonn. DOI: 10.5281/zenodo.3553458.

Jakubcsiková, M., Čerevková, A. and Renčo, M. (2021) Influence of *Asclepias syriaca* on soil nematode communities. *Folia Oecologica* 48(1), 73–81. DOI: 10.2478/foecol-2021-0009.

Jamin, J., Diehl, D., Meyer, M., David, J., Schaumann, G.E. *et al*. (2022) Physico-chemical soil properties affected by invasive plants in Southwest Germany (Rhineland-Palatinate)—a case study. *Soil Systems* 6(4), 93. DOI: 10.3390/soilsystems6040093.

Jones, G.L., Scullion, J., Worgan, H. and Gwynn-Jones, D. (2019) Litter of the invasive shrub *Rhododendron ponticum* (Ericaceae) modifies the decomposition rate of native UK woodland litter. *Ecological Indicators* 107, 105597. DOI: 10.1016/j.ecolind.2019.105597.

Jurová, J., Renčo, M., Gömöryová, E. and Čerevková, A. (2020) Effects of the invasive common milkweed (*Asclepias syriaca*) on nematode communities in natural grasslands. *Nematology* 22(4), 423–438. DOI: 10.1163/15685411-00003314.

Kato-Noguchi, H. and Kato, M. (2023) Evolution of the secondary metabolites in invasive plant species *Chromolaena odorata* for the defense and allelopathic functions. *Plants (Basel, Switzerland)* 12(3), 521. DOI: 10.3390/plants12030521.

Keiser, A.D. and Bradford, M.A. (2017) Climate masks decomposer influence in a cross-site litter decomposition study. *Soil Biology and Biochemistry* 107, 180–187. DOI: 10.1016/j.soilbio.2016.12.022.

Kourtev, P.S., Ehrenfeld, J.G. and Häggblom, M. (2003) Experimental analysis of the effect of exotic and native plant species on the structure and function of soil microbial communities. *Soil Biology and Biochemistry* 35(7), 895–905. DOI: 10.1016/S0038-0717(03)00120-2.

Koutika, L.-S., Vanderhoeven, S., Chapuis-Lardy, L., Dassonville, N. and Meerts, P. (2007) Assessment of changes in soil organic matter after invasion by exotic plant species. *Biology and Fertility of Soils* 44(2), 331–341. DOI: 10.1007/s00374-007-0210-1.

Kulmatiski, A., Beard, K.H., Stevens, J.R. and Cobbold, S.M. (2008) Plant-soil feedbacks: A meta-analytical review. *Ecology Letters* 11, 980–992. DOI: 10.1111/j.1461-0248.2008.01209.x.

Li, F.-F., Hao, Q., Cui, X., Lin, R.-Z., Luo, B.-S. *et al*. (2025) Global invasive alien plant management lists: Assessing current practices and adapting to new demands. *Plant Diversity* 47, 666–680. DOI: 10.1016/j.pld.2024.11.002.

Li, Y., Feng, Y., Kang, Z., Zheng, Y., Zhang, J. *et al*. (2017) Changes in soil microbial communities due to biological invasions can reduce allelopathic effects. *Journal of Applied Ecology* 54(5), 1281–1290. DOI: 10.1111/1365-2664.12878.

Li, Y.-P., Feng, Y.-L., Chen, Y.J. and Tian, Y.-H. (2015) Soil microbes alleviate allelopathy of invasive plants. *Science Bulletin* 60(12), 1083–1091. DOI: 10.1007/s11434-015-0819-7.

Liao, C., Peng, R., Luo, Y., Zhou, X., Wu, X. *et al*. (2008) Altered ecosystem carbon and nitrogen cycles by plant invasion: A meta-analysis. *The New Phytologist* 177(3), 706–714. DOI: 10.1111/j.1469-8137.2007.02290.x.

Liu, J., Xu, G., Yin, L., Xu, X., Armitage, D.W. *et al*. (2020) Invasive plants exert disproportionately negative allelopathic effects on the growth and physiology of the earthworm *Eisenia fetida*. *Science of the Total Environment* 747, 141534. DOI: 10.1016/j.scitotenv.2020.141534.

Liu, J., Wang, Y., Li, Y., Peñuelas, J., Zhao, Y. *et al*. (2023) Soil ecological stoichiometry synchronously regulates stream nitrogen and phosphorus concentrations and ratios. *Catena* 231, 107357. DOI: 10.1016/j.catena.2023.107357.

Lobe, J.W., Callaham Jr, M.A., Hendrix, P.F. and Hanula, J.L. (2014) Removal of an invasive shrub (Chinese privet: *Ligustrum sinense* Lour) reduces exotic earthworm abundance and promotes recovery of Native North American earthworms. *Applied Soil Ecology* 83, 133–139. DOI: 10.1016/j.apsoil.2014.03.020.

Loranger-Merciris, G., Imbert, D., Bernhard-Reversat, F., Lavelle, P. and Ponge, J.F. (2008) Litter N-content influences soil millipede abundance, species richness and feeding preferences in a semi-evergreen dry forest of Guadeloupe (lesser Antilles). *Biology and Fertility of Soils* 45(1), 93–98. DOI: 10.1007/s00374-008-0321-3.

Mack, R.N., Simberloff, D., Mark Lonsdale, W., Evans, H., Clout, M. *et al*. (2000) Biotic invasions: Causes, epidemiology, global consequences, and control. *Ecological Applications* 10(3), 689–710. DOI: 10.1890/1051-0761(2000)010[0689:BICEGC]2.0.CO;2.

Mayer, P.M., Tunnell, S.J., Engle, D.M., Jorgensen, E.E. and Nunn, P. (2005) Invasive grass alters litter decomposition by influencing macrodetritivores. *Ecosystems* 8(2), 200–209. DOI: 10.1007/s10021-004-0018-x.

Mbaabu, P.R., Ng, W.-T., Schaffner, U., Gichaba, M., Olago, D. *et al.* (2019) Spatial evolution of prosopis invasion and its effects on LULC and livelihoods in Baringo, Kenya. *Remote Sensing* 11(10), 1217. DOI: 10.3390/rs11101217.

McCary, M.A. and Wise, D.H. (2019) Plant invader alters soil food web via changes to fungal resources. *Oecologia* 191(3), 587–599. DOI: 10.1007/s00442-019-04510-0.

McCary, M.A., Mores, R., Farfan, M.A. and Wise, D.H. (2016) Invasive plants have different effects on trophic structure of green and brown food webs in terrestrial ecosystems: A meta-analysis. *Ecology Letters* 19(3), 328–335. DOI: 10.1111/ele.12562.

McKown, A.D., Guy, R.D., Azam, M.S., Drewes, E.C. and Quamme, L.K. (2013) Seasonality and phenology alter functional leaf traits. *Oecologia* 172(3), 653–665. DOI: 10.1007/s00442-012-2531-5.

McTee, M.R., Lekberg, Y., Mummey, D., Rummel, A. and Ramsey, P.W. (2017) Do invasive plants structure microbial communities to accelerate decomposition in intermountain grasslands? *Ecology and Evolution* 7(24), 11227–11235. DOI: 10.1002/ece3.3608.

Meier, C.L. and Bowman, W.D. (2008) Links between plant litter chemistry, species diversity, and below-ground ecosystem function. *Proceedings of the National Academy of Sciences USA* 105(50), 19780–19785. DOI: 10.1073/pnas.0805600105.

Merilä, J. and Hendry, A.P. (2014) Climate change, adaptation, and phenotypic plasticity: The problem and the evidence. *Evolutionary Applications* 7, 1–14. DOI: 10.1111/eva.12137.

Mincheva, T., Barni, E., Varese, G.C., Brusa, G., Cerabolini, B. *et al.* (2014) Litter quality, decomposition rates and saprotrophic mycoflora in *Fallopia japonica* (Houtt.) Ronse Decraene and in adjacent native grassland vegetation. *Acta Oecologica* 54, 29–35. DOI: 10.1016/j.actao.2013.03.010.

Peralta, G., Schon, N.L., Dickie, I.A., St John, M.G., Orwin, K.H. *et al.* (2019) Contrasting responses of soil nematode communities to native and non-native woody plant expansion. *Oecologia* 190(4), 891–899. DOI: 10.1007/s00442-019-04456-3.

Phillips, H.R.P., Cameron, E.K., Eisenhauer, N., Burton, V.J., Ferlian, O. *et al.* (2024) Global changes and their environmental stressors have a significant impact on soil biodiversity—a meta-analysis. *iScience* 27(9), 110540. DOI: 10.1016/j.isci.2024.110540.

Potts, D.L., Harpole, W.S., Goulden, M.L. and Suding, K.N. (2008) The impact of invasion and subsequent removal of an exotic thistle, *Cynara cardunculus*, on CO_2 and H_2O vapor exchange in a coastal California grassland. *Biological Invasions* 10(7), 1073–1084. DOI: 10.1007/s10530-007-9185-y.

Prescott, C.E. and Zukswert, J.M. (2016) Invasive plant species and litter decomposition: Time to challenge assumptions. *New Phytologist* 209(1), 5–7. DOI: 10.1111/nph.13741.

Pritchard, S.G. (2011) Soil organisms and global climate change. *Plant Pathology* 60(1), 82–99. DOI: 10.1111/j.1365-3059.2010.02405.x.

Pyšek, P., Jarošík, V., Hulme, P.E., Pergl, J., Hejda, M. *et al.* (2012) A global assessment of invasive plant impacts on resident species, communities and ecosystems: The interaction of impact measures, invading species' traits and environment. *Global Change Biology* 18(5), 1725–1737. DOI: 10.1111/j.1365-2486.2011.02636.x.

Qin, R., Zheng, Y., Valiente-Banuet, A., Callaway, R.M., Barclay, G.F. *et al.* (2013) The evolution of increased competitive ability, innate competitive advantages, and novel biochemical weapons act in concert for a tropical invader. *New Phytologist* 197(3), 979–988. DOI: 10.1111/nph.12071.

Qiu, S., Yang, H., Zhang, S., Huang, S., Zhao, S. *et al.* (2023) Carbon storage in an arable soil combining field measurements, aggregate turnover modeling and climate scenarios. *Catena* 220, 106708. DOI: 10.1016/j.catena.2022.106708.

Quist, C.W., Vervoort, M.T.W., Van Megen, H., Gort, G., Bakker, J. *et al.* (2014) Selective alteration of soil food web components by invasive giant goldenrod *Solidago gigantea* in two distinct habitat types. *Oikos* 123(7), 837–845. DOI: 10.1111/oik.01067.

Raheem, A., Yohanna, P., Li, G., Jin Noh, N., Iqbal, B. *et al.* (2024) Unraveling the ecological threads: How invasive alien plants influence soil carbon dynamics. *Journal of Environmental Management* 356, 120556. DOI: 10.1016/j.jenvman.2024.120556.

Rasmann, S., Erwin, A.C., Halitschke, R. and Agrawal, A.A. (2011) Direct and indirect root defences of milkweed (*Asclepias syriaca*): Trophic cascades, trade-offs and novel methods for studying subterranean herbivory. *Journal of Ecology* 99(1), 16–25. DOI: 10.1111/j.1365-2745.2010.01713.x.

Renčo, M., Čerevková, A. and Homolová, Z. (2021) Nematode communities indicate the negative impact of *Reynoutria japonica* invasion on soil fauna in ruderal habitats of tatra national park in Slovakia. *Global Ecology and Conservation* 26, e01470. DOI: 10.1016/j.gecco.2021.e01470.

Robinson, D.A., Thomas, A., Reinsch, S., Lebron, I., Feeney, C.J. *et al.* (2022) Analytical modelling of soil porosity and bulk density across the soil organic matter and land-use continuum. *Scientific Reports* 12(1), 7085. DOI: 10.1038/s41598-022-11099-7.

Rodgers, V.L., Wolfe, B.E., Werden, L.K. and Finzi, A.C. (2008) The invasive species *Alliaria petiolata* (garlic mustard) increases soil nutrient availability in northern hardwood-conifer forests. *Oecologia* 157, 459–471. DOI: 10.1007/s00442-008-1089-8.

Sackett, T.E., Classen, A.T. and Sanders, N.J. (2010) Linking soil food web structure to above- and belowground ecosystem processes: A meta-analysis. *Oikos* 119(12), 1984–1992. DOI: 10.1111/j.1600-0706.2010.18728.x.

Scharfy, D., Eggenschwiler, H., Olde Venterink, H., Edwards, P.J. and Güsewell, S. (2009) The invasive alien plant species *Solidago gigantea* alters ecosystem properties across habitats with differing fertility. *Journal of Vegetation Science* 20(6), 1072–1085. DOI: 10.1111/j.1654-1103.2009.01105.x.

Scharfy, D., Güsewell, S., Gessner, M.O. and Venterink, H.O. (2010) Invasion of *Solidago gigantea* in contrasting experimental plant communities: Effects on soil microbes, nutrients and plant–soil feedbacks. *Journal of Ecology* 98(6), 1379–1388. DOI: 10.1111/j.1365-2745.2010.01722.x.

Shah, M.A., Reshi, Z. and Rashid, I. (2008) Mycorrhizal source and neighbour identity differently influence *Anthemis cotula* L. invasion in the Kashmir Himalaya, India. *Applied Soil Ecology* 40(2), 330–337. DOI: 10.1016/j.apsoil.2008.06.002.

Stefanowicz, A.M., Stanek, M., Nobis, M. and Zubek, S. (2017) Few effects of invasive plants *Reynoutria japonica, Rudbeckia laciniata* and *Solidago gigantea* on soil physical and chemical properties. *Science of the Total Environment* 574, 938–946. DOI: 10.1016/j.scitotenv.2016.09.120.

Stefanowicz, A.M., Majewska, M.L., Stanek, M., Nobis, M. and Zubek, S. (2018) Differential influence of four invasive plant species on soil physicochemical properties in a pot experiment. *Journal of Soils and Sediments* 18(4), 1409–1423. DOI: 10.1007/s11368-017-1873-3.

Sun, J., Rutherford, S., Saif Ullah, M., Ullah, I., Javed, Q. *et al.* (2022) Plant–soil feedback during biological invasions: Effect of litter decomposition from an invasive plant (*Sphagneticola trilobata*) on its native congener (*S. calendulacea*). *Journal of Plant Ecology* 15(3), 610–624. DOI: 10.1093/jpe/rtab095.

Syngkli, R.B., Rai, P.K. and Lalnuntluanga (2025) Expanding horizon of invasive alien plants under the interacting effects of global climate change: Multifaceted impacts and management prospects. *Climate Change Ecology* 9, 100092. DOI: 10.1016/j.ecochg.2025.100092.

Talbot, J.M., Yelle, D.J., Nowick, J. and Treseder, K.K. (2012) Litter decay rates are determined by lignin chemistry. *Biogeochemistry* 108(1–3), 279–295. DOI: 10.1007/s10533-011-9599-6.

Tamura, M., Suseela, V. and Tharayil, N. (2017) Plant litter chemistry alters the content and composition of organic carbon associated with soil mineral and aggregate tions in invaded ecosystems. *Global Change Biology* 23, 4002–4018. DOI: 10.1111/gcb.13751.

Tang, Y., Li, X., Song, Y., Alabbosh, K.F., Li, K. *et al.* (2023) Nitrogen deposition enhanced the effect of *Solidago canadensis* invasion on soil microbial metabolic limitation and carbon use efficiency. *Polish Journal of Environmental Studies* 32, 1–12. DOI: 10.15244/pjoes/170766.

Tanner, R.A., Varia, S., Eschen, R., Wood, S., Murphy, S.T. *et al.* (2013) Impacts of an invasive non-native annual weed, *Impatiens glandulifera*, on above- and below-ground invertebrate communities in the United Kingdom. *PLoS ONE* 8, e67271. DOI: 10.1371/journal.pone.0067271.

Thorpe, A.S., Thelen, G.C., Diaconu, A. and Callaway, R.M. (2009) Root exudate is allelopathic in invaded community but not in native community: Field evidence for the novel weapons hypothesis. *Journal of Ecology* 97(4), 641–645. DOI: 10.1111/j.1365-2745.2009.01520.x.

Unger, I.M., Kremer, R.J., Veum, K.S. and Goyne, K.W. (2022) Immediate and long-term effects of invasive plant species on soil characteristics. *Soil Ecology Letters* 4, 276–288. DOI: 10.1007/s42832-021-0104-4.

Vanderhoeven, S., Dassonville, N., Chapuis-Lardy, L., Hayez, M. and Meerts, P. (2006) Impact of the invasive alien plant *Solidago giganteaon* primary productivity, plant nutrient content and soil mineral nutrient concentrations. *Plant and Soil* 286(1–2), 259–268. DOI: 10.1007/s11104-006-9042-2.

Varughese, A.S. and Joseph, S. (2023) Influence of invasive plant species on soil properties in tropical riparian zones. *Pollution Research* 42(01), 105–109. DOI: 10.53550/PR.2023.v42i01.016.

Veen, G.F.C., Freschet, G.T., Ordonez, A. and Wardle, D.A. (2014) Litter quality and environmental controls of home-field advantage effects on litter decomposition. *Oikos* 124(2), 187–195. DOI: 10.1111/oik.01374.

Vilà, M., Espinar, J.L., Hejda, M., Hulme, P.E., Jarošík, V. *et al*. (2011) Ecological impacts of invasive alien plants: A meta-analysis of their effects on species, communities and ecosystems. *Ecology Letters* 14(7), 702–708. DOI: 10.1111/j.1461-0248.2011.01628.x.

Vitousek, P.M., D'Antonio, C.M., Loope, L.L., Rejmánek, M. and Westbrooks, R. (1997) Introduced species: A significant component of human-caused global change. *New Zealand Journal of Ecology* 21, 1–16.

Vujanović, D., Losapio, G., Milić, S. and Milić, D. (2022) The impact of multiple species invasion on soil and plant communities increases with invasive species co-occurrence. *Frontiers in Plant Science* 13, 875824. DOI: 10.3389/fpls.2022.875824.

Williams, D.G. and Baruch, Z. (2000) African grass invasion in the Americas: Ecosystem consequences and the role of ecophysiology. *Biological Invasions* 2(2), 123–140. DOI: 10.1023/A:1010040524588.

Wolf, J.J., Beatty, S.W. and Seastedt, T.R. (2004) Soil characteristics of Rocky Mountain National Park grasslands invaded by *Melilotus officinalis* and *M. alba*. *Journal of Biogeography* 31(3), 415–424. DOI: 10.1046/j.0305-0270.2003.00983.x.

Xiao, H.F., Schaefer, D.A., Lei, Y.B., Zhen, Y.L., Li, Y.P. *et al*. (2013) Influence of invasive plants on nematode communities under simulated CO_2 enrichment. *European Journal of Soil Biology* 58, 91–97. DOI: 10.1016/j.ejsobi.2013.07.002.

Xu, G., Long, Z., Ren, P., Ren, C., Cao, Y. *et al*. (2020) Differential responses of soil hydrolytic and oxidative enzyme activities to the natural forest conversion. *Science of the Total Environment* 716, 136414. DOI: 10.1016/j.scitotenv.2019.136414.

Yadav, R.K.P., Karamanoli, K. and Vokou, D. (2005) Bacterial colonization of the phyllosphere of mediterranean perennial species as influenced by leaf structural and chemical features. *Microbial Ecology* 50(2), 185–196. DOI: 10.1007/s00248-004-0171-y.

Zhang, L., Wang, H., Zou, J., Rogers, W.E. and Siemann, E. (2014) Non-native plant litter enhances soil carbon dioxide emissions in an invaded annual grassland. *PLoS ONE* 9(3), e92301. DOI: 10.1371/journal.pone.0092301.

Zheng, Y.L., Feng, Y.L., Zhang, L.K., Callaway, R.M., Valiente-Banuet, A. *et al*. (2015) Integrating novel chemical weapons and evolutionarily increased competitive ability in success of a tropical invader. *New Phytologist* 205, 1350–1359. DOI: 10.1111/nph.13135.

5 Ecological Impacts of Invasive Plants: Community Transformation and Ecosystem Disruption

Mirjana Šipek* and Nina Šajna
Department of Biology, Faculty of Natural Sciences and Mathematics, University of Maribor, Maribor, Slovenia

Introduction

Anthropogenic disturbances and invasive non-native species are recognized as major global drivers of ecosystem change (IPBES, 2023). Invasive species are a consequence of globalization and the increasing exchange of goods across regions (Meyerson and Mooney, 2007). By overcoming natural geographical barriers, these species establish themselves in new environments and induce significant ecosystem alterations. Ecosystem functioning may be threatened or altered through changes in both above- and belowground communities that follow invasion (Wolfe and Klironomos, 2005).

Biological invasions can be classified according to the part of the ecosystem being invaded. Aboveground invasions include both floral (non-native plants) and faunal (non-native animals and microorganisms) species, whereas belowground invasions mainly involve fauna (Bohlen, 2006) and microbiota, including fungi, which remain underexplored.

In this chapter, we examine how invasive non-native plants affect aboveground communities and modify soil environmental conditions. We specifically explore how invaders alter plant community richness, diversity and composition, and how they influence both biotic and abiotic soil properties, thereby modifying ecosystem processes. These changes, in turn, shape the structure and functioning of soil organism communities.

Effects of Non-native Plant Invasions on Plant Communities

Non-native plants influence ecosystems both directly and indirectly, often resulting in profound ecological changes that can be perceived as positive, neutral or negative. Figure 5.1 presents a conceptual framework illustrating how invasive plants affect ecosystems. Invasive plants influence native plant communities directly through competition and indirectly by disrupting mutualistic interactions between plants and other organisms. Altered plant communities, in turn, affect soil communities by modifying ecosystem processes and resource availability. Such changes may disadvantage specialists dependent on specific resources, while favouring generalists that can exploit increased biomass or nutrient inputs. Invasive plants also directly and

*Corresponding author: mirjana.sipek1@um.si

DOI: 10.1079/9781836991298.0005

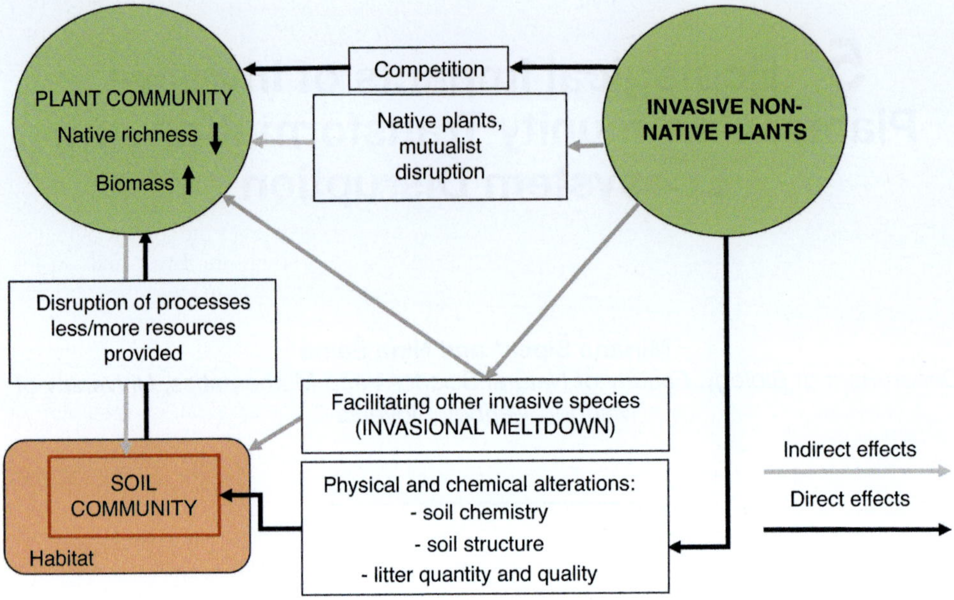

Fig. 5.1. Conceptual framework of direct and indirect effects of invasive plants on plant communities through mechanisms such as competition, disruption of mutualistic interactions, facilitation of other invasive species and alterations of physical and chemical soil properties, all of which influence soil as a habitat and the communities it supports.

indirectly impact soil communities by facilitating other non-native species and altering physical and chemical soil properties (e.g. soil chemistry, structure, litter quality and quantity). These belowground changes can feed back to further modify plant communities. Additionally, invasive plants may promote invasional meltdown by facilitating the establishment of other invaders, such as the introduction of specialist herbivores that follow their invasive host plants (Fig. 5.1).

Invasive non-native plants alter the composition (plant species present in the community and their abundance) and structure (vertical stratification, functional traits, chemical exudates) of plant communities (Šipek *et al.*, 2022; Kortz *et al.*, 2023; Šipek and Šajna, 2024). The effect of non-native invaders on relative species abundances often include displacement of native plants (Šipek *et al.*, 2022; Kortz *et al.*, 2023) and a reduction in diversity (Vilà *et al.*, 2006; Stinson *et al.*, 2007; Gaertner *et al.*, 2009; Hejda *et al.*, 2009; Vujanović *et al.*, 2022; De Almeida *et al.*, 2022). One meta-analysis confirmed that non-native plant invasions negatively affect abundance, fitness and diversity of native species; however, some ecosystem processes, including primary production, may be enhanced following invasion (Vilà *et al.*, 2011).

Implementation of the theory of island biogeography to predict plant invasions in heavily fragmented landscapes

In a rapidly changing world, vast areas of natural habitats are being transformed into arable or urbanized land, creating opportunities for invasive non-native plant species to establish within the remaining native plant communities. Heavily fragmented landscapes are considered especially prone to invasions because of the increased edge contact with adjacent human-disturbed areas (Ohlemüller *et al.*, 2006; Vilà and Ibáñez, 2011). In such landscapes, the theory of island biogeography can be applied to predict extinction and immigration dynamics within habitat 'islands' that remain suitable for wildlife (MacArthur and Wilson, 1967). Native plant communities in smaller and more isolated habitat fragments are expected to be more vulnerable to invasions by non-native species. Consequently, the overall impact of invasions should be greater in small, isolated fragments. Some species may go extinct while others may colonize, with the magnitude of these processes largely determined by habitat size, degree of isolation and connectivity with other fragments of the same type.

We hypothesized that this framework could be applied to lowland landscapes, where most forests have been cleared and converted into intensive agricultural fields (Šipek *et al.*, 2022). In this context, arable land represents a 'sea', while the remaining forest patches can be regarded as 'islands of forest habitat'. Accordingly, we expected that larger forest fragments would support higher species richness due to lower extinction rates and reduced isolation effects. In contrast, smaller fragments were expected to experience higher extinction rates of forest specialist plants and an increased influx of non-native species. This shift would alter both the composition and structure of plant communities, as native specialist herbs are gradually replaced by tall, invasive herbaceous species.

Surprisingly, the smallest fragments contained relatively few non-native and invasive plant species. Their number increased in small and medium-sized fragments with increasing fragment area, but in the largest fragments there was a decline in non-native species richness towards levels comparable to reference forest: large, unfragmented lowland forest stands (Šipek *et al.*, 2022). However, a stronger predictor of non-native plant species richness in forest fragments proved to be the type of land use adjacent to the forest, rather than fragment size or isolation itself (Šipek and Šajna, 2024).

Effects of invasive plants on the richness and diversity of native plant species

One of the central questions in invasion ecology is how non-native invasive plants influence native plant communities. In lowland oak/hornbeam forest fragments in Slovenia, a positive correlation was observed between the richness of non-native (including invasive) and native plant species. This pattern was evident at both the plot scale and the scale of entire forest fragments, with the correlation being even stronger at the latter level (Šipek, unpublished data). In practical terms, larger forest fragments supported greater numbers of both non-native and native species (Fig. 5.2). Such a relationship may reflect higher habitat heterogeneity and resource availability in larger fragments, which simultaneously promote native diversity and provide opportunities for establishment of non-native plants.

However, a vast number of case studies reveal a wide range of outcomes of interactions between non-native and native plant species. Moreover, even a single invader can be associated with different effects depending on the landscape context in which it occurs. For example, *Prunus serotina* Ehr. has been shown to exert positive effects on forest understorey herbaceous species (Dyderski and Jagodziński, 2021; Bury *et al.*, 2024), no detectable effects (Chabrerie *et al.*, 2010) or negative effects (Gentili *et al.*, 2019).

Even within individual studies, outcomes may vary depending on the ecological indicator considered. Chabrerie *et al.* (2010) found that functional diversity of the herbaceous layer was significantly lower in invaded plots compared to uninvaded plots, suggesting that *P. serotina* caused trait convergence among understorey species. Species richness, however, was similar between invaded and uninvaded sites, although with high variability among locations. Importantly, invaded plots supported fewer forest specialist species, while the community specialization index showed the opposite trend, indicating that canopy invasion by *P. serotina* created new environmental conditions that favoured species with a specific set of traits.

Another study examining the effects of three invasive trees – *P. serotina*, *Quercus rubra* L. and *Robinia pseudoacacia* L. – confirmed that all invaders altered species composition, but their impacts on understorey diversity (within-site diversity and differences between sites) were species-specific. Species richness was higher in forests invaded by *P. serotina* compared to uninvaded sites, and in *R. pseudoacacia* stands compared to native forest stands. In contrast, *Q. rubra* stands supported lower species richness than mature native forests (Dyderski and Jagodziński, 2021). Similarly, Bury *et al.* (2024) reported a positive effect of *P. serotina* on plant

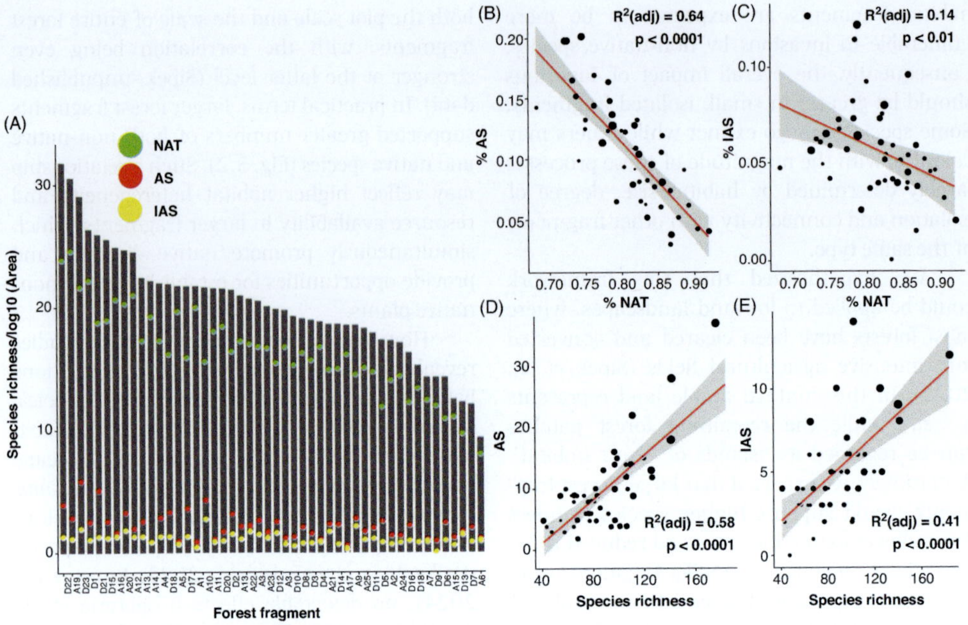

Fig. 5.2. (A) Overall plant species richness in lowland oak/hornbeam forest fragments plotted per $\log_{10}$ of fragment size. Green circles indicate native species (NAT), red circles non-native species (AS) and yellow circles invasive non-native species (IAS). Linear regressions were applied to the percentage of (B) non-native (%AS) and (C) invasive non-native (%IAS) species relative to native species (%NAT), and to (D) non-native (AS) and (E) invasive non-native (IAS) species relative to overall plant species richness in lowland forest fragments.

species richness and functional richness, likely driven by two correlated factors: greater light availability that facilitated a richer understorey and enhanced growth of the invader itself.

Gentili *et al.* (2019) further investigated the same three woody invaders across a wide range of ecosystem quality indicators. Their results showed that *R. pseudoacacia* stands had the highest plant species richness and diversity, while *P. serotina* stands did not differ in species richness from uninvaded sites; however, these stands exhibited lower diversity. *Quercus rubra* stands showed the lowest richness and diversity overall. Interestingly, *R. pseudoacacia* stands were also the most distinct in terms of community composition, characterized by the highest abundance and cover of non-native plant species.

There is evidence that the negative impacts of invasive plants on native plant communities intensify with increasing numbers of invaders (Vujanović *et al.*, 2022). For example, in a riparian ecosystem, native plant diversity declined more strongly in the presence of a higher number of invasive plant species (Vujanović *et al.*, 2022). However, decreases in plant richness following invasion are not universal (Kortz and Magurran, 2019; Šipek *et al.*, 2022). In some cases, invasions may even increase within-site species diversity, although this is often accompanied by a homogenization of community composition, as observed in the Brazilian biodiversity hotspot Cerrado (Kortz and Magurran, 2019).

Thus, effects of plant invasions on recipient communities cannot be generalized as a single, uniform outcome. When considering different invasive species, habitat types or even localities, their impacts emerge as multiple, context-dependent and often difficult to predict, ranging from negative to neutral to positive.

A meta-analysis of 1041 case studies on ecological impacts of invasive plants confirmed this variability (Vilà *et al.*, 2011). Plant production after invasion was, on average, enhanced, although this pattern was based on 58 studies, while 32 reported reduced production. In contrast, plant growth, abundance and diversity were generally reduced, although positive effects were also observed in some cases. The only parameter consistently negatively affected by non-native plant invasion was plant fitness, although this conclusion was drawn from a relatively small number of studies (n=25; Vilà *et al.*, 2011).

Diverse Effects of Invasive Plants on Soils

Invasive plants, by reshaping plant community composition and structure, alter litter quality, root exudates and rhizosphere processes, as well as abiotic soil properties such as pH (Vilà *et al.*, 2011) and soil microclimate (Jo *et al.*, 2017). These changes can drive substantial shifts in nutrient cycling dynamics, influencing rates of nutrient turnover (Bohlen, 2006). The rhizosphere – the soil zone directly affected by root exudates and their associated microorganisms – is particularly sensitive, as both above- and belowground processes are modified following plant invasion.

Many invasive species also directly alter soil chemistry and nutrient availability (Jo *et al.*, 2017; Jelinčić *et al.*, 2022), disadvantaging native plants while reinforcing their own dominance. Beyond soil interactions, plant invasions may alter phenology and even shift fire regimes, resulting in broader ecosystem consequences. Such changes in the biological, physical and chemical properties of soil can cascade through entire food webs, ultimately affecting herbivores, pollinators, higher trophic levels and soil microbial communities (Fig. 5.3; Bohlen, 2006).

Soil microbiota

The impacts of non-native species on invaded ecosystems have become a major focus of ecological research. Invasive organisms can influence ecosystems through multiple pathways, including effects on belowground communities such as soil microbes (Medina-Villar *et al.*, 2016; Pattison *et al.*, 2016; Gaggini *et al.*, 2018; Wang *et al.*, 2018). Changes in soil microbial communities can, in turn, shape plant and soil fauna composition through interactions with key groups such as mycorrhizal fungi, nitrifying bacteria and pathogens.

Aboveground plant traits and communities are much better understood than belowground traits and soil microbial communities. A

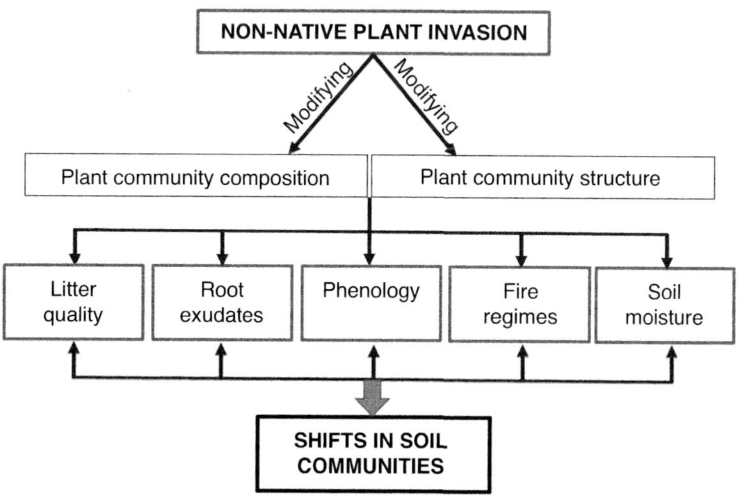

Fig. 5.3. Conceptual framework illustrating the impact of non-native plant invasions on ecosystems. Invasions modify plant community composition and structure directly and induce shifts in soil communities through various indirect mechanisms. Adapted from Bohlen (2006).

Table 5.1. The impact of selected invasive non-native plant species on the composition of soil bacterial and fungal communities, as well as on species richness and diversity.

Invader	Effect on community composition		Species richness, microbiota diversity	Method	Reference
	Bacteria	Fungi			
Ailanthus altissima Mill.	+	Not studied	No effect	Electrophoresis	Medina-Villar *et al.* (2016)
Impatiens glandulifera Royle	+	+	–, +	Biolog EcoPlates (microbes), T-RFLP (fungi), phospholipid profiling (microbes)	Pattison *et al.* (2016), Gaggini *et al.* (2018)
Prunus laurocerasus L.	+	Not studied	+ (– plants)	Biolog EcoPlates	Rusterholz *et al.* (2018)
Robinia pseudoacacia	+	Not studied	No effect	Electrophoresis	Medina-Villar *et al.* (2016)
Solidago canadensis L.	No effect	Not studied	No effect	High-throughput sequencing	Wang *et al.* (2018)

+, Indicates an effect on species composition and an increase in richness and diversity; –, indicates a decline in richness and diversity. T-RFLP, terminal restriction fragment length polymorphism.

macroecological study of invasive plant root traits across the USA addressed this gap, showing that invasive roots differ from those of native plants and explain twice as much variation in microbial community composition as native plant root traits. Additionally, roots of non-native invasive plants were associated with more homogeneous microbial communities, a pattern frequently observed in plant invasions (Nunez-Mir and McCary, 2024). These findings underscore the role of invasive plants as ecosystem transformers with substantial impacts on soil microbial communities.

Most studies compare microbial communities between invaded and uninvaded reference sites, whereas fewer investigate effects at the level of individual microbial taxa. The following section reviews research on invasive plants common in Central Europe, including *Ailanthus altissima* Mill., *Impatiens glandulifera* Royle, *Prunus laurocerasus* L., *Robinia pseudoacacia* and *Solidago canadensis* L. and their effects on soil microbiota (Table 5.1).

Bacterial communities

Invasive plants not only alter plant communities but also strongly influence soil microbial assemblages, which play a central role in nutrient cycling and ecosystem functioning. Their impacts on microbial diversity and composition, however, vary across species and environmental contexts. Medina-Villar *et al.* (2016) investigated the effects of *R. pseudoacacia* and *A. altissima* on soil microbial communities. They compared invaded sites to areas dominated by the native *Populus alba* L. Microbial composition differed markedly between *A. altissima* and *P. alba* stands, whereas differences associated with *R. pseudoacacia* were weaker. Bacterial richness and diversity did not differ significantly between invaded and control sites for either species, although both were higher in spring than in autumn. Soils under *R. pseudoacacia* contained more mineral nitrogen, likely due to symbiotic nitrogen-fixing bacteria in the roots (Medina-Villar *et al.*, 2016).

Soil nitrifying bacteria play a crucial role in the nitrogen cycle, converting atmospheric nitrogen into nitrogen ions that are available for plants. This process regulates plant growth by influencing soil fertility, making soil nitrifying bacteria communities critical indicators of ecosystem functioning and environmental change.

Solidago canadensis is one of the most widespread invasive plants in Central Europe, capable of forming dense monospecific stands. Soil nitrifying bacterial communities were examined in nine urban ecosystems in eastern China, comparing control (uninvaded) soils with soils invaded by *S. canadensis* (Wang *et al.*, 2018). Beta-diversity varied among ecosystem types and between invaded and control sites, whereas alpha-diversity differed only among ecosystem types. Several soil nitrifying bacterial families, including Micromonosporaceae, Oscillatoriaceae and Bacillaceae, exhibited significant shifts in abundance. Overall, *S. canadensis* did not significantly affect soil nitrifying bacterial diversity, although allelopathic compounds may have influenced specific taxa.

To assess the impact of invasive plants on soil bacterial community composition, metabolic diversity approaches are often employed. *Prunus laurocerasus*, an invasive shrub/small tree, is increasingly spreading, particularly adjacent to urban areas, where it invades nearby forests from residential ornamental gardens (Šipek and Šajna, 2020). Rusterholz *et al.* (2018) assessed the impact of *P. laurocerasus* on microbial metabolic diversity using Biolog EcoPlates™. Average well colour development (AWCD), an indication of metabolic activity, did not differ during the first 96 hours of incubation; however, after 120–168 hours, soils invaded by *P. laurocerasus* exhibited significantly higher AWCD, metabolic richness and Shannon diversity index. Principal component analysis revealed moderate separation in carbon substrate utilization, whereas enzyme activities (β−1,4-glucosidase and acid phosphatase) remained unchanged. The authors concluded that 10 years of *P. laurocerasus* presence increased the catabolic potential and functional diversity of soil microbial communities (Rusterholz *et al.*, 2018).

A study by Gaggini *et al.* (2018) investigated the seasonal effects of *I. glandulifera*, an invasive non-native plant of concern in central Europe, on bacterial metabolic activity in temperate forests. In June, invaded soils exhibited lower carbon substrate utilization compared to controls in both coniferous and deciduous forests, with activity strongly correlated with soil moisture and pH. By December, no invasion effect was detected, although differences between forest types persisted. These findings suggest that the impact of *I. glandulifera* on soil microbial communities may be seasonally variable, potentially due to changes in environmental conditions and microbial activity over time. Pattison *et al.* (2016) examined microbial communities in soils under *I. glandulifera* using phospholipid fatty acid (PLFA) analysis. Total and bacterial fatty acid biomass were higher in invaded soils, indicating enhanced microbial abundance and metabolic activity in invaded areas. PLFA profiles clearly separated invaded and control communities, with invaded samples showing lower variability but partial overlap, indicating compositional shifts.

Fungal communities

Soil fungal communities were investigated in uninvaded and invaded soils with *I. glandulifera* using internal transcribed spacer (ITS) sequencing. Across six sampling sites, 153 operational taxonomic units (OTUs) were identified. OTU numbers differed significantly between invaded and uninvaded sites across all sampling months and between coniferous and deciduous forests, with higher OTU numbers in invaded areas, especially in coniferous forests. OTU diversity also increased with higher soil moisture (June–July) and higher pH (July–September).

Across these studies, invasive plants consistently altered soil microbial communities, though the specific effects varied by species, ecosystem type, invasion duration and methodology. While plant invasions typically reduce plant diversity, microbial richness and functional diversity often increase or remain unchanged. For example, *P. laurocerasus* decreased understorey plant diversity but enhanced microbial functional diversity after a decade of invasion (Rusterholz *et al.*, 2018). *Ailanthus altissima* altered microbial composition in natural sites but not in seedling experiments, suggesting time-dependent effects (Medina-Villar *et al.*, 2016).

Results for *I. glandulifera* were mixed: Pattison *et al.* (2016) reported increased microbial diversity but reduced mycorrhizal colonization, whereas Gaggini *et al.* (2018) observed reduced bacterial activity but increased fungal diversity. Differences in methods (PLFA vs EcoPlates, sterilized seeds vs natural forest soils) likely explain the discrepancies. A meta-analysis also confirmed increased microbial activity in soils after non-native plant invasion. Microbial activity was on average 32.3% higher, with a positive effect observed in nine out of 14 case studies (Vilà *et al.*, 2011).

Overall, invasive plants influence soil microbial communities in species-specific and context-dependent ways. Shifts in microbial composition, activity and functional diversity have important implications for nutrient cycling and plant–microbe interactions in invaded ecosystems (Table 5.1).

Litter traits and decomposition

Litter decomposition is defined as the breakdown of organic matter into carbon dioxide and smaller, simpler organic molecules (nutrients) through physical, biological and chemical degradation (Krishna and Mohan, 2017). Organic material such as leaves and branches is first physically fragmented into smaller pieces by mechanical forces (e.g. trampling) and soil fauna. These fragments are then transformed by microorganisms into simpler organic compounds that re-enter nutrient cycles.

Biological invasions can alter decomposition primarily through changes in litter quality and litter composition. Different plant species produce litter of varying quality, determined by traits such as leaf toughness, nitrogen content, lignin concentration and the presence of secondary metabolites (e.g. polyphenols; Krishna and Mohan, 2017). These traits strongly influence the rate and pathway of decomposition. Consequently, under the same environmental conditions, decomposition rates can vary substantially among plant species (Krishna and Mohan, 2017). Among litter traits, nitrogen and lignin content are recognized as the most influential determinants of the decomposition rate. Based on such traits, decomposition speed can be predicted under given environmental conditions, and biogeochemical models can be developed (Krishna and Mohan, 2017).

Invasive plants can substantially alter ecosystem processes (Liao *et al.*, 2008). Many invasive species are fast-growing, produce large amounts of aboveground biomass and reproduce clonally. However, their leaves often lack durable structural compounds, making their litter highly decomposable. This is considered a key trait why invasive plants, compared with native species, are associated with different rates of litter decomposition. Decomposition is the primary process responsible for releasing molecules into nutrient cycles; therefore changes in decomposition directly affect nutrient dynamics. As a result, nutrient cycling is accelerated, and soils may accumulate higher levels of organic matter (Fig. 5.4).

Meta-analyses have revealed contrasting findings regarding the effects of invasive plants on decomposition, highlighting the strong context dependency that characterizes all aspects of biological invasions (Liao *et al.*, 2008; Vilà *et al.*, 2011). Vilà *et al.* (2011) reported that non-native plant invasion decreased litter decomposition by 15.6%, although this effect was not statistically significant due to high variability among studies. By contrast, Liao *et al.* (2008), after synthesizing 94 experimental studies, found that litter decomposition rates, net nitrogen mineralization and nitrification were more than 50% higher in invaded ecosystems compared to those dominated by native species. Additionally, soil concentrations of NH_4^+ and NO_3^- were 30% and 17% higher, respectively, in invaded ecosystems (Liao *et al.*, 2008).

Nutrient cycles

Many non-native plants increase primary production (Vilà *et al.*, 2011), producing large amounts of biomass that are often low in structural compounds and therefore highly decomposable. This accelerates decomposition and nutrient release, particularly nitrogen, often increasing soil fertility in invaded areas. We should not forget that one of the main sources of particulate organic matter in soils is that derived from decomposing roots (Coleman *et al.*, 2018). Meta-analyses support consistent differences in soil nutrient properties between invaded and uninvaded ecosystems (Liao *et al.*, 2008). Furthermore, positive feedback loops between plant invasion and soil nutrient cycling appear to occur regardless of habitat type (Liao *et al.*, 2008).

To investigate nitrogen dynamics following plant invasion, Jo *et al.* (2017) conducted a 3 year monoculture experiment in a common garden in New York, USA, using five pairs of understorey woody species. Each pair consisted of a native species – *Celastrus scandens* L., *Frangula caroliniana* (Walt.) A. Gray, *Lonicera canadensis* (L.) Roth, *Lonicera sempervirens* L. and *Lonicera villosa* Michx. – and its non-native invasive congener: *Celastrus orbiculatus* Thunb., *Frangula alnus* Mill.,

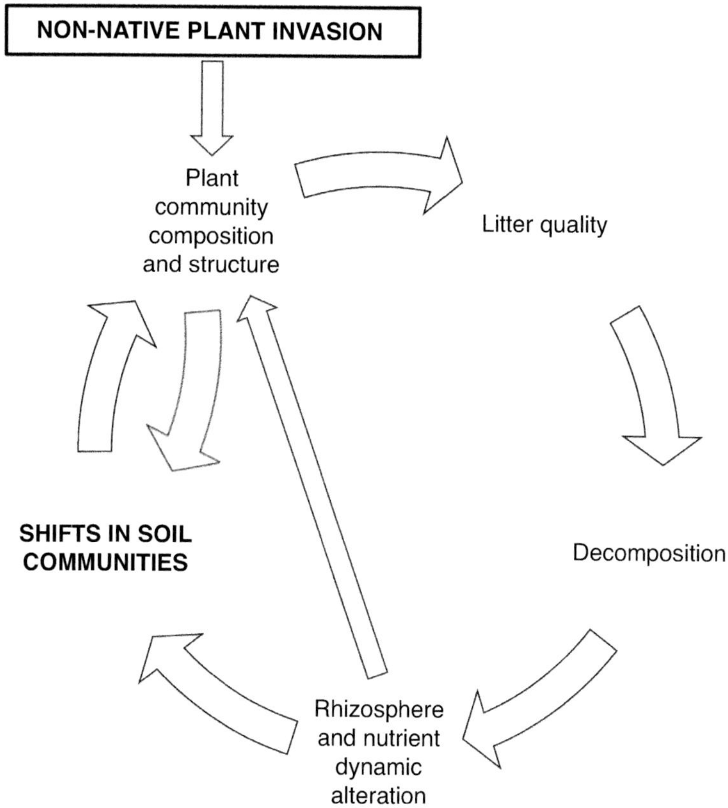

Fig. 5.4. Circle of ecosystem change following non-native plant invasion. Non-native plant invasions alter plant community composition and structure. This modifies litter quality and decomposition rates, which in turn influence the rhizosphere and nutrient dynamics. These changes feed back to affect plant communities and drive shifts in soil communities.

Lonicera fragrantissima Lindl., *Lonicera japonica* Thunb. and *Lonicera morrowii* A. Gray. After 3 years, both aboveground and belowground biomass were greater in the invasive species. The nitrogen pool was higher in all plant compartments, including leaf litter, stems, branches and roots. However, there were no significant differences between invasive and native species in soil inorganic nitrogen or total soil nitrogen concentration. These results suggest that nitrogen cycling was enhanced through increased production of nitrogen-rich litter by the invaders and greater uptake of soil nitrogen by native and non-native plants, facilitated by root adaptations such as accelerated growth of finer roots.

Non-native plants capable of atmospheric nitrogen fixation such as *R. pseudoacacia* and other legumes have particularly strong effects on nitrogen-cycling parameters, whereas impacts of other invaders are more variable (Vilà *et al.*, 2011). Liao *et al.* (2008) also concluded in their meta-analysis that the most pronounced ecosystem impacts were caused by woody and nitrogen-fixing invasive species. Some soil parameters strongly depend on whether the plant invader is capable of nitrogen fixation or not. For example, nitrogen-fixing invaders significantly increase soil nitrogen pools and nitrification, whereas non-nitrogen-fixing invaders have smaller effects (Vilà *et al.*, 2011). In contrast, the soil C:N ratio tends to decrease under nitrogen-fixing invaders but increase under non-nitrogen-fixing invaders (Vilà *et al.*, 2011). On average, non-native plant invasions increase available soil nitrogen by 53.7% (positive effect in 32 out of 47 cases), phosphorus pools by 19.7% (positive effect in 31 out of 50 cases) and carbon pools by 11.6% (positive effect in 35 out of 63 cases). Effects on soil organic matter, nitrogen mineralization and salinity were generally not significant (Vilà *et al.*, 2011). Invaded ecosystems showed increases ranging from a 5% increase in root carbon stocks to a 133% increase in shoot carbon stocks (Liao *et al.*, 2008).

Gentili *et al.* (2019) demonstrated alteration of soil properties for three invasive tree species in Central Europe: *R. pseudoacacia*, *P. serotina* and *Q. rubra*. All three species were associated with changes in soil characteristics, including humus forms, properties of the organic horizon and the mineral layer. However, these changes were species-specific rather than following a uniform pattern. Interestingly, soils under *R. pseudoacacia* stands were characterized by higher pH values compared to the other species.

Riparian ecosystems are among the most invasion-prone habitats in Central Europe. The presence of three invasive riparian species – *Amorpha fruticosa* L., *Fraxinus pennsylvanica* Marshall and *Acer negundo* L. – has been shown to negatively affect native plant diversity, soil properties and soil functioning (Vujanović *et al.*, 2022). In this study, a comparison of invaded sites and uninvaded sites showed that soils in invaded areas contained higher levels of organic matter, nitrogen and carbon, greater abundances of nitrifying bacteria and a lower C:N ratio. The impact of invasive species on soil chemical properties was even more pronounced when multiple invaders co-occurred in the same plant community, highlighting the interactive and potentially synergistic effects of multiple invasions (Vujanović *et al.*, 2022).

Similar patterns in altered soil chemical properties have been documented for *Aster squamatus* (Sprengel) Hieron., a hemicryptophyte native to central South America that is invading coastal grasslands in several Mediterranean countries. On the Slovenian coast, soil invaded by *A. squamatus* contained higher levels of potassium, humus, organic carbon, total nitrogen and zinc compared to uninvaded soils, where statistically significant differences were found for higher phosphorus and lower C:N ratio (Šajna *et al.*, 2014). Because of the increased soil nitrogen, *A. squamatus* is considered an 'engineering species' capable of modifying its environment through habitat fertilization resulting from rapidly decomposing litter (Šajna *et al.*, 2014). The increased nutrient availability in invaded plots on nutrient-poor soils has been shown for several other invasive species, some of them studied extensively: the notorious invasive species in Europe, *Solidago gigantea* Aiton. (Goossens *et al.*, 2024), *Carpobrotus edulis* (L.) N.E.Br., a South African succulent chamaephyte invading Mediterranean coastal habitats globally (Novoa *et al.*, 2014) and the exotic nitrogen-fixing *Myrica faya Ait.* fertilizing Hawai'ian ecosystems with nitrogen (ammonium + nitrate; Vitousek *et al.*, 1987). Because differences were observed also in soil concentrations of metal pollutants, physiological differences between invasive and native plants may also be implicated (Ehrenfeld, 2003). Increased nutrient cycling rates could cause positive feedback for invasion (Allison and Vitousek, 2004), especially when invasive plants are invading grass-dominated habitats with low biomass productivity (Scharfy *et al.*, 2011).

Root biomass and architecture

Roots not only facilitate water and nutrient uptake, but they can also alter physical and chemical soil conditions through their biomass and architecture (Nunez-Mir and McCary, 2024). Invasive plants can affect the quantity and allocation of belowground biomass in addition to often accelerated aboveground plant productivity following invasions (Vilà *et al.*, 2011). They frequently possess distinctive root traits, such as higher root biomass (mass or volume of roots), finer root morphology, deeper root penetration or denser rhizome networks than native plants. These traits allow invasive plants to occupy larger soil volumes for their benefit.

Invasive plants are frequently characterized by lower root investment and greater allocation to rapid aboveground growth, traits that help them outcompete native species (Schoenfelder *et al.*, 2010; Lin *et al.*, 2019). Nevertheless, experimental evidence suggests that this generalization does not apply to all invasive species. A pot experiment demonstrated that *Duchesnea indica* (Andrews) Teschem. (syn. *Potentilla indica* (Andrews) Th.Wolf), an invasive plant increasingly common in Europe, invests heavily in its root system under specific conditions. When exposed to full sunlight, optimal nutrient availability and interspecific competition with the native species *Glechoma hederacea* L., *D. indica* nearly doubled its root investment compared to the no-competition scenario and increased root allocation by about 60% relative to intraspecific competition (Kreća *et al.*, 2025). This case highlights that invasive plants can employ diverse strategies, with some relying on rapid aboveground growth while others, such as *D. indica*, succeed by allocating substantial resources below ground to enhance competitiveness.

A high belowground/aboveground biomass ratio (root/shoot ratio), indicating greater plant investment in roots relative to shoots, reflects enhanced competitiveness for water and nutrients. This strategy is often characteristic of plants in arid or nutrient-poor environments, where it promotes survival under stressful conditions (Bloom *et al.*, 1985). By contrast, a high aboveground/belowground biomass ratio represents greater investment in photosynthetic tissue, facilitating rapid biomass accumulation. Such a strategy is typical of fertile habitats with sufficient water and nutrient supply, where competition for light is a dominant factor (Bloom *et al.*, 1985). This ratio is thus a key indicator of plant ecological strategies, reflecting adaptation to different habitat conditions and shaping plant impacts on carbon and nutrient cycling. Greater root biomass contributes to soil organic matter accumulation (Yang *et al.*, 2023), whereas greater shoot biomass results in higher litter input and faster nutrient turnover (Tang *et al.*, 2010).

The impacts of greater root biomass can also be physical as they change soil structure and aggregation. How the invasive plants change soil properties physically depends on their root architecture. Nunez-Mir and McCary (2024) showed that invasive plants have longer roots and can outcompete native plant species because their roots are fast-growing. Deep-rooting invasive plants also benefit from nutrient uplift (Jobbágy and Jackson, 2004). Extensive rhizomes, on the other hand, can increase soil aggregation, causing compact soil, reduced porosity, lower aeration and weaker water infiltration. *Fallopia japonica* (Houtt.) Ronse Decr., invasive in Europe and North America, develops spreading rhizome systems up to 2m deep with high biomass and long-term legacy effects on soil structure (Child and Wade, 2000). Similarly, fine root networks can form dense mats that enhance water and nutrient uptake, depleting resources for natives. Invasive plants therefore affect soil health and can have a profound effect on both above- and belowground biodiversity.

Allelopathic potential of invasive plants in soil

Many invasive plants produce and release allelopathic chemical compounds that are new for the introduced ecosystems. Allelopathic interactions thus have received increased attention (Callaway and Aschehoug, 2000; Ridenour and Callaway, 2001), resulting in the so-called novel weapons hypothesis (Bais *et al.*, 2003) and homeland security hypothesis (Cummings *et al.*, 2012) proposing that allelopathy is an important trait for predicting which plant species could become successful invader. The analysis of Kalisz *et al.* (2021) showed that allelopathy was present in 72% of the 113 plant families surveyed, confirming that allelopathic potential for invasive plants is rather high. Globally important invasive plants like *A. altissima*, *Alliaria petiolata* (M.Bieb.) Cavara and Grande, *Imperata cylindrica* (L.) P. Beauv., *Fallopia japonica* and *Mikania micrantha* Kunth exemplify the novel weapons hypothesis.

Depending on the plant organ from which allelochemicals are released and the form they take (e.g. volatiles, exudates, leachates), as well as whether they enter the environment directly or through processes such as decomposition or volatilization, their effects on above- and belowground systems differ. A meta-analysis of 384 studies showed that leachates from belowground organs had 13.6% stronger negative effects on plant performance than those from aboveground organs (Zhang *et al.*, 2021). Aboveground allelopathic effects between non-native and native plants are often masked by other forms of interspecific competition. For example, allelopathy reduces plant performance by an average of 25% (Zhang *et al.*, 2021), whereas competition reduces performance by about 43% (Lekberg *et al.*, 2018). Importantly, native plant species respond more negatively to leachates from invasive plants than to those from other native species (Zhang *et al.*, 2021).

When allelochemicals enter the soil, either directly through root exudates or indirectly via decomposition products (and their derivatives influenced by environmental conditions; e.g. juglone vs hydrojuglone), they impact soil chemistry and soil biota. The effects, particularly negative ones, on plants can manifest as reduced seed germination. Laboratory experiments with *A. petiolata* root and shoot extracts and volatile compounds showed that germination of the test plant *Lepidium sativum* L. was reduced and that both *L. sativum* seedlings and radicles were shorter when compared to the control for each material tested. In contrast, no difference was found between the tested root or shoot materials (Šajna, 2017). Another effect on seeds in soil is an altered composition of dormant seeds in the soil seed bank (Gioria and Pyšek, 2016). Not only was impaired root growth of seedlings frequently observed, caused by invasive plants, but also roots of fully grown competing plants were impaired. Root suppression can happen because of alterations in auxin transport and root gravitropism: inhibited roots experience oxidative stress and cell division in root tips is inhibited (Šoln and Dolenc Koce, 2021).

The negative effects of allelochemicals from invasive plant species on soil biota are less well understood. Experiments in the field are often limited by insufficient knowledge of allelochemical pathways in the soil, while laboratory experiments often show significant inhibition effects; however, because they do not include effects of microbiota, these results poorly describe the actual situation in natural soils or only indirectly. For example, in the laboratory, *A. petiolata* volatile compounds decreased germination, while at the same time seeds were less likely to be attacked by fungal overgrowth (Šajna, 2017). This effect indirectly confirms the understanding that allelochemicals produced by *A. petiolata* inhibit root fungal symbionts and the soil hyphal network of mycorrhizal fungi (Hale *et al.*, 2016).

Overall, research has focused largely on studying the negative impacts of invasive plant allelochemicals on symbiotic root organisms and the rhizosphere microbiome. When the roots of invasive plants produce secondary metabolites that affect the colonization of native plants by arbuscular mycorrhizal fungi, the negative effects are exhibited in several ways. Allelochemicals can lower the abundance of arbuscular mycorrhizal fungi spores (propagules) or decrease the abundance of root structures (vesicles) (Wolfe and Klironomos, 2005).

Allelochemicals from *A. petiolata*, a non-mycorrhizal species itself, decrease the rate of root colonization by arbuscular mycorrhizal fungi in the invaded range. For example, seedling roots of *Acer saccharum* Marshall, native to southern Ontario, Canada, were not colonized by arbuscular mycorrhizal fungi when soils were invaded by *A. petiolata* (Wolfe and Klironomos, 2005).

The effects of allelopathy in the soil may persist for a long time even after the removal of invasive plants. Possible explanations include that some allelochemicals are relatively stable compounds and that dead roots of invasive plants represent a substantial portion of the necromass, which decomposes and releases allelochemicals slowly over time. Allelopathic exudates from bracken ferns may also transform vegetation composition (de Jesus Jatoba *et al.*, 2016). For example, vegetation succession after clearing bracken (*Pteridium Gled. ex Scop.*) may not occur as expected and/or happen very slowly (Gil da Costa *et al.*, 2024). During secondary succession, seral stages dominated allelopathic monodominant stands, frequently observed after the abandonment of pastures and meadows with low biomass productivity, showing unexpected patterns of succession. For example, secondary succession of abandoned pastures with low biomass productivity exhibited a seral stage dominated by allelopathic tall plants belonging to the Apiaceae family. They formed a monodominant seral stage which was later invaded by *Pinus sylvestris* L. (a species tolerating Apiaceae allelochemicals) instead of expected *Ostrya carpinifolia* Scop. forest (Kaligarič *et al.*, 2011).

Above- and Belowground Biodiversity Are Interdependent

Aboveground and belowground biotic systems are inextricably linked and mutually dependent. Both components function through continuous interactions, with direct and indirect effects

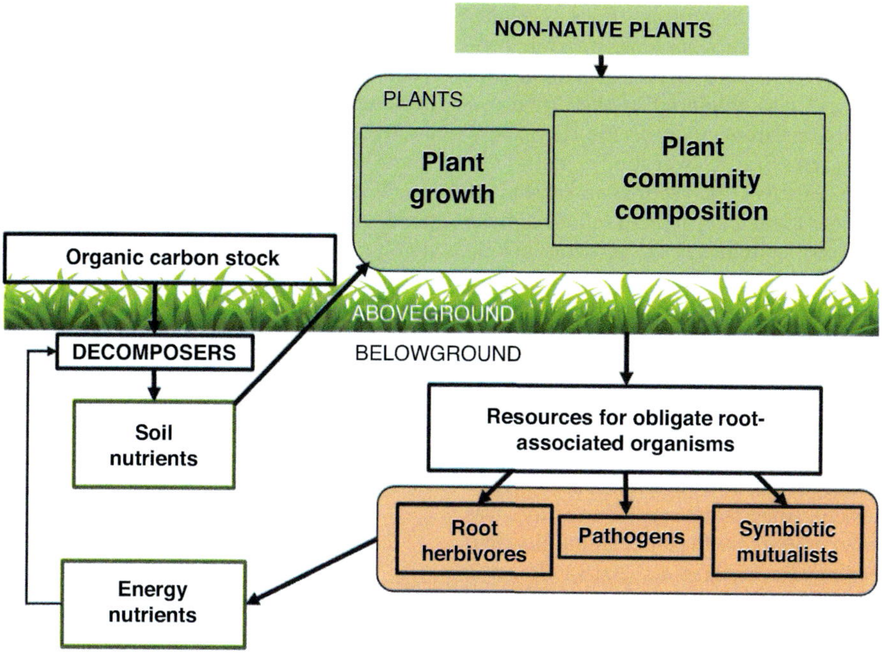

Fig. 5.5. Invasive non-native plants reshape aboveground communities by altering plant composition and growth patterns. Since plants are the main source of organic carbon above and below ground, these shifts cascade to decomposers and root-associated organisms such as herbivores, pathogens and mutualists. The resulting changes in nutrient cycling and energy flow feed back to further modify aboveground communities. Adapted from Wardle *et al.* (2004).

on each other. Plants, as primary producers, provide the main source of organic carbon for decomposers and resources for root-associated organisms such as root herbivores, pathogenic microbes and symbiotic mutualists (e.g. mycorrhizal fungi). In turn, these belowground communities influence plant performance and growth by regulating the flow of energy and nutrients. When an invasive plant enters a biotic system, it can trigger cascading effects throughout the entire ecosystem. These effects begin with aboveground changes, such as alterations in plant community composition and structure and the suppression of native plant growth, and extend to belowground processes, including shifts in microbial communities, changes in litter decomposition rates and alterations in nutrient cycling (Wardle *et al.*, 2004; Fig. 5.5).

The impact of an invasive plant on aboveground–belowground interactions is greater when the invader differs markedly from the pool of native plants in morphological and physiological traits, such as root architecture and root absorption rate, photosynthetic efficiency, the capability of nitrogen fixation and water stress regulation (Wardle *et al.*, 2004).

It is relatively well understood how invaded aboveground communities drive changes in belowground communities through alterations in nutrient cycling and related processes. By contrast, the effects of invasion on belowground communities affecting aboveground organisms are far less explored.

Klironomos (2002) provided important insights into the factors shaping the relative abundance of plant species and highlighted the critical role of plant–soil microbiota interactions. In a prairie community in Canada, he demonstrated that interactions between plants and soil organisms strongly influenced plant growth rate. Specifically, rare native plants growing in their 'home soil' exhibited lower relative growth when compared to growth in soils from other plants, due to the accumulation of soil pathogens. In contrast, invasive plants gained a competitive advantage through beneficial interactions with mycorrhizal fungi, and their growth was enhanced in their own soil compared to soil from other species. This experiment illustrates that belowground communities shape the growth of plant species and that plant species differ in their capacity to influence their own relative abundance by altering belowground communities.

Several studies have confirmed a strong positive correlation between plant species richness and the richness of soil fauna. More diverse aboveground ecosystems tend to support more diverse, complex, and stable belowground communities. For instance, in traditional agroforestry systems in Tabasco (humid tropics of Mexico), macro-arthropod diversity was positively correlated with plant community richness (Villanueva-López *et al.*, 2019). However, both plant species richness and soil arthropod diversity varied considerably among habitats. In this study, Hymenoptera and Coleoptera were the most abundant soil arthropods, while Dermaptera were the rarest (Villanueva-López *et al.*, 2019). These results suggest that changes in plant species composition and abundance – such as those caused by invasive plants – can disrupt soil ecosystems and negatively affect soil organisms.

A substantial body of evidence shows that aboveground and belowground biodiversity is strongly interdependent, including in ecosystems dominated by invasive plants. A notable example is *Fallopia* spp., a highly competitive invasive plant in Central Europe that forms dense monocultures, especially in riparian zones (Drazan *et al.*, 2021). By outcompeting native vegetation, *Fallopia* markedly reduces plant species richness and abundance, which in turn negatively affects belowground biodiversity. For instance, *Fallopia* stands have been reported to reduce soil macroinvertebrate biomass by up to 60% (Lavoie, 2017; De Almeida *et al.*, 2022).

In northern France, invaded riparian forests and meadows had similar total plant cover to uninvaded sites but experienced near-complete loss of native species, accumulation of thicker litter layers and shifts in soil fauna communities. Effects varied by trophic group and habitat: abundances of total Collembola, hemiedaphic Collembola and plant-feeding nematodes decreased in invaded plots, while epedaphic Collembola were more abundant in meadows than forests. Conversely, plant-feeding nematodes were higher in uninvaded forest sites, but were unaffected by invasion in meadows (De Almeida *et al.*, 2022).

These studies demonstrate that invasive plants not only dominate aboveground vegetation but also restructure belowground communities, altering nutrient cycling, litter dynamics and soil fauna composition. The interdependence of above- and belowground biodiversity means that invasive species can profoundly disrupt ecosystem functioning through both direct and indirect pathways.

Conclusion

Invasive non-native plants are powerful ecosystem engineers that reshape both aboveground and belowground communities through a combination of direct and indirect mechanisms. By altering plant community composition, structure and functional traits, they modify litter inputs, root traits, soil chemistry and microbial communities, ultimately affecting nutrient cycling, decomposition and energy flow. These changes create feedback loops that often reinforce invader dominance while reducing the abundance and fitness of native species. The impacts of invasions are highly context-dependent, varying with species identity, landscape configuration, habitat type and the presence of multiple invaders. Importantly, aboveground and belowground biodiversity are tightly interlinked: changes in plant communities cascade to soil organisms, while soil microbial and faunal communities influence plant performance and community dynamics. Understanding these complex, multi-level interactions is essential for predicting ecosystem responses to invasions and for developing effective management strategies. Overall, invasive plants act as ecosystem transformers, capable of fundamentally altering biodiversity, nutrient dynamics and ecosystem functioning in both natural and human-modified landscapes.

In conclusion, we would like to emphasize the high context-dependency of invasive plant effects on both above- and belowground communities, as well as on the physical environment in which organisms live. The most prominent variables influencing the outcomes of invasions include differences in the invaded habitats, geographical variation such as climatic conditions, topography and biogeographical context, the edaphic variation including soil type, physical and chemical properties of the soil and the effects of species co-occurrence.

References

Allison, S.D. and Vitousek, P.M. (2004) Rapid nutrient cycling in leaf litter from invasive plants in Hawai'i. *Oecologia* 141(4), 612–619. DOI: 10.1007/s00442-004-1679-z.

Bais, H.P., Vepachedu, R., Gilroy, S., Callaway, R.M. and Vivanco, J.M. (2003) Allelopathy and exotic plant invasion: From molecules and genes to species interactions. *Science* 301, 1377–1380.

Bloom, A.J., Chapin III, F.S. and Mooney, H.A. (1985) Resource limitation in plants: An economic analogy. *Annual Review of Ecology and Systematics* 16(1), 363–392. DOI: 10.1146/annurev.es.16.110185.002051.

Bohlen, P.J. (2006) Biological invasions: Linking the aboveground and belowground consequences. *Applied Soil Ecology* 32(1), 1–5. DOI: 10.1016/j.apsoil.2005.10.001.

Bury, S., Jagodziński, A.M. and Dyderski, M.K. (2024) In search of per capita effects of *Prunus serotina* Ehrh. invasion on temperate forest understory alpha diversity. *Biologia* 79(10), 3011–3025. DOI: 10.1007/s11756-024-01766-7.

Callaway, R.M. and Aschehoug, E.T. (2000) Invasive plants versus their new and old neighbors: A mechanism for exotic invasion. *Science (New York, N.Y.)* 290(5491), 521–523. DOI: 10.1126/science.290.5491.521.

Chabrerie, O., Loinard, J., Perrin, S., Saguez, R. and Decocq, G. (2010) Impact of *Prunus serotina* invasion on understory functional diversity in a European temperate forest. *Biological Invasions* 12(6), 1891–1907. DOI: 10.1007/s10530-009-9599-9.

Child, L.E. and Wade, P.M. (2000) *The Japanese Knotweed Manual*. Packard Publishing, Sussex.

Coleman, D.C., Callaham, M.A. and Crossley, D.A. (2018) *Fundamentals of Soil Ecology*, 3rd edn. Academic Press. DOI: 10.1016/C2015-0-04083-7.

Cummings, J.A., Parker, I.M. and Gilbert, G.S. (2012) Allelopathy: A tool for weed management in forest restoration. *Plant Ecology* 213, 1975–1989. DOI: 10.1007/s11258-012-0154-x.

De Almeida, T., Forey, E. and Chauvat, M. (2022) Non-native invasive plant effect on soil fauna is habitat dependent. *Diversity* 14(2), 61. DOI: 10.3390/d14020061.

de Jesus Jatoba, L., Varela, R.M., Molinillo, J.M.G., Ud Din, Z., Juliano Gualtieri, S.C. *et al.* (2016) Allelopathy of bracken fern (*Pteridium arachnoideum*): New evidence from green fronds, litter, and soil. *PLoS ONE* 11(8), e0161670. DOI: 10.1371/journal.pone.0161670.

Drazan, D., Smith, A.G., Anderson, N.O., Becker, R., Clark, M. *et al.* (2021) History of knotweed (*Fallopia* spp.) invasiveness. *Weed Science* 69, 617–623. DOI: 10.1017/wsc.2021.62.

Dyderski, M.K. and Jagodziński, A.M. (2021) Impacts of invasive trees on alpha and beta diversity of temperate forest understories. *Biological Invasions* 23(1), 235–252. DOI: 10.1007/s10530-020-02367-6.

Ehrenfeld, J.G. (2003) Effects of exotic plant invasions on soil nutrient cycling processes. *Ecosystems* 6(6), 503–523. DOI: 10.1007/s10021-002-0151-3.

Gaertner, M., Breeyen, A.D., Hui, C. and Richardson, D.M. (2009) Impacts of alien plant invasions on species richness in Mediterranean-type ecosystems: A meta-analysis. *Progress in Physical Geography* 33(3), 319–338. DOI: 10.1177/0309133309341607.

Gaggini, L., Rusterholz, H.P. and Baur, B. (2018) The invasive plant *Impatiens glandulifera* affects soil fungal diversity and the bacterial community in forests. *Applied Soil Ecology* 124, 335–343. DOI: 10.1016/j.apsoil.2017.11.032.

Gentili, R., Ferrè, C., Cardarelli, E., Montagnani, C., Bogliani, G. *et al.* (2019) Comparing negative impacts of *Prunus serotina*, *Quercus rubra* and *Robinia pseudoacacia* on native forest ecosystems. *Forests* 10(10), 842. DOI: 10.3390/f10100842.

Gil da Costa, R.M., Povey, A., Medeiros-Fonseca, B., Ramwell, C., O'Driscoll, C. *et al.* (2024) Sixty years of research on bracken fern (*Pteridium* spp.) toxins: Environmental exposure, health risks and recommendations for bracken fern control. *Environmental Research* 257, 119274. DOI: 10.1016/j.envres.2024.119274.

Gioria, M. and Pyšek, P. (2016) The legacy of plant invasions: Changes in the soil seed bank of invaded plant communities. *BioScience* 66(1), 40–53. DOI: 10.1093/biosci/biv165.

Goossens, E.P., Mertens, W. and Olde Venterink, H. (2024) *Solidago gigantea* invasion homogenizes soil properties and native plant communities. *Biological Invasions* 26(10), 3315–3327. DOI: 10.1007/s10530-024-03374-7.

Hale, A.N., Lapointe, L. and Kalisz, S. (2016) Invader disruption of belowground plant mutualisms reduces carbon acquisition and alters allocation patterns in a native forest herb. *New Phytologist* 209, 542–549. DOI: 10.1111/nph.13610.

Hejda, M., Pyšek, P. and Jarošík, V. (2009) Impact of invasive plants on the species richness, diversity and composition of invaded communities. *Journal of Ecology* 97(3), 393–403. DOI: 10.1111/j.1365-2745.2009.01480.x.

IPBES (2023) In: Roy, H.E., Pauchard, A., Stoett, P., Renard Truong, T., Bacher, S. *et al.* (eds) *Summary for Policymakers of the Thematic Assessment Report on Invasive Non-Native Species and Their Control of the Intergovernmental Science-Policy Platform on Biodiversity and Ecosystem Services*. IPBES Secretariat, Bonn. DOI: 10.5281/zenodo.7430692.

Jo, I., Fridley, J.D. and Frank, D.A. (2017) Invasive plants accelerate nitrogen cycling: Evidence from experimental woody monocultures. *Journal of Ecology* 105(4), 1105–1110. DOI: 10.1111/1365-2745.12732.

Jelinčić, A., Šajna, N., Zgorelec, Ž. and Perčin, A. (2022) Bracken-induced increase in soil P availability, along with its high P acquisition efficiency, enables it to invade P-deficient meadows. *Journal of Plant Ecology* 15(4), 783–794. DOI: 10.1093/jpe/rtab114.

Jobbágy, E.G. and Jackson, R.B. (2004) The uplift of soil nutrients by plants: Biogeochemical consequences across scales. *Ecology* 85(9), 2380–2389. DOI: 10.1890/03-0245.

Kaligarič, M., Meister, M.H., Škornik, S., Šajna, N., Kramberger, B. *et al.* (2011) Grassland succession is mediated by umbelliferous colonizers showing allelopathic potential. *Plant Biosystems* 145, 688–698. DOI: 10.1080/11263504.2011.601340.

Kalisz, S., Kivlin, S.N. and Bialic-Murphy, L. (2021) Allelopathy is pervasive in invasive plants. *Biological Invasions* 23(2), 367–371. DOI: 10.1007/s10530-020-02383-6.

Klironomos, J.N. (2002) Feedback with soil biota contributes to plant rarity and invasiveness in communities. *Nature* 417(6884), 67–70. DOI: 10.1038/417067a.

Kortz, A.R. and Magurran, A.E. (2019) Increases in local richness (α-diversity) following invasion are offset by biotic homogenization in a biodiversity hotspot. *Biology Letters* 15(5), 20190133. DOI: 10.1098/rsbl.2019.0133.

Kortz, A.R., Moyes, F., Pivello, V.R., Pyšek, P., Dornelas, M. *et al.* (2023) Elevated compositional change in plant assemblages linked to invasion. *Proceedings of the Royal Society B Biological Sciences* 290(1998), 20222450. DOI: 10.1098/rspb.2022.2450.

Kreća, M., Šajna, N. and Šipek, M. (2025) Response of the invasive alien plant *Duchesnea indica* (Andrews) Teschem. to different environmental and competitive settings 14(11), 1563. DOI: 10.3390/plants14111563.

Krishna, M.P. and Mohan, M. (2017) Litter decomposition in forest ecosystems: A review. *Energy, Ecology and Environment* 2(4), 236–249. DOI: 10.1007/s40974-017-0064-9.

Lavoie, C. (2017) The impact of invasive knotweed species (*Reynoutria* spp.) on the environment: Review and research perspectives. *Biological Invasions* 19(8), 2319–2337. DOI: 10.1007/s10530-017-1444-y.

Lekberg, Y., Bever, J.D., Bunn, R.A., Callaway, R.M., Hart, M.M. *et al.* (2018) Relative importance of competition and plant-soil feedback, their synergy, context dependency and implications for coexistence. *Ecology Letters* 21(8), 1268–1281. DOI: 10.1111/ele.13093.

Liao, C., Peng, R., Luo, Y., Zhou, X., Wu, X. *et al.* (2008) Altered ecosystem carbon and nitrogen cycles by plant invasion: A meta-analysis. *The New Phytologist* 177(3), 706–714. DOI: 10.1111/j.1469-8137.2007.02290.x.

Lin, T., Klinkhamer, P.G.L., Pons, T.L., Mulder, P.P.J. and Vrieling, K. (2019) Evolution of increased photosynthetic capacity and its underlying traits in invasive *Jacobaea vulgaris*. *Frontiers in Plant Science* 10, 1016. DOI: 10.3389/fpls.2019.01016.

MacArthur, R.H. and Wilson, E.O. (1967) *The Theory of Island Biogeography*. Princeton University Press. DOI: 10.1515/9781400881376.

Medina-Villar, S., Rodríguez-Echeverría, S., Lorenzo, P., Alonso, A., Perez-Corona, E. *et al.* (2016) Impacts of the alien trees *Ailanthus altissima* (Mill.) Swingle and *Robinia pseudoacacia* L. on soil nutrients and microbial communities. *Soil Biology and Biochemistry* 96, 65–73. DOI: 10.1016/j.soilbio.2016.01.015.

Meyerson, L.A. and Mooney, H.A. (2007) Invasive alien species in an era of globalization. *Frontiers in Ecology and the Environment* 5(4), 199–208. DOI: 10.1890/1540-9295(2007)5[199:IASIAE]2.0.CO;2.

Novoa, A., Rodríguez, R., Richardson, D. and González, L. (2014) Soil quality: A key factor in understanding plant invasion? The case of *Carpobrotus edulis* (L.) N.E.Br. *Biological Invasions* 16(2), 429–443. DOI: 10.1007/s10530-013-0531-y.

Nunez-Mir, G.C. and McCary, M.A. (2024) Invasive plants and their root traits are linked to the homogenization of soil microbial communities across the United States. *Proceedings of the National Academy of Sciences USA* 121(44), e2418632121. DOI: 10.1073/pnas.2418632121.

Ohlemüller, R., Walker, S. and Wilson, J.B. (2006) Local vs. regional factors as determinants of the invasibility of indigenous forest fragments by alien plant species. *Oikos* 112, 493–501. DOI: 10.1111/j.0030-1299.2006.14115.x.

Pattison, Z., Rumble, H., Tanner, R.A., Jin, L. and Gange, A.C. (2016) Positive plant–soil feedbacks of the invasive *Impatiens glandulifera* and their effects on above-ground microbial communities. *Weed Research* 56, 198–207. DOI: 10.1111/wre.12202.

Ridenour, W.M. and Callaway, R.M. (2001) The relative importance of allelopathy in interference: The effects of an invasive weed on a native bunchgrass. *Oecologia* 126(3), 444–450. DOI: 10.1007/s004420000533.

Rusterholz, H.-P., Schneuwly, J. and Baur, B. (2018) Invasion of the alien shrub *Prunus laurocerasus* in suburban deciduous forests: Effects on native vegetation and soil properties. *Acta Oecologica* 92, 44–51. DOI: 10.1016/j.actao.2018.07.003.

Šajna, N. (2017) Habitat preference within its native range and allelopathy of garlic mustard (*Alliaria petiolata*). *Polish Journal of Ecology* 65, 46–56. DOI: 10.3161/15052249PJE2017.65.1.004.

Šajna, N., Kaligarič, M. and Ivajnšič, D. (2014) Reproduction biology of an alien invasive plant: A case of drought-tolerant *Aster squamatus* on the Northern Adriatic Seacoast, Slovenia. In: Rannow, S. and Neubert, M. (eds) *Managing Protected Areas in Central and Eastern Europe under Climate Change*. Springer, Dordrecht, pp. 297–314. DOI: 10.1007/978-94-007-7960-0_19.

Scharfy, D., Funk, A., Olde Venterink, H. and Güsewell, S. (2011) Invasive forbs differ functionally from native graminoids, but are similar to native forbs. *The New Phytologist* 189(3), 818–828. DOI: 10.1111/j.1469-8137.2010.03531.x.

Schoenfelder, A.C., Bishop, J.G., Martinson, H.M. and Fagan, W.F. (2010) Resource use efficiency and community effects of invasive *Hypochaeris radicata* (Asteraceae) during primary succession. *American Journal of Botany* 97(11), 1772–1779. DOI: 10.3732/ajb.0900383.

Šipek, M. and Šajna, N. (2020) Vrstna pestrost tujerodnih rastlin v Miklavškem gozdu na Dravskem polju [Alien plant diversity in urban forest Miklavški gozd]. *Hladnikia* 46, 53–65.

Šipek, M. and Šajna, N. (2024) Lowland forest fragment characteristics and anthropogenic disturbances determine alien plant species richness and composition. *Biological Invasions* 26(5), 1595–1614. DOI: 10.1007/s10530-024-03269-7.

Šipek, M., Kutnar, L., Marinšek, A. and Šajna, N. (2022) Contrasting responses of alien and ancient forest indicator plant species to fragmentation process in the temperate lowland forests. *Plants (Basel, Switzerland)* 11(23), 3392. DOI: 10.3390/plants11233392.

Šoln, K. and Dolenc Koce, J. (2021) Allelopathic root inhibition and its mechanisms. *Allelopathy Journal* 52(2), 181–198. DOI: 10.26651/allelo.j/2021-52-2-1315.

Stinson, K., Kaufman, S., Durbin, L. and Lowenstein, F. (2007) Impacts of garlic mustard invasion on a forest understory community. *Northeastern Naturalist* 14(10), 73–88. DOI: 10.1656/10926194(2007)14[73:IOGMIO]2.0.CO;2.

Tang, J.W., Cao, M., Zhang, J.H. and Li, M.H. (2010) Litterfall production, decomposition and nutrient use efficiency varies with tropical forest types in Xishuangbanna, SW China: A 10-year study. *Plant and Soil* 335(1–2), 271–288. DOI: 10.1007/s11104-010-0414-2.

Vilà, M. and Ibáñez, I. (2011) Plant invasions in the landscape. *Landscape Ecology* 26(4), 461–472. DOI: 10.1007/s10980-011-9585-3.

Vilà, M., Tessier, M., Suehs, C.M., Brundu, G., Carta, L. *et al.* (2006) Local and regional assessments of the impacts of plant invaders on vegetation structure and soil properties of Mediterranean islands. *Journal of Biogeography* 33(5), 853–861. DOI: 10.1111/j.1365-2699.2005.01430.x.

Vilà, M., Espinar, J.L., Hejda, M., Hulme, P.E., Jarošík, V. *et al.* (2011) Ecological impacts of invasive alien plants: A meta-analysis of their effects on species, communities and ecosystems. *Ecology Letters* 14(7), 702–708. DOI: 10.1111/j.1461-0248.2011.01628.x.

Villanueva-López, G., Lara-Pérez, L.A., Oros-Ortega, I., Ramírez-Barajas, P.J., Casanova-Lugo, F. *et al.* (2019) Diversity of soil macro-arthropods correlates to the richness of plant species in traditional agroforestry systems in the humid tropics of Mexico. *Agriculture, Ecosystems & Environment* 286, 106658. DOI: 10.1016/j.agee.2019.106658.

Vitousek, P.M., Walker, L.R., Whiteaker, L.D., Mueller-Dombois, D. and Matson, P.A. (1987) Biological invasion by *Myrica faya* alters ecosystem development in Hawaii. *Science* 238(4828), 802–804. DOI: 10.1126/science.238.4828.802.

Vujanović, D., Losapio, G., Milić, S. and Milić, D. (2022) The impact of multiple species invasion on soil and plant communities increases with invasive species co-occurrence. *Frontiers in Plant Science* 13, 875824. DOI: 10.3389/fpls.2022.875824.

Wang, C., Jiang, K., Zhou, J. and Wu, B. (2018) *Solidago canadensis* invasion affects soil N-fixing bacterial communities in heterogeneous landscapes in urban ecosystems in East China. *Science of the Total Environment* 631–632, 702–713. DOI: 10.1016/j.scitotenv.2018.03.061.

Wardle, D.A., Bardgett, R.D., Klironomos, J.N., Setälä, H., van der Putten, W.H. *et al.* (2004) Ecological linkages between aboveground and belowground biota. *Science* 304(5677), 1629–1633. DOI: 10.1126/science.1094875.

Wolfe, B.E. and Klironomos, J.N. (2005) Breaking new ground: Soil communities and exotic plant invasion. *BioScience* 55(6), 477. DOI: 10.1641/0006-3568(2005)055[0477:BNGSCA]2.0.CO;2.

Yang, X., Wang, B., Fakher, A., An, S. and Kuzyakov, Y. (2023) Contribution of roots to soil organic carbon: From growth to decomposition experiment. *Catena* 231, 107317. DOI: 10.1016/j.catena.2023.107317.

Zhang, Z., Liu, Y., Yuan, L., Weber, E. and van Kleunen, M. (2021) Effect of allelopathy on plant performance: A meta-analysis. *Ecology Letters* 24(2), 348–362. DOI: 10.1111/ele.13627.

6 Invasive Plants as Soil Engineers: Their Effects on Organic Matter Decomposition and Nutrient Dynamics

Erika Gömöryová* and Dušan Gömöry
Technical University in Zvolen, Faculty of Forestry, Zvolen, Slovakia

Introduction

In recent decades, invasive plants have become an increasingly significant ecological and economic problem worldwide (Chiu *et al.*, 2023). As global trade and mobility accelerate, many invasive plant species have spread into both terrestrial and aquatic environments. While some may have beneficial effects, they more often have extensive negative impacts on the ecosystems they invade (Kourantidou *et al.*, 2022; Raheem *et al.*, 2024). Their contribution to biodiversity loss is well documented, especially in ecosystems under pressure from land use and climate change (Bellard *et al.*, 2013; Sage, 2020). However, invasive plants can significantly affect not only the diversity of understorey flora and fauna but also forest regeneration as they compete with native trees for water, nutrients and light (Pile Knapp *et al.*, 2023). They are also known to affect the fire regimes (Underwood *et al.*, 2019).

Traditionally, research has focused on the aboveground effects of invasive plant species. However, there is now a growing interest in their belowground impact (Marchante *et al.*, 2008; Weidenhamer and Callaway, 2010; Zhang *et al.*, 2019; Unger *et al.*, 2022). It is well known that plants and soils are closely linked. Soil influences the growth and production of plants, while plants influence the soil physical, chemical and biological properties directly and/or indirectly. Invasive plants can act as ecosystem engineers in soil systems. For example, they can influence bulk density and porosity (Weidenhamer and Callaway, 2010), as well as soil chemistry through litter input, root exudates and nutrient uptake and release (Ehrenfeld, 2003). Invasive plants also influence biological communities by shaping microbial and invertebrate communities (Reinhart and Callaway, 2006; Landsman *et al.*, 2021). Their impact on soil systems is often profound and lasting. Invasive plants often introduce novel traits into ecosystems as soil engineers, such as nitrogen fixation and allelopathy (Qu *et al.*, 2021; Ortiz and Wolf, 2024). This modifies the quantity and chemical composition of litter and alters the processes of decomposition, the functional groups of microorganisms and the patterns of nutrient retention or leaching. This role extends their ecological impact below ground, making them powerful agents of change in the soil system, an area which often receives relatively little attention in invasion ecology.

The decomposition of organic matter and nutrient cycling belong to the key soil ecosystem processes that influence soil fertility

*Corresponding author: gomoryova@tuzvo.sk

DOI: 10.1079/9781836991298.0006

and productivity. These processes also regulate carbon storage and greenhouse gas emissions, and support plant–microbe interactions (Gautam and Berg, 2025). Invasive plants can alter these processes, resulting in long-term consequences for ecosystem resilience and restoration potential. Since future climate change is expected to increase plant invasions, and the patterns of their impact and ecosystems at greatest risk are unclear, more research is needed to develop effective control measures in high-risk areas (Seebens *et al.*, 2015; Deeley and Petrovskaya, 2022).

In this chapter, we summarize the current knowledge of the role of invasive plant species as soil engineers, and their effect on the organic matter decomposition and nutrient dynamics. We explore the mechanisms through which invasive plants alter soil processes, evaluate empirical evidence from various ecosystems and assess the implications for soil ecosystem functions.

Mechanisms of Soil Engineering by Invasive Plants

Like other plants, invasive plants can directly or indirectly affect the decomposition of organic matter and nutrient cycling. For example, they can influence these processes by adding organic matter through litter and roots and by altering the microclimate (Fig. 6.1). Furthermore, characteristics such as plant size, leaf area and canopy cover can affect photosynthesis, transpiration and nutrient utilization efficiency, thereby influencing the soil water regime and nutrient cycling (Allison and Vitousek, 2004; Weidenhamer and Callaway, 2010).

Litter quality and quantity

The quantity and chemical composition of plant biomass are critical factors that influence

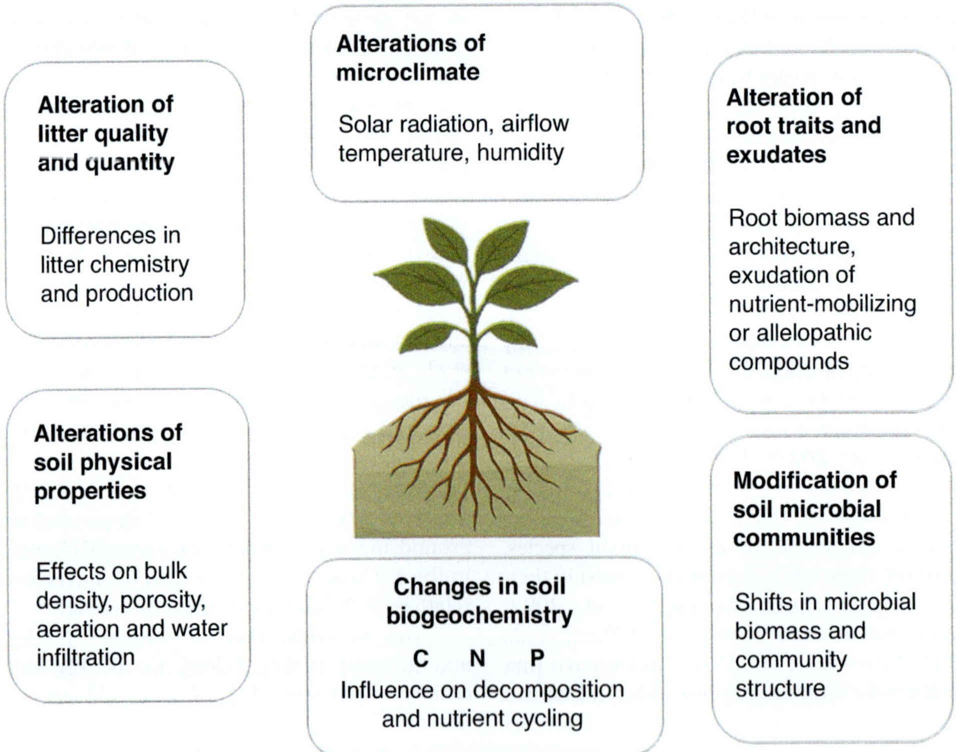

Fig. 6.1. Mechanisms of soil engineering by invasive plants.

decomposition rates and nutrient cycling in ecosystems. These factors determine how quickly and efficiently decomposers break down organic matter and release essential nutrients back into the environment. The long-term impact of litter can modify soil nutrient pools. Evidence suggests that invasive plant species may alter nutrient cycling differently to native species (Weidenhamer and Callaway, 2010).

The size and leaf area of a plant influence its ability to conduct photosynthesis. Larger plants with greater leaf areas have a larger surface area and can capture more sunlight. This results in increased carbon assimilation and higher organic matter production. Consequently, more organic matter is input into the soil. Although studies on invasive plant biomass are less common than studies on plant chemistry, several studies suggest that some invasive plant species have higher specific leaf areas, growth rates and photosynthetic rates than native species in the same locations (Allison and Vitousek, 2004; van Kleunen *et al.*, 2010). Japanese knotweed (*Fallopia japonica*) is an example of an invasive plant with large leaves and a perennial life cycle. It builds up a dense canopy and produces leaves up to 15 cm long (Fennell *et al.*, 2018). This enables the plant to release significant amounts of carbon into the soil over an extended period (Weidenhamer and Callaway, 2010). Other invasive species, such as *Reynoutria japonica*, *Reynoutria sachalinensis*, and *Reynoutria* × *bohemica* were also found to produce large amounts of litter. *Aegilops triuncialis* was found to have an aboveground biomass that was twice as high as that of native plants at the end of the season (Drenovsky and Batten, 2007; Lavoie, 2017). Wang *et al.* (2018) analysed several characteristics of plants, including height, number of branches, number of reproductive branches, belowground-to-aboveground biomass ratio and reproductive allocation coefficient. They found that all of these characteristics were notably higher in invasive plants than in native species. However, the specific leaf area values of the invasive plants were lower. The plasticity indices of specific leaf area, maximum branch angle and branch number were each remarkably lower for invasive plants than for native species.

Litter chemistry also strongly affects the rate of decomposition and ecosystem biogeochemistry (Tamura *et al.*, 2017). Invasive plant species often produce litter with biochemical traits that differ from those of native species. Many invasive plant species have been found to have higher nitrogen content in their tissues than their native counterparts (Jo *et al.*, 2017; Huang *et al.*, 2020). These plants produce nutrient-rich litter with a lower C:N ratio that decomposes rapidly. Nitrogen-fixing plants, such as *Acacia*, *Albizia* and *Leucaena* species, also exhibit high nitrogen content in their litter (Adams *et al.*, 2016; Xiao *et al.*, 2024). Hu *et al.* (2022) compared six native and two invasive plant species: *Alternanthera philoxeroides* and *Bidens pilosa*. According to the initial elemental content of the litter, the carbon content of all species was similar; however, *A. philoxeroides* and *B. pilosa* had significantly higher nitrogen content (up to 33.1 $mg \cdot g^{-1}$). *Alternanthera philoxeroides* had a C:N ratio of 12.1, which was low compared to the other species.

Similarly, Jo *et al.* (2015) compared the chemistry of ten native and 14 non-native invasive forest shrubs and lianas in the eastern USA. They found that non-native species had higher leaf production and leaf nitrogen concentration but lower leaf nitrogen resorption rates than native species. In a study of Mediterranean ecosystems, Incerti *et al.* (2018) found significant differences in litter chemical composition between native and invasive species, with invaders generally exhibiting higher nitrogen concentrations and lower lignin:N ratios. The researchers conducted a 1 year litterbag experiment with eight invasive and 24 autochthonous plant species in four different plant communities (grassland, sand dune, riparian and mixed forests). Conversely, Drenovsky and Batten (2007) found that aboveground plant litter from *Aegilops triuncialis* plots had significantly higher lignin:N and C:N ratios, as well as lower total nitrogen, phosphorus and potassium, than litter from native plant plots. The decomposition rate is strongly influenced by the presence of lignin and polyphenols. Lignin, a complex organic polymer, is more resistant to microbial breakdown, thus slowing the decomposition process (Talbot and Treseder, 2012; Talbot *et al.*, 2012).

Invasive trees can alter soil properties and ecosystem function by producing phenolic compounds (Stanek *et al.*, 2021). These compounds can regulate the availability of carbon and nitrogen, influence enzymatic activity and function as signalling molecules. In plant–microorganism interactions, species-specific phenolic compounds may serve as nutrient sources or exert toxic effects. Depending on their properties, these compounds can have antimicrobial or stimulatory effects, thereby shaping soil microbial communities and modulating enzymatic processes in distinct ways.

Root architecture and exudates

Invasive plant species have a profound impact on ecosystems through both their above- and belowground traits (Núñez-Mir and McCary, 2024). Their root architecture and exudates alter the soil physical, chemical and biological properties. These changes often persist long after plant removal, influencing the regeneration of native plants, nutrient cycling and microbial communities (Weidenhamer and Callaway, 2010; Maclean *et al.*, 2018; Zhang *et al.*, 2019; Afzal *et al.*, 2023). However, unlike more easily measurable aboveground plant traits, the belowground traits of invasive plants have largely been overlooked in studies of their impact.

According to Núñez-Mir and McCary (2024), native and invasive plants differ significantly in their root characteristics. Invasive species generally have higher specific root length, while native species tend to have greater root tissue density. These trait patterns in invasive species are associated with rapid root growth, low C:N ratios and enhanced nutrient foraging: attributes often observed in invasive plants. These root characteristics can influence soil conditions and affect microbial growth. Therefore, understanding the impact of invasive plant root traits on soil microbial communities is essential to comprehending ecosystem functioning and health. Root traits commonly found in invasive species, such as increased specific root length and reduced tissue density, may inhibit microbial activity and alter the composition of soil microbial communities.

Root systems of invasive species may differ from those of native species in terms of depth, density and branching patterns. Plants with deeper and denser root networks can have greater access to soil water and nutrients. This increases soil turnover and modifies soil porosity, affecting water and oxygen availability. Deep or aggressive rooting (e.g. *Tamarix*) can redistribute nutrients and moisture from lower soil layers to the surface (Zhang *et al.*, 2002). In their study of the root systems of ten native and 14 non-native invasive forest shrubs and lianas in the eastern USA Jo *et al.* (2015) found that the non-native species had greater root production but lower root nitrogen concentration than the native species. The soil nitrogen concentration associated with non-native species was significantly lower than that associated with native species. Their results also suggest that increased aboveground productivity in invasive forest species is linked to greater fine root production, which may enhance their ability to uptake soil resources.

Root growth and turnover can affect the physical properties of soil. For example, some invasive species, such as *Solidago canadensis*, have been shown to disrupt soil aggregates by increasing bulk density and decreasing aggregate stability (Zhang *et al.*, 2009). Changes in soil bulk density and compaction affect the penetration of the roots of native species and alter their susceptibility to erosion (Kyle *et al.*, 2007). Invasive species with high root turnover rates (i.e. rapid decomposition) can reduce stable root-derived organic matter, which may negatively impact aggregate formation (Baloglanni *et al.*, 2015). While several studies have examined how invasive species disrupt soil structure, the promotion of aggregation by invasive species is not well documented.

Invasive roots often release organic acids, enzymes or allelopathic chemicals that can directly influence the growth and development of neighbouring plants. These chemicals can also interact with soil microorganisms, altering their composition and activity and, consequently, plant performance (Elgersma *et al.*, 2011; Tian *et al.*, 2021; Yu *et al.*, 2022; Rathore *et al.*, 2025). For example, Yu *et al.* (2022) found that invaders exhibited greater arbuscular mycorrhizal colonization and biomass compared to natives. Their root exudates contained higher concentrations of two common chemical signals, quercetin and strigolactones. These signals stimulate arbuscular mycorrhizal fungal growth and root colonization. An exudate exchange experiment showed that root exudates from invaders increased root colonization more than exudates from natives. However, Rathore *et al.* (2025) found no significant differences in root morphology or exudate metabolome composition, although certain metabolites differed between the two groups.

Some invasive species are highly effective at nitrogen fixation. These nitrogen-fixing invasive plants profoundly impact on ecosystems by increasing soil nitrogen availability. This can disrupt native plant communities adapted to low-nitrogen soils, accelerate nutrient cycling, alter microbial communities and facilitate further invasions (Ortiz and Wolf, 2024). These species play a dual role: although they can improve degraded lands and boost soil fertility in agriculture or restoration, they disrupt nutrient balances, promote further invasions and threaten native biodiversity in wild ecosystems (Hiltbrunner *et al.*, 2014; Ortiz and Wolf, 2024; Zhu *et al.*, 2025). Managing them requires carefully evaluating their above- and belowground impacts, especially in ecosystems that evolved under low-nutrient conditions.

Many studies have examined the allelopathic compounds released by invasive species. Studies have shown that the success of an invasive species depends on the presence of harmful plant biochemicals, including allelopathic antimicrobial root exudates, which directly inhibit plant growth and soil microbial activity. Qu *et al.* (2021) investigated the allelopathic effects of the invasive plant species *Rhus typhina* on the cultivated plant *Tagetes erecta*, as well as on soil properties and microbial communities. They found that *R. typhina* root extracts inhibited both plant growth and soil microbial activity. The root extracts significantly reduced the utilization of carbohydrate, carboxylic acid and polymer carbon sources, but enhanced phenolic acid utilization.

Microclimate and soil water regime alteration

Much research on invasive plant species has focused on the modification of aboveground litter input. However, their role in modifying microclimatic conditions has received less attention. Microclimate is essential for regulating biological processes, including seed germination, species interactions, microbial activity and decomposition (Kemppinen *et al.*, 2024). Plants, including invasive species, can significantly alter the local microclimate by changing in vegetation structure, ground cover, litter amount and water dynamics. Typically, air and soil temperatures in vegetated areas are more stable than in open areas. Canopies can also affect light distribution, relative humidity and vapour pressure deficit (Chen *et al.*, 2024). These changes can give invasive species a competitive advantage and make ecosystem restoration more difficult, especially in systems already vulnerable to climate variability.

One of the ways invasive plants alter microclimates is by changing the structure of the vegetation canopy. Invasive plants often form denser canopies than native vegetation, reducing incoming solar radiation and increasing shading. These changes can lead to lower daytime temperatures near the soil surface, reduced diurnal temperature fluctuations and higher relative humidity. For example, *Lonicera maackii* creates a dense canopy that significantly decreases light availability and soil temperatures in the understory (McNeish and McEwan, 2016). Conversely, Carter *et al.* (2019) found slightly higher average soil temperatures and lower soil surface photosynthetically active radiation in areas with Scotch broom (*Cytisus scoparius* (L.) Link).

Furthermore, plants with a large, dense canopy often transpire more water, which affects the microclimate and soil water content in the vicinity. However, the effect of plants on the microclimate can differ depending on the type of vegetation. For instance, Kozhoridze *et al.* (2025) illustrate the differences in microclimate between two invasive species: *Acacia saligna*, a densely growing tree, and *Heterotheca subaxillaris*, a resilient herb. *H. subaxillaris* increased surface temperatures throughout its growing season by 0.6°C in spring, 1.8°C in summer and 2.19°C in autumn compared to native vegetation. It intensified water stress both directly by reducing soil moisture due to poor ground shading and high potential evapotranspiration, and indirectly by raising temperatures and increasing water vapour loss. Unlike *H. subaxillaris*, the dense canopy of *A. saligna* provided substantial soil shading and cooled ambient temperatures. However, it had high assimilation and transpiration rates, which increased water vapour and potential evapotranspiration levels and led to indirect moisture depletion despite its shading.

Invasive plants could also potentially moderate wind conditions. Dense, tall stands may reduce near-ground airflow. This limits evapotranspiration, increases soil humidity and affects seed dispersal and plant–pollinator interactions. However, studies on this topic are lacking.

Many invasive species alter soil surface properties by introducing dense ground cover or by changing litter quantity. Dense ground cover has been shown to affect surface albedo, soil temperature dynamics and moisture retention (Chen *et al.*, 2024). Thick or persistent litter layers can insulate the soil, keeping it cooler in the summer and warmer in the winter. It reduces diurnal soil temperature fluctuations and traps moisture, maintaining higher humidity near the soil surface.

Some invasive plants have higher transpiration rates, especially woody species. This reduces soil moisture, uses water more aggressively and depletes groundwater or accesses deep water tables unavailable to shallow-rooted native plants (Le Maitre *et al.*, 2020; Shiferaw *et al.*, 2021).

Alteration of the structure of microbial communities

Changes in litter inputs, root exudates and altered microclimate conditions lead to altered habitats for soil organisms, including microbiota. Invasive plants can affect the structure and function of these communities, which play a key role in decomposition and nutrient cycling (Hawkes *et al.*, 2005). Depending on the type of carbon input, the stimulation or inhibition of different microbial taxa can shift community composition and alter the efficiency of organic matter decomposition and nutrient turnover. For instance, cheatgrass invasion in the northern Great Basin of the western USA reduces the capacity of the fungal nitrogen consumption sink, enhances the capacity of a cycloheximide-resistant nitrogen sink and alters the contributions of bacteria and fungi to gross nitrogen mineralization (DeCrappeo *et al.*, 2017). According to the meta-analyses performed by Torres *et al.* (2021), invasive plants affect soil biota differentially. Bacterial diversity in soil increases with the presence of invasive plants, specifically herbaceous plants that produce allelopathic substances in temperate forest ecosystems. Invasive plants may also disrupt mycorrhizal associations. Arbuscular mycorrhizal fungi, which are crucial for nutrient uptake, often decline in abundance in invaded soils (Stinson *et al.*, 2006), negatively impacting the nutrient acquisition of native plants.

Effects of Invasive Plants on Organic Matter Decomposition and Nutrient Cycling

Organic matter decomposition

Litter decomposition is a key ecological process that regulates the cycling of nutrients such as nitrogen, phosphorus and carbon. This process is influenced by a variety of direct and indirect mechanisms. Invasive plant species can alter decomposition rates by changing organic matter inputs and climate conditions. They can also influence abiotic soil conditions, such as pH, soil structure, moisture content and temperature regimes, as well as soil microbial communities. These changes can persist even after the removal of invaders, continuing to influence nutrient cycling (Ehrenfeld *et al.*, 2001; Strickland *et al.*, 2009; Ehrenfeld, 2010).

Invasive plant species can have different chemical functions and traits to native species, which may influence litter decomposition. Table 6.1 provides examples of various invasive plant species and their effects on litter decomposition. Some studies have shown that plant invasion accelerates litter decomposition, while others have demonstrated neutral or retarding effects in comparison to native species. One of the primary pathways through which invasive species affect decomposition is via their litter inputs.

The litter of nitrogen-rich invasive plants decomposes more rapidly than that of native plants (Hu *et al.*, 2022). The higher nitrogen content and lower C:N ratio of their litter favour

Table 6.1. Invasive plant species and their effects on litter decomposition.

Invasive plant species	Litter decomposition effect	Country	Reference
Acacia longifolia	Accelerated	Portugal	Marchante *et al.* (2009)
Acer negundo	Accelerated	Lithuania	Krevš and Kučinskiene (2017)
Aegilops triuncialis	Suppressed	USA	Drenovsky and Batten (2007)
Ailanthus altissima	Accelerated	Slovakia, Spain	Halabuk and Gerhátová (2011), Castro-Díez *et al.* (2012)
Alliaria petiolata	Accelerated	USA	Rodgers *et al.* (2008)
Alternanthera philoxeroides	Accelerated	China	Hu *et al.* (2022)
Bidens pilosa	Accelerated	China	Hu *et al.* (2022)
Celastrus orbiculatus	Accelerated	USA	Leicht-Young *et al.* (2009)
Centaurea maculosa	Not consistent	USA	Hook *et al.* (2004)
Centaurea stoebe	Not consistent	USA, Romania	Thorpe and Callaway (2011)
Conyza canadensis	Suppressed	China	Liu *et al.* (2025)
Eucalyptus spp.	Accelerated	Brazil	Silva *et al.* (2023)
Eucalyptus globulus	Suppressed	Spain	Castro-Díez *et al.* (2012)
Fallopia japonica	Suppressed	Italy	Mincheva *et al.* (2014)
Impatiens parviflora	Accelerated	Slovakia	Halabuk and Gerhátová (2011)
Lonicera maackii	Accelerated	USA	Poulette and Arthur (2012)
Melaleuca quinquenervia	No significant change	USA	Martin *et al.* (2010)
Phragmites australis	Variable	Austria	Hietz (1992)
Pueraria montana	Accelerated	USA	Hickman *et al.* (2010)
Quercus rubra	Suppressed	Poland	Stanek *et al.* (2021)
Rhamnus cathartica	Accelerated	USA	Heneghan *et al.* (2007)
Rhododendron ponticum	Suppressed	UK	Jones *et al.* (2019)
Solidago canadensis	Suppressed	China	Liu *et al.* (2025)
Tamarix spp.	Suppressed	Argentina	Bedano *et al.* (2014)
Ulmus pumila	Suppressed	Spain	Castro-Díez *et al.* (2012)

invertebrates and microorganisms. Therefore, litter from invasive plants likely breaks down more quickly, releasing nitrogen and other nutrients into the soil and changing the environment for the plants that have been overtaken. For instance, *Acacia longifolia* produces high-quality litter with low C:N ratios, which often accelerates decomposition and leads to rapid nutrient mineralization. Conversely, the litter of some invasive species is rich in lignin or phenolic compounds, which slows the decomposition rate. One of the most aggressive invasive species is *F. japonica*. Studies have shown that *F. japonica* litter decomposes three to four times more slowly than native grassland litter, primarily due to its low nitrogen content and consequently high C:N ratio in both leaves and stems (Mincheva *et al.*, 2014). However, as decomposition proceeds, the C:N ratio of *F. japonica* litter decreases to values approaching those of grassland litter.

Some invasive plants decompose at notably slow rates, possibly due to high concentrations of secondary metabolites. For instance, the leaves of *Melaleuca* species contain essential oils (monoterpenes) that significantly reduce the decomposition rate, resulting in less than a 10% mass loss during the first year (Boon and Johnstone, 1997, as cited in Ehrenfeld, 2003).

In general, the litter of Leguminosae such as *Robinia pseudoacacia*, is characterized by a high nitrogen content due to symbiotic nitrogen fixation and moderate to low levels of recalcitrant compounds. This results in faster decomposition rates compared to non-leguminous native trees at least in the initial stage (Stogner *et al*., 2024). However, the long-term effects are context-dependent. In dry ecosystems or where microbial communities are not adapted to high nitrogen availability, the shift in decomposition dynamics may lead to imbalances in microbial nutrient demand. This favours fast-growing bacteria over more functionally diverse communities (Liu *et al*., 2024). According to Castro-Díez *et al*. (2012), the litter of exotic nitrogen-fixing trees (*Acacia dealbata* and *R. pseudoacacia*) does not increase soil nitrogen pools due to the inhibition of microbial activity by secondary compounds. Therefore, the secondary metabolites in the litter play a significant role in explaining the impact of exotic litter on soil properties. For example, the litter of invasive species such as *Rhododendron ponticum* is rich in phenolic compounds or lignin, which can suppress microbial activity and slow decomposition (Jones *et al*., 2019).

Similarly, *S. canadensis* litter is relatively low in nitrogen and high in polyphenolic compounds, such as flavonoids and phenolic acids (Zhang *et al*., 2009). These compounds contribute to the slow decomposition rates of litter. These secondary products can inhibit microbial activity, thereby suppressing extracellular enzyme activity (Triebwasser *et al*., 2012). However, this effect depends on the ecosystem and varies based on microbial community adaptation and local soil conditions.

Nutrient cycling

The consequences of altered decomposition rates can vary. Generally, accelerated decomposition increases nutrient availability. However, when litter decomposition slows, nutrients become immobilized, which can lead to nutrient limitation for native species.

Carbon cycling

Biological invasions can significantly alter the carbon cycle, consequently affecting soil organic carbon (SOC) levels. Depending on species traits and context, invasions can increase, decrease or redistribute carbon between vegetation and soil. Martin *et al*. (2017) emphasize that the impact of invasive species on an ecosystem may depend on differences in the traits of invasive and native species rather than solely the traits of invasive species, and that climate can alter the influence of these trait differences. Invasions often alter the allocation of organic matter above and below ground, change litter quality and decomposition rates and cause shifts in soil microbial community composition. These changes collectively affect SOC stocks and CO_2 fluxes (Martin *et al*., 2017; Torres *et al*., 2021). High-quality litter (low in lignin and high in nitrogen) speeds decomposition and can increase CO_2 release. In contrast, low-quality litter slows decomposition and can build SOC, at least initially (Marschner *et al*., 2008).

Invaders alter the composition and activity of soil microbial communities and change microbial biomass, which is one of the key fractions of SOC. He *et al*. (2025) conducted a meta-analysis of 445 observations from 61 published reports and found that plant invasion leads to a 29% increase in microbial biomass carbon (MBC). The magnitude of these effects depends on the type of invasive species, the type of ecosystem and the depth of the soil layer, which highlights the complex role of biological invasion in the global carbon cycle. Specifically, plant invasion caused MBC increases of 59% in estuaries, 59% in alluvial lands, 53% in wetlands and 80% in orchards. Interestingly, invasion by Asteraceae plants resulted in a 33% MBC increase, whereas Lythraceae invasion caused a 72% MBC increase.

Invasive plants alter the substrate availability for extracellular enzymes by changing litter inputs and soil chemistry. This shifts decomposition pathways and the balance between labile and recalcitrant carbon pools. Invasions often exhibit altered enzyme activities linked to observed changes in SOC and CO_2 fluxes (Kourtev *et al*., 2002). However, Zhou and Staver's (2019) meta-analysis of 60 case studies indicates that enzymes that act on/break down carbon compounds are not consistently or significantly affected by invasions. This may be because soil microbes in invaded sites do not need to invest heavily in oxidizing lignin and other recalcitrant organic compounds. Interestingly, they found that sites invaded by nitrogen-fixing species had significantly higher impacts on the activity of β-glucosidase compared to those invaded by non-nitrogen-fixing species, which suggests that nitrogen-fixing species produce more abundant carbohydrate substrates (e.g. cellulose) for soil microbes to hydrolyse, increasing the quantity of substrate for β-glucosidase. Similarly, a meta-analysis by Negesse *et al.* (2025) showed that despite higher enzymatic activities in soils of invasive plants, the impact of invasive plants on carbon-decomposing enzymes was inconsistent.

According to Tamura and Tharayil (2014), invasive plants have contrasting effects on soil organic matter despite higher net primary productivity. While most studies reported an increase in soil carbon stocks in invaded areas, others documented a decline in soil carbon concentrations following the invasion of exotic species. Based on their research, the authors observed that species replacement influences the total amount and quality of soil organic matter in ecosystems. Furthermore, they emphasize that the most significant soil organic matter alterations occur when the tissue quality of invading species differs substantially from that of the historically dominant species. Their research clearly demonstrates that species invasions can significantly affect both the quantity and stability of the soil organic matter stored in ecosystems, thereby influencing our ability to use natural and managed landscapes to mitigate anthropogenic carbon emissions.

Nitrogen cycling

The nitrogen cycle is one of the most sensitive to invasion due to its close relationship with plant inputs and microbial transformations (Liao *et al.*, 2008). Invasive plants often restructure nitrogen pools and fluxes by altering inputs (e.g. fixation, deposition and capture), transformations (e.g. mineralization, nitrification and denitrification) and losses (e.g. leaching and gaseous emissions) (Lee *et al.*, 2017). In contrast to carbon, where directionality is ambiguous, invasions accelerate nitrogen cycling and often lead to soil eutrophication.

Some invasive plants, such as nitrogen-fixing legumes (e.g. *Acacia*, *Robinia*, *Lespedeza*, *Mimosa*), directly enrich the nitrogen pool by biologically fixing atmospheric N_2, which leads to higher levels of soil nitrogen that can persist long after the plant is removed. This shifts ecosystems from nitrogen-limited to nitrogen-enriched states (Vitousek and Walker, 1989; Marchante *et al.*, 2008; Hellmann *et al.*, 2011). For instance, *Robinia* can add tens to hundreds of kilograms of nitrogen per hectare per year, thereby increasing soil inorganic nitrogen and nitrification (Lindström and Mousavi, 2020). Its high-nitrogen litter decomposes rapidly, increasing nitrogen availability and potentially promoting nitrate leaching. Nodulation responds dynamically to background nitrogen status (Jo *et al.*, 2017). These changes can fundamentally alter competitive dynamics in nutrient-poor ecosystems and enable the establishment of other non-native species that are adapted to high-nutrient conditions (Vitousek and Walker, 1989). Nitrogen-fixing plants have a stronger impact on nitrogen pools when they become invasive, especially in communities lacking nitrogen-fixers (Castro-Díez *et al.*, 2014).

However, non-nitrogen-fixing invasive species have also been associated with increases in soil nitrogen pools and fluxes. This is due to their fast nutrient uptake and release, which promotes fast nitrogen cycling, or their slow nutrient uptake and release, which promotes slow nitrogen cycling (Liao *et al.*, 2008; Lee *et al.*, 2017). Invader litter often has a low C:N ratio and a high nutrient content, which leads to faster decomposition and increased net nitrogen mineralization (Lee *et al.*, 2017). For instance, *Alliaria petiolata* increases the availability of inorganic nitrogen by directly or indirectly promoting microbial nitrogen-transforming processes (Rodgers *et al.*, 2008). Non-nitrogen-fixing invasive plants may increase nitrogen pools in the ecosystem by utilizing them differently than native species (e.g. by taking up nitrogen at different soil depths or times or by accessing different forms of nitrogen). This can increase total nutrient use in the ecosystem and reduce losses of mineral nitrogen by leaching (Castro-Díez *et al.*, 2014).

Elevated nitrogen availability under invasions stimulates nitrification; however, there are also signs of contrasting nitrification patterns in terms of magnitude and direction (Negesse *et al.*, 2025). Increasing NO_3^- concentrations in soil and its subsequent leaching often leads to soil acidification and downstream eutrophication (Ehrenfeld, 2003; Liao *et al.*, 2008; Castro-Díez *et al.*, 2014).

Soil microbial community composition is altered by invasions, which increase the abundance of nitrifiers and change enzyme activities associated with nitrogen cycling (Zhao *et al.*, 2023). Nitrogen-acquiring enzymes often exhibit increased activity during invasions, reflecting microbial responses to altered nitrogen inputs. A meta-analysis by Negesse *et al.* (2025) showed that plant invasions significantly increased soil enzymes related to nitrogen metabolism compared to soils with native plants. Specifically, invasion increased levels of N-acetyl-glucosaminidase, catalase, aminopeptidase, urease and protease by 15% to 27%. In general, plant or ecosystem type did not influence nitrogen-releasing enzymatic activities. However, soils invaded by allelopathic and woody invasive plant species exhibited significantly higher urease activity than soils invaded by non-allelopathic and non-woody invasive plant species.

Denitrification may also be altered by invasive plants. For instance, Yu *et al.* (2022) found that the invasive plant *Micania micrantha* increases denitrification as well as the abundance of denitrifying microorganisms. However, other studies suggest that invasive plants reduce the abundance of denitrifying soil bacteria and denitrification (Dassonville *et al.*, 2011).

A meta-analysis by Lee *et al.* (2017) showed that invaded areas had significantly larger ammonium pools (+49%), faster nitrification (+371%), faster mineralization rates (+105%), more total soil nitrogen (+65%) and more soil organic matter (+ 37%) than reference areas. Soil nitrogen increases more when the invasive species has high leaf nitrogen content. However, the direction and magnitude of the effects depend on the functional traits of the invader and the site conditions. For instance, Castro-Díez *et al.* (2014) found that moist and warm environments are more susceptible to alterations in the nitrogen cycle following exotic plants invasions, likely due to these conditions exacerbating the acceleration of nitrogen fluxes.

Phosphorus cycling

Although less commonly reported, invasive plants also affect phosphorus cycling (Sun *et al.*, 2022; Luo *et al.*, 2024; Tang *et al.*, 2025). Some invasive species release root exudates that enhance phosphorus solubilization, leading to increased phosphorus availability and shifts in microbial phosphorus limitation. Luo *et al.* (2024), for example, found that soil phosphorus availability increased after plant invasion, which was associated with elevated root carboxylate release and increased microbial activities involved in hydrochloric acid-extractable phosphorus (HCl-P) and organic phosphorus mineralization. According to Tang *et al.* (2025), a significant number of invasive plants release substantial amounts of root carboxylates to enhance phosphorus acquisition. Mycorrhizal invasive plants likely rely on their mycorrhizal fungal partners to acquire phosphorus in moderately poor soils and primarily depend on root carboxylates to acquire phosphorus in severely phosphorus-impoverished habitats. However, different invasive species demonstrate different invasion strategies.

Soil phosphatase enzymes, which are produced by plant roots and microorganisms, play a crucial role in the phosphorus cycling (Margalef *et al.*, 2021). Negesse *et al.* (2025) demonstrated that acid and alkaline phosphatase activities were significantly higher in soils under invasive plants than in soils under native plants, at 8.05% and 21.19%, respectively. Changes in enzyme activities often reflect shifts in nutrient demand and turnover rates. Luo *et al.* (2024) found that following the invasion of *Wollastonia biflora* and *Chromolaena odorata*, the proportion of soil organic phosphorus decreased, while soil alkaline phosphatase activity and the proportion of labile phosphorus increased. These changes allow the invasive plants to meet their high demand for phosphorus to sustain their rapid growth and luxury phosphorus uptake.

However, phosphorus availability is highly substrate-dependent. On tropical coral islands with calcareous soils, invasive species enhanced soil phosphorus cycling. Root carboxylates mobilize otherwise inaccessible calcium-bound phosphorus (Ca-P), and invaders invest more phosphorus in leaves. This indicates a strong biogeochemical effect in phosphorus-limited soils (Luo *et al.*, 2024). Conversely, in nutrient-poor tropical forests, some invasive trees exhibited limited capacity to alter phosphorus dynamics, demonstrating that the impact of invasive species on nutrient cycling is more complex and unpredictable in such ecosystems, where multiple nutrients may be co-limiting (Kueffer *et al.*, 2008).

Conclusion

When discussing plant invasions, the primary concern is their effect on the aboveground biodiversity. This is partly because the belowground living world largely escapes our attention and interest and is largely unknown to us. While we are concerned about the loss of rare plant and animal species caused by biological invasions, the loss of belowground biodiversity may in fact be even more dramatic. Since the advent of molecular methods, the number of newly described fungal and protist taxa has steadily increased (Blackwell, 2011; Burki *et al.*, 2021). Changes of the soil environment caused by invasive plants may push prokaryotic strains or species of fungi and unicellular eukaryotes to extinction, and we may not yet be aware of them.

As shown in this chapter, plant invasions do not necessarily disturb soil processes; sometimes even the opposite may be true. The influence of invasive plants on decomposition and nutrient cycling varies greatly and is shaped by context-dependent factors such as ecosystem type, climate, soil properties and the functional traits of the invading species. While many invasive species generally enhance nutrient availability and decomposition rates, others can reduce these processes, particularly in nutrient-poor ecosystems. These inconsistencies highlight the importance of considering both biotic and abiotic factors when evaluating the effects of invasions on soil processes. Additionally, the temporal dynamics must also be considered, as the impacts of newly established invasions may differ substantially from those of long-term, well-established ones. Changes to the soil environment induced by invasions also push the composition and abundance of microbial communities far from their natural state, leading to changes in soil food webs and subsequent impacts on soil fauna.

References

Adams, M.A., Turnbull, T.L., Sprent, J.I. and Buchmann, N. (2016) Legumes are different: Leaf nitrogen, photosynthesis, and water use efficiency. *Proceedings of the National Academy of Sciences USA* 113(15), 4098–4103. DOI: 10.1073/pnas.1523936113.

Afzal, M.R., Naz, M., Ullah, R. and Du, D. (2023) Persistence of root exudates of *Sorghum bicolor* and *Solidago canadensis*: Impacts on invasive and native species. *Plants* 13(1), 58. DOI: 10.3390/plants13010058.

Allison, S.D. and Vitousek, P.M. (2004) Rapid nutrient cycling in leaf litter from invasive plants in Hawai'i. *Oecologia* 141(4), 612–619. DOI: 10.1007/s00442-004-1679-z.

Balogianni, V.G., Wilson, S.D., Farrell, R.E. and MacDougall, A.S. (2015) Rapid root decomposition decouples root length from increased soil C following grassland invasion. *Ecosystems* 18(8), 1307–1318. DOI: 10.1007/s10021-015-9900-y.

Bedano, J.C., Sacchi, L., Natale, E. and Reinoso, H. (2014) Saltcedar (*Tamarix ramosissima*) invasion alters decomposer fauna and plant litter decomposition in a temperate xerophytic deciduous forest. *Advances in Ecology* 2014(1), 1–8. DOI: 10.1155/2014/519297.

Bellard, C., Thuiller, W., Leroy, B., Genovesi, P., Bakkenes, M. *et al.* (2013) Will climate change promote future invasions. *Global Change Biology* 19(12), 3740–3748. DOI: 10.1111/gcb.12344.

Blackwell, M. (2011) The fungi: 1, 2, 3 ... 5.1 million species? *American Journal of Botany* 98(3), 426–438. DOI: 10.3732/ajb.1000298.

Burki, F., Sandin, M.M. and Jamy, M. (2021) Diversity and ecology of protists revealed by metabarcoding. *Current Biology* 31(19), R1267–R1280. DOI: 10.1016/j.cub.2021.07.0.

Carter, D.R., Slesak, R.A., Harrington, T.B., Peter, D.H. and D'Amato, A.W. (2019) Scotch broom (*Cytisus scoparius*) modifies microenvironment to promote nonnative plant communities. *Biological Invasions* 21(4), 1055–1073. DOI: 10.1007/s10530-018-1885-y.

Castro-Díez, P., Godoy, O., Alonso, A., Gallardo, A. and Saldaña, A. (2014) What explains variation in the impacts of exotic plant invasions on the nitrogen cycle? A meta-analysis. *Ecology Letters* 17(1), 1–12. DOI: 10.1111/ele.12197.

Castro-Díez, P., Fierro-Brunnenmeister, N., González-Muñoz, N. and Gallardo, A. (2012) Effects of exotic and native tree leaf litter on soil properties of two contrasting sites in the Iberian Peninsula. *Plant and Soil* 350(1), 179–191. DOI: 10.1007/s11104-011-0893-9.

Chen, J., North, M. and Franklin, J.F. (2024) The contributions of microclimatic information in advancing ecosystem science. *Agricultural and Forest Meteorology* 355, 110105. DOI: 10.1016/j.agrformet.2024.110105.

Chiu, J.H., Chong, K.Y., Lum, S.K. and Wardle, D.A. (2023) Trends in the direction of global plant invasion biology research over the past two decades. *Ecology and Evolution* 13(1), e9690. DOI: 10.1002/ece3.9690.

Dassonville, N., Guillaumaud, N., Piola, F., Meerts, P. and Poly, F. (2011) Niche construction by the invasive Asian knotweeds (species complex Fallopia): Impact on activity, abundance and community structure of denitrifiers and nitrifiers. *Biological Invasions* 13(5), 1115–1133. DOI: 10.1007/s10530-011-9954-5.

DeCrappeo, N.M., DeLorenze, E.J., Giguere, A.T., Pyke, D.A. and Bottomley, P.J. (2017) Fungal and bacterial contributions to nitrogen cycling in cheatgrass-invaded and uninvaded native sagebrush soils of the western USA. *Plant and Soil* 416(1–2), 271–281. DOI: 10.1007/s11104-017-3209-x.

Deeley, B. and Petrovskaya, N. (2022) Propagation of invasive plant species in the presence of a road. *Journal of Theoretical Biology* 548, 111196. DOI: 10.1016/j.jtbi.2022.111196.

Drenovsky, R.E. and Batten, K.M. (2007) Invasion by *Aegilops triuncialis* (Barb goatgrass) slows carbon and nutrient cycling in a serpentine grassland. *Biological Invasions* 9(2), 107–116. DOI: 10.1007/s10530-006-0007-4.

Ehrenfeld, J.G. (2003) Effects of exotic plant invasions on soil nutrient cycling processes. *Ecosystems* 6(6), 503–523. DOI: 10.1007/s10021-002-0151-3.

Ehrenfeld, J.G. (2010) Ecosystem consequences of biological invasions. *Annual Review of Ecology, Evolution, and Systematics* 41(1), 59–80. DOI: 10.1146/annurev-ecolsys-102209-144650.

Ehrenfeld, J.G., Kourtev, P. and Huang, W. (2001) Changes in soil functions following invasions of exotic understory plants in deciduous forests. *Ecological Applications* 11(5), 1287–1300. DOI: 10.1890/0761(2001)011[1287:CISFFI]2.0.CO;2.

Elgersma, K.J., Ehrenfeld, J.G., Yu, S. and Vor, T. (2011) Legacy effects overwhelm the short-term effects of exotic plant invasion and restoration on soil microbial community structure, enzyme activities, and nitrogen cycling. *Oecologia* 167(3), 733–745. DOI: 10.1007/s00442-011-2022-0.

Fennell, M., Wade, M. and Bacon, K.L. (2018) Japanese knotweed (*Fallopia japonica*): An analysis of capacity to cause structural damage (compared to other plants) and typical rhizome extension. *PeerJ* 6, e5246. DOI: 10.7717/peerj.5246.

Gautam, M.K. and Berg, B. (2025) Editorial: Plant litter decomposition: Patterns, processes, and element cycling. *Frontiers in Environmental Science* 13, 1562030. DOI: 10.3389/fenvs.2025.1562030.

Halabuk, A. and Gerhátová, K. (2011) Comparative study of leaf litter decomposition of exotic and native species in an ecotop of the hornbeam-oak forest near Báb village, SW Slovakia. *Folia Oecologica* 38(1), 17–27.

Hawkes, C.V., Wren, I.F., Herman, D.J. and Firestone, M.K. (2005) Plant invasion alters nitrogen cycling by modifying the soil nitrifying community. *Ecology Letters* 8(9), 976–985. DOI: 10.1111/j.1461-0248.2005.00802.x.

He, Y., Li, J., Siemann, E., Li, B., Xu, Y. *et al.* (2025) Plant invasion increases soil microbial biomass carbon: Meta-analysis and empirical tests. *Global Change Biology* 31(3), e70109. DOI: 10.1111/gcb.70109.

Hellmann, C., Sutter, R., Rascher, K.G., Máguas, C., Correia, O. *et al.* (2011) Impact of an exotic N_2-fixing acacia on composition and N status of a native Mediterranean community. *Acta Oecologica* 37(1), 43–50. DOI: 10.1016/j.actao.2010.11.005.

Heneghan, L., Steffen, J. and Fagen, K. (2007) Interactions of an introduced shrub and introduced earthworms in an Illinois urban woodland: Impact on leaf litter decomposition. *Pedobiologia* 50(6), 543–551. DOI: 10.1016/j.pedobi.2006.10.002.

Hickman, J.E., Wu, S., Mickley, L.J. and Lerdau, M.T. (2010) Kudzu (*Pueraria montana*) invasion doubles emissions of nitric oxide and increases ozone pollution. *Proceedings of the National Academy of Sciences USA* 107(22), 10115–10119. DOI: 10.1073/pnas.0912279107.

Hietz, P. (1992) Decomposition and nutrient dynamics of reed (*Phragmites australis* (Cav.) Trin. ex Steud.) litter in Lake Neusiedl, Austria. *Aquatic Botany* 43(3), 211–230. DOI: 10.1016/0304-3770(92)90068-T.

Hiltbrunner, E., Aerts, R., Bühlmann, T., Huss-Danell, K., Magnusson, B. *et al.* (2014) Ecological consequences of the expansion of N_2-fixing plants in cold biomes. *Oecologia* 176(1), 11–24. DOI: 10.1007/s00442-014-2991-x.

Hook, P.B., Olson, B.E. and Wraith, J.M. (2004) Effects of the invasive forb *Centaurea maculosa* on grassland carbon and nitrogen pools in Montana, USA. *Ecosystems* 7(6), 686–694. DOI: 10.1007/s10021-004-0006-1.

Hu, X., Arif, M., Ding, D., Li, J., He, X. *et al.* (2022) Invasive plants and species richness impact litter decomposition in riparian zones. *Frontiers in Plant Science* 13, 955656. DOI: 10.3389/fpls.2022.955656.

Huang, K., Kong, D.L., Lu, X.R., Feng, W.W., Liu, M.C. *et al.* (2020) Lesser leaf herbivore damage and structural defense and greater nutrient concentrations for invasive alien plants: Evidence from 47 pairs of invasive and non-invasive plants. *The Science of the Total Environment* 723, 137829. DOI: 10.1016/j.scitotenv.2020.137829.

Incerti, G., Cartenì, F., Cesarano, G., Sarker, T.C., Abd El-Gawad, A.M. *et al.* (2018) Faster N release, but not C loss, from leaf litter of invasives compared to native species in Mediterranean ecosystems. *Frontiers in Plant Science* 9, 534. DOI: 10.3389/fpls.2018.00534.

Jo, I., Fridley, J.D. and Frank, D.A. (2015) Linking above- and belowground resource use strategies for native and invasive species of temperate deciduous forests. *Biological Invasions* 17(5), 1545–1554. DOI: 10.1007/s10530-014-0814-y.

Jo, I., Fridley, J.D. and Frank, D.A. (2017) Invasive plants accelerate nitrogen cycling: Evidence from experimental woody monocultures. *Journal of Ecology* 105(4), 1105–1110. DOI: 10.1111/1365-2745.12732.

Jones, G.L., Scullion, J., Allison, G., Stott, H., Johnson, D. *et al.* (2019) Reduced soil respiration beneath invasive *Rhododendron ponticum* persists after cutting and is related to substrate quality rather than microbial community. *Soil Biology and Biochemistry* 138, 107577. DOI: 10.1016/j.soilbio.2019.107577.

Kemppinen, J., Lembrechts, J.J., Van Meerbeek, K., Carnicer, J., Chardon, N.I. *et al.* (2024) Microclimate, an important part of ecology and biogeography. *Global Ecology and Biogeography* 33(6), e13834. DOI: 10.1111/geb.13834.

Kourantidou, M., Haubrock, P.J., Cuthbert, R.N., Bodey, T.W., Lenzner, B. *et al.* (2022) Invasive alien species as simultaneous benefits and burdens: Trends, stakeholder perceptions and management. *Biological Invasions* 24(7), 1905–1926. DOI: 10.1007/s10530-021-02727-w.

Kourtev, P.S., Ehrenfeld, J.G. and Häggblom, M. (2002) Exotic plant species alter the microbial community structure and function in the soil. *Ecology* 83(11), 3152–3166. DOI: 10.1890/0012-9658(2002)083[3152:EPSATM]2.0.CO;2.

Kozhoridze, G., Ben-Dor, E., Moudrý, V. and Sternberg, M. (2025) Remote sensing assessment of invasive plant species impacts on microclimate and water stress in Mediterranean coastal ecosystems. *Agricultural and Forest Meteorology* 371, 110606. DOI: 10.1016/j.agrformet.2025.110606.

Krevš, A. and Kučinskiene, A. (2017) Influence of invasive *Acer negundo* leaf litter on benthic microbial abundance and activity in the littoral zone of a temperate river in Lithuania. *Knowledge & Management of Aquatic Ecosystems* 418(418), 26. DOI: 10.1051/kmae/2017015.

Kueffer, C., Klingler, G., Zirfass, K., Schumacher, E., Edwards, P.J. *et al.* (2008) Invasive trees show only weak potential to impact nutrient dynamics in phosphorus-poor tropical forests in the Seychelles. *Functional Ecology* 22(2), 359–366. DOI: 10.1111/j.1365-2435.2007.01373.x.

Kyle, G.P., Beard, K.H. and Kulmatiski, A. (2007) Reduced soil compaction enhances establishment of non-native plant species. *Plant Ecology* 193(2), 223–232. DOI: 10.1007/s11258-006-9260-y.

Landsman, A.P., Schmit, J.P. and Matthews, E.R. (2021) Invasive plants differentially impact forest invertebrates, providing taxon-specific benefits by enhancing structural complexity. *Frontiers in Ecology and Evolution* 9, 682140. DOI: 10.3389/fevo.2021.682140.

Lavoie, C. (2017) The impact of invasive knotweed species (*Reynoutria* spp.) on the environment: Review and research perspectives. *Biological Invasions* 19(8), 2319–2337. DOI: 10.1007/s10530-017-1444-y.

Le Maitre, D.C., Blignaut, J.N., Clulow, A., Dzikiti, S., Everson, C.S. *et al.* (2020) Impacts of plant invasions on terrestrial water flows in South Africa. In: van Wilgen, B.W., Measey, J., Richardson, D.M., Wilson,

J.R. and Zengeya, T.A. (eds) *Biological Invasions in South Africa*. Springer International Publishing, Richardson, D.M., Wilson, pp. 431–457. DOI: 10.1007/978-3-030-32394-3.

Lee, M.R., Bernhardt, E.S., van Bodegom, P.M., Cornelissen, J.H.C., Kattge, J. *et al.* (2017) Invasive species' leaf traits and dissimilarity from natives shape their impact on nitrogen cycling: A meta-analysis. *New Phytologist* 213(1), 128–139. DOI: 10.1111/nph.14115.

Leicht-Young, S.A., O'Donnell, H., Latimer, A.M. and Silander, J.A., Jr. (2009) Effects of an invasive plant species, *Celastrus orbiculatus*, on soil composition and processes. *The American Midland Naturalist* 161(2), 219–231. DOI: 10.1674/0003-0031-161.2.219.

Liao, C., Peng, R., Luo, Y., Zhou, X., Wu, X. *et al.* (2008) Altered ecosystem carbon and nitrogen cycles by plant invasion: A meta-analysis. *New Phytologist* 177(3), 706–714. DOI: 10.1111/j.1469-8137.2007.02290.x.

Lindström, K. and Mousavi, S.A. (2020) Effectiveness of nitrogen fixation in Rhizobia. *Microbial Biotechnology* 13(5), 1314–1335. DOI: 10.1111/1751-7915.13517.

Liu, J., Zhu, M., Shi, X., Hui, C., Sun, Y. *et al.* (2024) Cascading impacts of nitrogen deposition on soil microbiome and herbivore communities in desert steppes. *Science of The Total Environment* 955, 176892. DOI: 10.1016/j.scitotenv.2024.176892.

Liu, Y., Du, Y., Li, C., Li, Y., Xu, Z. *et al.* (2025) Invasive plants can modify the soil environment through litter decomposition. *Environmental Conservation* 1–6, 1–6. DOI: 10.1017/S0376892925100131.

Luo, X., Liu, N., Lambers, H., Cai, H., Hou, E. *et al.* (2024) Plant invasion alters soil phosphorus cycling on tropical coral islands: Insights from *Wollastonia biflora* and *Chromolaena odorata* invasions. *Soil Biology and Biochemistry* 193, 109412. DOI: 10.1016/j.soilbio.2024.109412.

Maclean, J.E., Mitchell, R.J., Burslem, D.F.R.P., Genney, D., Hall, J. *et al.* (2018) Understorey plant community composition reflects invasion history decades after invasive Rhododendron has been removed. *Journal of Applied Ecology* 55(2), 874–884. DOI: 10.1111/1365-2664.12973.

Marchante, E., Kjøller, A., Struwe, S. and Freitas, H. (2008) Short- and long-term impacts of *Acacia longifolia* invasion on the belowground processes of a Mediterranean coastal dune ecosystem. *Applied Soil Ecology* 40(2), 210–217. DOI: 10.1016/j.apsoil.2008.04.004.

Marchante, E., Kjøller, A., Struwe, S. and Freitas, H. (2009) Soil recovery after removal of the N_2-fixing invasive *Acacia longifolia*: Consequences for ecosystem restoration. *Biological Invasions* 11(4), 813–823. DOI: 10.1007/s10530-008-9295-1.

Margalef, O., Sardans, J., Maspons, J., Molowny-Horas, R., Fernández-Martínez, M. *et al.* (2021) The effect of global change on soil phosphatase activity. *Global Change Biology* 27(22), 5989–6003. DOI: 10.1111/gcb.15832.

Marschner, B., Brodowski, S., Dreves, A., Gleixner, G., Gude, A. *et al.* (2008) How relevant is recalcitrance for the stabilization of organic matter in soils? *Journal of Plant Nutrition and Soil Science* 171(1), 91–110. DOI: 10.1002/jpln.200700049.

Martin, M.R., Tipping, P.W. and Reddy, K.R. (2010) Comparing native and exotic litter decomposition and nutrient dynamics. *Journal of Aquatic Plant Management* 48, 72–79.

Martin, P.A., Newton, A.C. and Bullock, J.M. (2017) Impacts of invasive plants on carbon pools depend on both species' traits and local climate. *Ecology* 98(4), 1026–1035. DOI: 10.1002/ecy.1711.

McNeish, R.E. and McEwan, R.W. (2016) A review on the invasion ecology of Amur honeysuckle (*Lonicera maackii*, Caprifoliaceae) a case study of ecological impacts at multiple scales. *The Journal of the Torrey Botanical Society* 143(4), 367–385. DOI: 10.3159/TORREY-D-15-00049.1.

Mincheva, T., Barni, E., Varese, G.C., Brusa, G., Cerabolini, B. *et al.* (2014) Litter quality, decomposition rates and saprotrophic mycoflora in *Fallopia japonica* (Houtt.) Ronse decraene and in adjacent native grassland vegetation. *Acta Oecologica* 54, 29–35. DOI: 10.1016/j.actao.2013.03.010.

Negesse, Z., Pan, K., Guadie, A., Justine, M.F., Azene, B. *et al.* (2025) Plant invasions alter soil biota and microbial activities: A global meta-analysis. *Plant and Soil* 513(1), 1031–1050. DOI: 10.1007/s11104-025-07227-7.

Núñez-Mir, G.C. and McCary, M.A. (2024) Invasive plants and their root traits are linked to the homogenization of soil microbial communities across the United States. *Proceedings of the National Academy of Sciences USA* 121(44), e2418632121. DOI: 10.1073/pnas.2418632121.

Ortiz, S.K. and Wolf, A.A. (2024) Nitrogen-fixing plants increase soil nitrogen and neighbouring-plant biomass, but decrease community diversity: A meta-analysis reveals the mediating role of soil texture. *Journal of Ecology* 112(10), 2374–2385. DOI: 10.5061/dryad.4qrfj6qk1.

Pile Knapp, L.S., Coyle, D.R., Dey, D.C., Fraser, J.S., Hutchinson, T. *et al.* (2023) Invasive plant management in eastern North American forests: A systematic review. *Forest Ecology and Management* 550, 121517. DOI: 10.1016/j.foreco.2023.121517.

Poulette, M.M. and Arthur, M.A. (2012) The impact of the invasive shrub *Lonicera maackii* on the decomposition dynamics of a native plant community. *Ecological Applications* 22(2), 412–424. DOI: 10.1890/11-1105.1.

Qu, T., Du, X., Peng, Y., Guo, W., Zhao, C. *et al.* (2021) Invasive species allelopathy decreases plant growth and soil microbial activity. *PLoS ONE* 16(2), e0246685. DOI: 10.1371/journal.pone.0246685.

Raheem, A., Yohanna, P., Li, G., Noh, N.J., Iqbal, B. *et al.* (2024) Unraveling the ecological threads: How invasive alien plants influence soil carbon dynamics. *Journal of Environmental Management* 356, 120556. DOI: 10.1016/j.jenvman.2024.120556.

Rathore, N., Hanzelková, V., Dostálek, T., Thakur, D. and Münzbergová, Z. (2025) Invasive species are more homogenous in their root morphology and exudate metabolome than non-invasive alien species. *Environmental and Experimental Botany* 229, 106057. DOI: 10.1016/j.envexpbot.2024.106057.

Reinhart, K.O. and Callaway, R.M. (2006) Soil biota and invasive plants. *New Phytologist* 170(3), 445–457. DOI: 10.1111/j.1469-8137.2006.01715.x.

Rodgers, V.L., Wolfe, B.E., Werden, L.K. and Finzi, A.C. (2008) The invasive species *Alliaria petiolata* (garlic mustard) increases soil nutrient availability in northern hardwood-conifer forests. *Oecologia* 157(3), 459–471. DOI: 10.1007/s00442-008-1089-8.

Sage, R.F. (2020) Global change biology: A primer. *Global Change Biology* 26(1), 3–30. DOI: 10.1111/gcb.14893.

Seebens, H., Essl, F., Dawson, W., Fuentes, N., Moser, D. *et al.* (2015) Global trade will accelerate plant invasions in emerging economies under climate change. *Global Change Biology* 21(11), 4128–4140. DOI: 10.1111/gcb.13021.

Shiferaw, H., Alamirew, T., Dzikiti, S., Bewket, W., Zeleke, G. *et al.* (2021) Water use of *Prosopis juliflora* and its impacts on catchment water budget and rural livelihoods in Afar region, Ethiopia. *Scientific Reports* 11(1), 2688. DOI: 10.1038/s41598-021-81776-6.

Silva, S.C., Valadão, M.B., Inkotte, J., Ribeiro, F.P., Miguel, E.P. *et al.* (2023) Legal provision on the use of exotic species in forest restoration: Eucalyptus leaf decomposition in different Brazilian Savannah formations. *Australian Journal of Crop Science* 17(1), 14–19. DOI: 10.3316/informit.950536727736471.

Stanek, M., Zubek, S. and Stefanowicz, A.M. (2021) Differences in phenolics produced by invasive quercus rubra and native plant communities induced changes in soil microbial properties and enzymatic activity. *Forest Ecology and Management* 482, 118901. DOI: 10.1016/j.foreco.2020.118901.

Stinson, K.A., Campbell, S.A., Powell, J.R., Wolfe, B.E., Callaway, R.M. *et al.* (2006) Invasive plant suppresses the growth of native tree seedlings by disrupting belowground mutualisms. *PLoS Biology* 4(5), e140. DOI: 10.1371/journal.pbio.0040140.

Stogner, J.B., Hall, D.B., Yu, M. and Hendricks, J.J. (2024) Legume versus non-legume foliar litter decomposition in regularly burned loblolly pine forests. *Forest Ecology and Management* 551, 121500. DOI: 10.1016/j.foreco.2023.121500.

Strickland, M.S., Osburn, E., Lauber, C., Fierer, N. and Bradford, M.A. (2009) Litter quality is in the eye of the beholder: Initial decomposition rates as a function of inoculum characteristics. *Functional Ecology* 23(3), 627–636. DOI: 10.1111/j.1365-2435.2008.01515.x.

Sun, F., Zeng, L., Cai, M., Chauvat, M., Forey, E. *et al.* (2022) An invasive and native plant differ in their effects on the soil food-web and plant-soil phosphorus cycle. *Geoderma* 410, 115672. DOI: 10.1016/j.geoderma.2021.115672.

Talbot, J.M. and Treseder, K.K. (2012) Interactions among lignin, cellulose, and nitrogen drive litter chemistry-decay relationships. *Ecology* 93(2), 345–354. DOI: 10.1890/11-0843.1.

Talbot, J.M., Yelle, D.J., Nowick, J. and Treseder, K.K. (2012) Litter decay rates are determined by lignin chemistry. *Biogeochemistry* 108(1–3), 279–295. DOI: 10.1007/s10533-011-9599-6.

Tamura, M. and Tharayil, N. (2014) Plant litter chemistry and microbial priming regulate the accrual, composition and stability of soil carbon in invaded ecosystems. *The New Phytologist* 203(1), 110–124. DOI: 10.1111/nph.12795.

Tamura, M., Suseela, V., Simpson, M., Powell, B. and Tharayil, N. (2017) Plant litter chemistry alters the content and composition of organic carbon associated with soil mineral and aggregate fractions in invaded ecosystems. *Global Change Biology* 23(10), 4002–4018. DOI: 10.1111/gcb.13751.

Tang, D., Yan, L., Fisher, J.L., Kang, H., Dallongeville, P. *et al.* (2025) Why are some invasive plant species so successful in nutrient-impoverished habitats in South-Western Australia: A perspective based on their phosphorus-acquisition strategies. *Functional Ecology* 39(2), 635–652. DOI: 10.1111/1365-2435.14726.

Thorpe, A.S. and Callaway, R.M. (2011) Biogeographic differences in the effects of *Centaurea stoebe* on the soil nitrogen cycle: Novel weapons and soil microbes. *Biological Invasions* 13(6), 1435–1445. DOI: 10.1007/s10530-010-9902-9.

Tian, B., Pei, Y., Huang, W., Ding, J. and Siemann, E. (2021) Increasing flavonoid concentrations in root exudates enhance associations between arbuscular mycorrhizal fungi and an invasive plant. *The ISME Journal* 15(7), 1919–1930. DOI: 10.1038/s41396-021-00894-1.

Torres, N., Herrera, I., Fajardo, L. and Bustamante, R.O. (2021) Meta-analysis of the impact of plant invasions on soil microbial communities. *BMC Ecology and Evolution* 21(1), 172. DOI: 10.1186/s12862-021-01899-2.

Triebwasser, D.J., Tharayil, N., Preston, C.M. and Gerard, P.D. (2012) The susceptibility of soil enzymes to inhibition by leaf litter tannins is dependent on the tannin chemistry, enzyme class and vegetation history. *New Phytologist* 196(4), 1122–1132. DOI: 10.1111/j.1469-8137.2012.04346.x.

Underwood, E.C., Klinger, R.C. and Brooks, M.L. (2019) Effects of invasive plants on fire regimes and postfire vegetation diversity in an arid ecosystem. *Ecology and Evolution* 9(22), 12421–12435. DOI: 10.1002/ece3.5650.

Unger, I.M., Kremer, R.J., Veum, K.S. and Goyne, K.W. (2022) Immediate and long-term effects of invasive plant species on soil characteristics. *Soil Ecology Letters* 4(3), 276–288. DOI: 10.1007/s42832-021-0104-4.

van Kleunen, M., Weber, E. and Fischer, M. (2010) A meta-analysis of trait differences between invasive and non-invasive plant species. *Ecology Letters* 13(2), 235–245. DOI: 10.1111/j.1461-0248.2009.01418.x.

Vitousek, P.M. and Walker, L.R. (1989) Biological invasion by *Myrica faya* in Hawai'i: Plant demography, nitrogen fixation, ecosystem effects. *Ecological Monographs* 59(3), 247–265. DOI: 10.2307/1942601.

Wang, C., Zhou, J., Liu, J., Xiao, H. and Wang, L. (2018) Differences in functional traits and reproductive allocations between native and invasive plants. *Journal of Central South University* 25(3), 516–525. DOI: 10.1007/s11771-018-3756-1.

Weidenhamer, J.D. and Callaway, R.M. (2010) Direct and indirect effects of invasive plants on soil chemistry and ecosystem function. *Journal of Chemical Ecology* 36(1), 59–69. DOI: 10.1007/s10886-009-9735-0.

Xiao, Y., Yang, D., Zhang, S.-B., Mo, Y.-X., Dong, Y.-Y. *et al.* (2024) Nitrogen-fixing and non-nitrogen-fixing legume plants differ in leaf nutrient concentrations and relationships between photosynthetic and hydraulic traits. *Tree Physiology* 44(5), tpae048. DOI: 10.1093/treephys/tpae048

Yu, H., He, Y., Zhang, W., Chen, L., Zhang, J. *et al.* (2022) Greater chemical signaling in root exudates enhances soil mutualistic associations in invasive plants compared to natives. *The New Phytologist* 236(3), 1140–1153. DOI: 10.1111/nph.18289.

Zhang, C.B., Wang, J., Qian, B.Y. and Li, W.H. (2009) Effects of the invader solidago canadensis on soil properties. *Applied Soil Ecology* 43(2–3), 163–169. DOI: 10.1016/j.apsoil.2009.07.001.

Zhang, D., Yin, L. and Pan, B. (2002) Biological and ecological characteristics of *Tamarix* L. and its effect on the ecological environment. *Science in China Series D* 45(S1), 18–22. DOI: 10.1007/BF02878384.

Zhang, P., Li, B., Wu, J. and Hu, S. (2019) Invasive plants differentially affect soil biota through litter and rhizosphere pathways: A meta-analysis. *Ecology Letters* 22(1), 200–210. DOI: 10.1111/ele.13181.

Zhao, P., Liu, B., Zhao, H., Lei, Z. and Zhou, T. (2023) Significant changes in soil microbial community structure and metabolic function after *Mikania micrantha* invasion. *Scientific Reports* 13(1), 1141. DOI: 10.1038/s41598-023-27851-6.

Zhou, Y. and Staver, A.C. (2019) Enhanced activity of soil nutrient-releasing enzymes after plant invasion: A meta-analysis. *Ecology* 100, 11. DOI: 10.1002/ecy.2830.

Zhu, Y., Gao, Z., Liu, L., Li, J., Zhu, T. *et al.* (2025) Nitrogen-fixing plants enhance soil phosphorus availability by promoting transformations among phosphorus fractions in a subtropical karst forest. *Forests* 16(2), 360. DOI: 10.3390/f16020360.

7 Impacts of Invasive Plants on Soil Fauna and Ecosystem Restoration

Peliyagodage Chathura Dineth Perera[1]*, Iwona Gruss[2], Jayawanage Oshani Kaushalya[1] and Gamage Kavinda Dhananjaya Gamage[1]
[1]Department of Agricultural Biology, Faculty of Agriculture, University of Ruhuna, Mapalana, Kamburupitiya, Sri Lanka; [2]Department of Plant Protection, Wrocław University of Environmental and Life Sciences, Wrocław, Poland

Introduction

Soil is inhabited by a wide range of animals collectively referred to as the edaphon. These organisms are typically categorized by body size into microfauna (e.g. nematodes, protozoa, rotifers), mesofauna (e.g. springtails, mites, insect larvae), macrofauna (e.g. earthworms, enchytraeids, centipedes, insects) and megafauna (e.g. large arthropods, small mammals) (Menta, 2012; Mekonen, 2019; Potapov *et al.*, 2022) (Fig. 7.1). Fertile soils may contain extraordinarily high densities and taxonomic diversity of soil invertebrates: up to 100,000 individuals per square metre (Breure *et al.*, 2004; Fitter *et al.*, 2005). These organisms serve critical ecological functions, including soil structure formation, organic matter decomposition, nutrient cycling, suppression of pathogens and regulation of pest populations (Lavelle *et al.*, 2006; Culliney, 2013; Abramczyk and Gałązka, 2017).

Microfauna – typically smaller than 100 µm – include protozoa, rotifers and nematodes, which inhabit water films around soil particles. These organisms are crucial in regulating microbial populations and nutrient mineralization, directly influencing nutrient availability for plants (Sławska, 2004; Scheu *et al.*, 2005).

Mesofauna (0.1–2 mm), such as mites and springtails, play an essential role in shredding plant residues, facilitating microbial access and redistributing organic and mineral particles within the soil (Scheu *et al.*, 2005; dos Santos Dias *et al.*, 2022). Their activity shapes microbial community structure and contributes to aggregate formation, thereby improving soil physical properties (Reynolds, 1994).

Macrofauna (>2 mm), including earthworms, beetles and molluscs, significantly impact soil aeration, porosity and root penetration through burrowing and bioturbation (Blair *et al.*, 1997; Lavelle *et al.*, 2006; Prabhakar, 2023). Earthworms in particular enhance carbon and nitrogen mineralization, increasing nutrient bioavailability (Bhadauria and Saxena, 2010).

Together, these invertebrates play a role in broader ecological services such as long-term carbon sequestration, greenhouse gas regulation, detoxification processes and water retention (Cock *et al.*, 2012; Johnson and Jones, 2017). Their presence contributes to improved resistance against erosion, flooding and climate-related

*Corresponding author: chathura@agri.ruh.ac.lk

DOI: 10.1079/9781836991298.0007

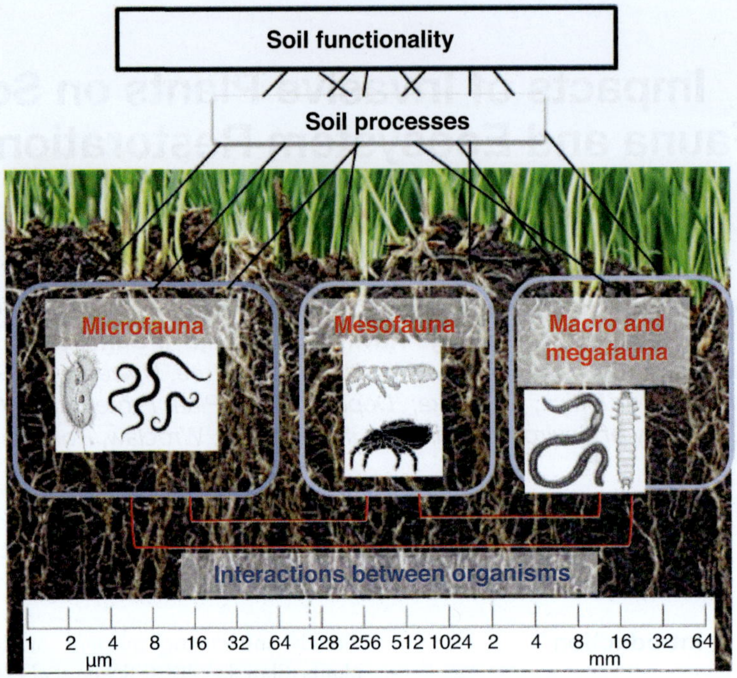

Fig. 7.1. Concept of soil functionality in relation to the presence of edaphic fauna.

stressors, while also supporting plant health and overall ecosystem resilience (dos Santos Dias *et al.*, 2022). The physical modifications of the soil matrix through their movements enhance gas exchange, water infiltration and root access, stabilizing the system and boosting fertility (Reynolds, 1994; Lavelle *et al.*, 2006).

Springtails (Collembola), belonging to the subphylum Hexapoda, are usually the most abundant and diverse arthropods in the edaphon (Lachnicht *et al.*, 2004). They are distinguished by specialized features such as the furcula (jumping organ), concealed mouthparts and a ventral tube. Morphological differentiation within Collembola reflects their stratified distribution in the soil: epigeic species are larger and pigmented, hemiedaphic forms are intermediate and live in the litter while euedaphic Collembola are small, unpigmented and adapted to life deep in the soil. These traits also reflect differences in dispersal capacity, developmental biology and environmental adaptations (Potapov *et al.*, 2022). Morphological classifications have led to the development of soil quality indices, such as QBS-C and QBS-Ar (the Biological Quality of Soil indices for Collembola and arthropods, respectively), which use the presence of adapted morphotypes as indicators of soil health (Parisi and Menta, 2008).

Due to their wide distribution, sensitivity to environmental stressors and tight integration with soil processes, soil invertebrates have high potential as bioindicators (Menta, 2012). Factors such as land-use intensity (Twardowski *et al.*, 2016), pollution by heavy metals or organic compounds (Gruss *et al.*, 2024), and climate change (Gruss *et al.*, 2023) can significantly affect their abundance, diversity and community structure. These disturbances may lead to the replacement of sensitive species with more tolerant but functionally limited ones, reducing the ecological integrity of the soil (Potapov *et al.*, 2020). The evaluation of invertebrate communities – both taxonomically and functionally – offers crucial insights into soil degradation and recovery potential.

Modern agriculture and environmental degradation have contributed to alarming declines in invertebrate abundance and diversity. Overuse of synthetic nitrogen fertilizers promotes acidification and microbial imbalances, often favouring fungi, including toxin-producing species, while suppressing beneficial bacteria (Natywa *et al.*, 2014). Contaminants such as pesticides, antibiotics, polycyclic aromatic hydrocarbons, heavy metals and microplastics have direct and indirect toxic effects on soil organisms, reducing biomass, enzymatic activity and reproductive output (de Souza *et al.*, 2018; Klimkowicz-Pawlas, 2021). Even at low concentrations, chronic exposure to pollutants can induce physiological stress, impair development and lead to altered life cycles and community shifts (Kammenga *et al.*, 2001).

Changes in Soil Characteristics after Plant Invasion

Invasive plant species are powerful agents of soil transformation, often modifying belowground environments in ways that both reinforce their persistence and undermining native biodiversity and ecosystem function. These impacts are mediated by several interlinked mechanisms, including allelopathic interference, shifts in nutrient cycling, disruption of microbial and faunal communities and changes to physical and chemical properties of soil (Weidenhamer and Callaway, 2010; Gibbons *et al.*, 2017; Chen *et al.*, 2018).

One of the most consistent patterns observed across invasive species is an increase in aboveground biomass and nutrient stocks relative to native vegetation, driven by traits such as rapid growth, extended phenology and efficient resource acquisition (Dassonville *et al.*, 2008; Gibbons *et al.*, 2017). *Fallopia japonica* (Japanese knotweed), for example, enhances nitrogen and carbon concentrations in surface soils, while *Solidago canadensis* promotes microbial activity and organic matter accumulation in disturbed habitats (Dassonville *et al.*, 2008). However, these effects depend on initial soil fertility: in nutrient-poor or degraded soils invaders often act as nutrient pumps, whereas in fertile systems they may disrupt existing balances or cause depleted resources (Dassonville *et al.*, 2008). These soil changes are closely linked to alterations in vegetation structure and species composition. Invasive species tend to form dense, often monospecific, stands that outcompete native plants through shading, rapid vegetative spread and suppression of regeneration. As native diversity declines, the complexity of root systems and functional traits in the community also diminishes, leading to more homogeneous and shallower root penetration. This reduces rhizosphere heterogeneity and limits microbial niche diversity, further reinforcing microbial shifts initiated by the invader. For instance, *Solidago gigantea* forms dense clonal patches that exclude native forbs and grasses, simplifying plant community structure and reducing aboveground biodiversity. The shift from diverse native assemblages to single-species dominance leads to reduced litter heterogeneity, which affects litter decomposition dynamics and microbial substrate availability. Over time, this transition results in decreased functional redundancy in microbial communities and impairs resilience to environmental stress (Gibbons *et al.*, 2017; Marjańska *et al.*, 2023).

On the other hand, species such as *Robinia pseudoacacia* not only fix atmospheric nitrogen but also drive successional shifts by facilitating the establishment of nitrophilic and ruderal species. This results in positive feedback loops: nutrient-enriched soils favour fast-growing, opportunistic plants (including other invasives), while oligotrophic specialists are excluded. As the vegetation continues to change, root exudation patterns, mycorrhizal associations and litter chemistry are altered, reinforcing the new soil conditions. These vegetation-driven feedbacks further exacerbate soil degradation processes. The loss of deep-rooted native species reduces bioturbation and water infiltration. At the same time, changes in litter input – such as lignin-rich material from *Eucalyptus* or waxy leaves from *Carpobrotus edulis* – lead to slower decomposition rates and organic matter accumulation in undecomposed layers. These layers may suppress seedling emergence and promote physical soil crusting, increasing runoff and erosion risks (Walker and Smith, 1997; Hobbie, 2015). Moreover, the simplification of plant cover alters microclimatic conditions at the soil surface. Denser canopies reduce temperature fluctuations but also limit sunlight and air circulation, fostering moist, anaerobic microsites that favour denitrification and the proliferation of facultative anaerobes. Such changes in redox conditions impact nitrogen cycling and greenhouse gas emissions (e.g. N_2O), especially in wetlands or riparian zones invaded by *Lythrum salicaria* or *Phalaris arundinacea* (Lindig-Cisneros and Zedler, 2002; Hawkes *et al.*, 2005; Mahaney *et al.*, 2006).

Over time, these transformations accumulate and lead to long-lasting soil legacies, which can persist even after invasive plant removal. Altered microbial communities, disrupted mutualistic networks (e.g. mycorrhizal fungi) and residual allelochemicals continue to influence nutrient availability and inhibit native plant recruitment (Hawkes *et al.*, 2005). Thus, the invasion of non-native plant species sets off a cascade of changes: from vegetation shifts and microbial reassembly to structural and functional transformation of soil. The resulting feedback loops often stabilize a new degraded state that is resistant to passive recovery. Effective ecological restoration in such contexts requires not only the removal of invasive biomass but also active interventions aimed at restoring soil biota, rebalancing nutrient pools and re-establishing functional vegetation–soil feedback (Pickett *et al.*, 2019) (Fig. 7.2).

Effects of invasive plants on soil macrofauna

Invasive plants can disturb mutualisms in the soil by decreasing microbial diversity and abundance, resulting in a reduction of macrofauna activity and soil health. The reduction of microfauna also decreases the decomposition of organic matter, resulting in diminished soil structure and fertility (Mekonen, 2019). It can be expected that invasive species will change the composition and structure of the soil food web. These modifications are predicted to influence the rates of soil processes, subsequently affecting soil nutrient accessibility. For example, in areas invaded by *Phragmites*, the abundance of certain native macrofauna (such as worms, insects and gastropods) was reduced in comparison to uninvaded marshes.

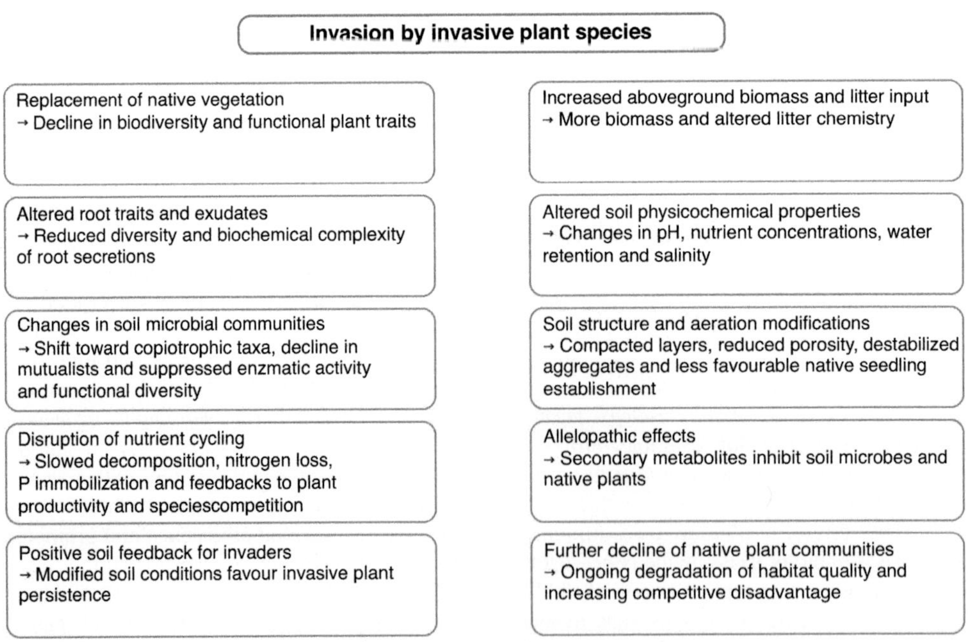

Fig. 7.2. Multidirectional impacts of invasive plants on vegetation and soil systems.

Organisms that feed on the surface, such as sabellid polychaetes and podurids, were more prevalent in *Phragmites* habitats (Talley and Levin, 2001). Land use and soil cover have a significant impact on soil macrofauna communities. Yearly cropping systems frequently decrease their variety and numbers because of soil disruption and insufficient permanent cover. Pastures may boost earthworm populations but could diminish overall diversity, whereas systems based on trees are more effective in promoting both high diversity and abundance. Generally, soil invertebrates abundance increase when land use closely mirrors the original natural ecosystem (Wu *et al.*, 2025).

Invasive plants combine with soil microorganisms such as fungi, bacteria and nematodes in ways that may either enhance or restrict their success. These interactions are frequently described by plant–soil interaction, in which plants affect the makeup of soil microbial communities, which later influence plant growth. In comparison to native species that generally face adverse feedback from accumulating soil pathogens, invasive species frequently avoid these threats or interact with advantageous microbes in their new environment, leading to positive feedback that boosts their growth and competitiveness. In certain instances, invasive plants establish strong connections with soil mutualists such as mycorrhizal fungi and nitrogen-fixing bacteria, enhancing their nutrient absorption and further supporting invasion (Reinhart and Callaway, 2006).

A key way invasive plants impact soil macrofauna is by altering litter input and its composition. Invasive species frequently generate litter that has a different chemical composition compared to that created by native plants, especially regarding lignin levels, C:N ratios and the occurrence of allelopathic substances (Hu *et al.*, 2022). Invasive plants can greatly change the soil microclimate by modifying the canopy structure and ground cover. Enhanced shading, decreased evapotranspiration and alterations in water infiltration can result in cooler, wetter soil conditions, potentially affecting the suitability for various macrofaunal taxa. For instance, *Lantana camara* can establish dense undergrowth conditions that benefit specific earthworm species while leaving out others. Certain invasive species emit allelopathic substances that influence soil macrofauna either directly or indirectly. These chemicals may inhibit microbial communities that act as nourishment for macrofauna or directly damage invertebrates (Ruwanza and Shackleton, 2016; De Almeida *et al.*, 2022).

Direct and Indirect Effects on Soil Fauna

Diversity and abundance shifts

Invasive plant species have profound and multifaceted impacts on soil fauna worldwide (Vilà *et al.*, 2011; Abgrall *et al.*, 2019; Rai and Singh, 2020). These impacts manifest through direct changes, such as shifts in litter quantity and quality, and indirect pathways, such as modifications to microclimate and soil physicochemical properties (Ehrenfeld, 2003; Pyšek *et al.*, 2012; Wardle and Peltzer, 2017). The specific impacts, including their direction and magnitude, are highly dependent on the particular invasive plant species involved (Grant and Paschke, 2012; Stefanowicz *et al.*, 2019).

Commonly invasive plants actively modify their subterranean environment through the release of root exudates, which can directly affect soil structure and the mobilization or chelation of nutrients. Native plants and soil biota in the invaded range are typically unaccustomed to these compounds. These allelochemicals can disrupt fundamental biogeochemical processes, leading to significant alterations in soil chemistry (Stinson *et al.*, 2006; Weidenhamer and Callaway, 2010; Xu *et al.*, 2022). For instance, garlic mustard (*Alliaria petiolata*), a widespread invader, releases sinigrin, a glucoside that not only inhibits the growth of other plant species but also suppresses arbuscular mycorrhizal fungi, which are crucial for tree root nutrient and water absorption. In invaded soils, this mechanism contributes to its invasion success (Tonole, 2025). Further, this shift in microbial communities has direct and significant consequences for macrofauna, particularly those that feed on specific microbial groups (Morales-Fonseca *et al.*, 2025). For example, the documented negative impact of garlic mustard on fungal abundance leads directly to declines in frugivorous arthropods, illustrating a clear bottom-up effect within the soil food web (McCary and Wise, 2019). Such disruptions can initiate trophic cascades, where changes at lower trophic levels propagate upwards, potentially affecting higher trophic levels and the overall stability of the ecosystem (Litt *et al.*, 2014; Heller, 2020; National Wildlife Federation, 2025). Another example in grasslands of tropical and subtropical regions invaded by *Imperata cylindrica* is that native beetle assemblages were significantly reduced, leading to local diversity losses (Jose *et al.*, 2002). Thus, invasions frequently induce taxon-specific shifts (Ricciardi and Cohen, 2007).

Conversely, numerous meta-analyses show that soil fauna abundance may shift under some invasions to promote the growth of specific microbial groups (Ehrenfeld, 2010; Rout and Callaway, 2012; Coyle *et al.*, 2017; Stefanowicz *et al.*, 2019). These invasive plants function as 'niche constructors' by significantly modifying soil microbial communities and altering carbon and nitrogen cycling within ecosystems (Wang *et al.*, 2015; Xu *et al.*, 2022; Xu *et al.*, 2023). For example, nitrogen-fixing invaders like *Acacia* spp. enhance soil nitrogen pools, favouring decomposer communities such as earthworms and Collembola (Suding *et al.*, 2004; Liao *et al.*, 2008). Similarly, *F. japonica* was found to support denser isopod and millipede populations due to its nutrient-rich litter (Gerber *et al.*, 2008). This phenomenon accelerates litter decomposition and nutrient cycling, leading to an increase in nitrogen availability but also potentially both increasing nutrient leaching from the soil profile (National Park Service, 2025) and leading to the release of stored carbon and a reduction in forest floor biodiversity (Wilson *et al.*, 2017).

The impact of invasive plants on the diversity and abundance of soil invertebrates is highly variable, demonstrating positive, negative or neutral effects depending on the specific context (Weidenhamer and Callaway, 2010; Pyšek *et al.*, 2012). Litt *et al.* (2014) reported that total arthropod abundance decreased in 62% of reviewed studies and increased in 15%, whereas taxonomic richness declined in 48% of cases and increased in 13%.

Changes in functional roles

Invasive plants can profoundly disrupt critical ecosystem functions. Changes in soil faunal functional roles reflect altered community composition and resource flows (Lavelle *et al.*, 2006; Wardle and Peltzer, 2017). Invasive plants, either directly or indirectly through associated invasive soil fauna, disrupt the critical ecosystem engineering functions performed by native soil fauna (Zhukov *et al.*, 2023). Decomposers often benefit from invasions that produce labile litter (Ehrenfeld, 2003). Invasive legumes like *Myrica faya* in Hawai'i have increased detritivore biomass by 30–40% while accelerating nitrogen mineralization (Vitousek and Walker, 1989). Earthworms, as ecosystem engineers, frequently expand under such invasions, modifying soil structure, porosity and organic matter distribution (Bohlen *et al.*, 2004). Such imbalances can affect soil nutrient regulation and nutrient retention (Didham *et al.*, 2007). In forest systems, the replacement of native leaf litter with invader litter changes decomposition pathways. Studies on *Rubus armeniacus* (Himalayan blackberry) in the Pacific northwest of the USA show elevated springtail and mite densities but reduced predator beetle diversity, indicating a functional simplification of the soil food web (Wolfe and Klironomos, 2005). These faunal shifts cascade into ecosystem processes such as carbon cycling, nutrient turnover and soil aggregation (Ehrenfeld, 2010; Wardle and Peltzer, 2017). Invasion-driven declines in soil biodiversity may reduce resilience against other stressors like drought or soil erosion (Pyšek *et al.*, 2012). For example, dense stands of *Phragmites australis* in wetlands change soil faunal assemblages, suppressing native detritivores and altering peat formation (Meyerson *et al.*, 2000). Additionally, altered macrofaunal activities can influence hydrological properties by changing infiltration rates and erosion risk (Ehrenfeld, 2003; Liao *et al.*, 2008). Understanding these belowground shifts is crucial for predicting long-term ecosystem trajectories. Integrating soil fauna into invasive species management enhances restoration outcomes by targeting processes like litter removal, soil amendment and reintroduction of native macrofauna where appropriate (Simberloff *et al.*, 2013) (Table 7.1).

Implications for Ecosystem Functioning

Ecosystem functioning refers to the collective ecological processes that maintain the condition and services of ecosystems, such as nutrient cycling, energy flow, decomposition and more. Various factors, including pollution, climate change, invasive plants and land use changes, can alter ecosystem functioning, sometimes irreversibly. One of the most significant and risky impacts comes from invasive plants (Dangles *et al.*, 2004). Invasive plants might enhance, diminish or have no impact on ecosystem functioning. They might occupy previously empty niches or outcompete indigenous species, thereby impacting growth and energy dynamics within ecosystems. Certain invaders, like nitrogen-fixing varieties, can enhance soil nutrients, while others might drain vital elements or yield nutrient-deficient debris (Schwartz *et al.*, 2000; Butler *et al.*, 2012). These alterations may influence the health and regrowth of native plants. Invasive species can modify soil moisture by utilizing more water than indigenous species, altering water infiltration patterns or tapping into deeper water supplies. Certain species also elevate soil salinity, leading to adverse conditions for indigenous species. Invasive species can alter the patterns of natural disturbances, including fire occurrence, erosion rates and grazing. For instance, grasses might heighten fire risk, whereas certain trees may mitigate it. Such disturbances frequently result in changes to community structure. Invasive species frequently diminish native biodiversity by dominating resources such as light, water and nutrients. They might also change successional pathways, resulting in enduring modifications to ecosystem structure.

Invasive plant species influence ecosystems in intricate, interconnected manners. Addressing these effects necessitates cohesive ecological understanding, ongoing observation and teamwork among ecologists and land management professionals (Walker and Smith, 1997). The implications of invasive plants for ecosystem functioning concern the ways that different factors, whether natural or human-made, influence the structure, processes and services of ecosystems (Tomimatsu *et al.*, 2013). The emergence and distribution of invasive plants present significant and extensive consequences for ecosystem operations. Ecosystems rely on a fragile equilibrium of biotic and abiotic interactions, including nutrient cycling, energy transfer, decomposition, pollination and water regulation. Invasive species can interfere with these essential processes, frequently in a permanent way, changing the structure, function and strength of ecosystems (Schläpfer *et al.*, 2005; Schmitz *et al.*, 2008).

A significant effect of invasive plants is their influence on nutrient cycling. Numerous invasive plants alter soil chemistry by generating litter that is low in nutrients and decomposes slowly or by producing allelopathic substances. For instance, *Lantana camara* and *Clidemia hirta* create thick litter layers that inhibit the germination of native seeds while transforming microbial communities that participate in nutrient cycling. These alterations may decrease nitrogen accessibility for native species, resulting in prolonged soil deterioration (Wijesundara, 2010).

Invasive species also impact primary productivity and energy transfer. Rapidly expanding species such as *Parthenium hysterophorus* and *Panicum maximum* surpass native plants in competition for light, water and nutrients, frequently creating monodominant stands that diminish plant diversity and structural complexity. This reduction constrains the variety of herbivores, pollinators and higher trophic levels, ultimately diminishing interactions and energy flow. Aquatic invasive species like *Eichhornia crassipes* (water hyacinth) and *Salvinia molesta* create thick layers on water surfaces, diminishing oxygen levels, boosting evapotranspiration and obstructing sunlight penetration. This interrupts aquatic photosynthesis, degrades water quality and impacts species dependent on open water environments. Invasive species with deep roots on land can change the patterns of groundwater recharge and surface runoff, leading to erosion or decreased stream flow. Certain invasive grasses and shrubs change

Table 7.1. Examples of invasive plant species connected to soil fauna, the observed shift in diversity and the primary mechanisms involved.

Invasive plant species	Affected macrofauna group(s)	Observed shift macrofauna diversity	Primary mechanism(s)	References
Hybrid *Spartina* (*Spartina alterniflora* and *Spartina foliosa*)	Oligochaetes, capitellid polychaetes	Decreased	Alters soil properties, changes oxygen level	Neira *et al.* (2007)
Centaurea maculosa	Soil nematodes and arthropods	Decreased or increased	Alters the soil macrofaunal community and soil nitrogen cycle	Reinhart and Callaway (2006)
Alternanthera philoxeroides	Mites, Diplura, springtails	Decreased	Alters soil microfaunal community, reduces soil diversity and abundance	Zhang *et al.* (2020)
Spartina alterniflora	Amphipods, crabs and polychaete worms	Decreased	Increases belowground biomass and litter buildup	Rejmanek (2000)
Phragmites australis	Invertebrates	Decreased	Changes nutrient cycling, sediment aeration and decomposition processes	Talley and Levin (2001)
Acacia dealbata	Soil invertebrates	Decreased	Changing litter chemistry and soil pH	Roy *et al.* (2014)
Euphorbia esula	Narive dentritivores, herbivores	Decreased	Produces allelopathic compounds in roots	Roy *et al.* (2014)
Bormus tectorum	Nematodes and microarthropods	Decreased	Alters soil food web structure	Belnap *et al.* (2005)
Fagus grandifalia and *Tsuga canadensis*	Earthworms	Increased	Changes soil chemical and physical properties (e.g. pH, nutrient availability, litter quality)	Davis (1987)
Soligado gigantea L	Collembola	Decreased	Increased soil pH and reduced soil humidity, conductivity and nutrient content.	Sterzyńska *et al.* (2017)
Alliaria petiolata	Fungivorous arthropods	Decreased	Changes the soil food web via to suppression of fungal resources	McCary and Wise (2019)

Continued

Table 7.1. Continued

Invasive plant species	Affected macrofauna group(s)	Observed shift macrofauna diversity	Primary mechanism(s)	References
Berberis thunbergii, *Rhamnus frangula*	Earthworms	Increased	Alters soil pH (reduced soil acidity)	National Park Service (2025)
Reynoutria spp.	Collembola	Decreased	Alters the soil organic matter inputs levels and microhabitat structure	De Almeida *et al.* (2022)
Pyrus calleryana	Millipedes	Increased	Alters the soil organic matter input levels and microhabitat structure	McGee (2023)
Solidago canadensis	Millipedes, symphylans	Increased	Alters habitat and increases soil moisture levels	Abgrall *et al.* (2019)
Gunnera tinctoria	Native earthworms	Increased	Increased inputs of higher-quality litter	Mantoani *et al.* (2025)
Robinia pseudoacacia	Collembola	Decreased	Reduces microbial biomass and alters nutrients	Zizolfi *et al.* (2025)

the frequency and intensity of fires by supplying an abundance of flammable biomass. For instance, *P. maximum* heightens the fire risk in arid grasslands, potentially damaging native seed banks and promoting additional invasions, a feedback cycle that speeds up ecosystem change. Moreover, it diminishes biodiversity and ecological stability. The replacement of invasive plants results in the reduction of related fauna, including soil macrofauna, pollinators and seed dispersers. The reduction in functional diversity increases the susceptibility of ecosystems to further stressors like climate change and disease outbreaks (Wijesundara, 2010).

Invasive plants threaten several ecosystem functions at once, resulting in cascading impacts that weaken ecological integrity and sustainability. Effective management and restoration must transcend species-specific control and address the systemic alterations induced by invasions, concentrating on sustaining ecological processes, fostering native biodiversity and improving the resilience of impacted ecosystems (Hector *et al.*, 2001).

References

Abgrall, C., Forey, E. and Chauvat, M. (2019) Soil fauna responses to invasive alien plants are determined by trophic groups and habitat structure: A global meta-analysis. *Oikos* 128(10), 1390–1401. DOI: 10.1111/oik.06493.

Abramczyk, K. and Gałązka, A. (2017) Różnorodność mikroorganizmów glebowych obszarów chronionych i oznaczających walory przyrodnicze [Diversity of soil microorganisms in areas under protection or with significant natural value]. *Sylwan* 161, 496–503.

Belnap, J., Phillips, S.L., Sherrod, S.K. and Moldenke, A. (2005) Soil biota can change after exotic plant invasion: Does this affect ecosystem processes? *Ecology* 86(11), 3007–3017. DOI: 10.1890/05-0333.

Bhadauria, T. and Saxena, K.G. (2010) Role of earthworms in soil fertility maintenance through the production of biogenic structures. *Applied and Environmental Soil Science* 2010, 1–7. DOI: 10.1155/2010/816073.

Blair, J.M., Bohlen, P.J. and Freckman, D.W. (1997) Soil invertebrates as indicators of soil quality. *Methods for Assessing Soil Quality* 49, 273–291. DOI: 10.2136/sssaspecpub49.c16.

Bohlen, P.J., Scheu, S., Hale, C.M., McLean, M.A., Migge, S. *et al.* (2004) Non-native invasive earthworms as agents of change in northern temperate forests. *Frontiers in Ecology and the Environment* 2(8), 427–435. DOI: 10.1890/1540-9295(2004)002[0427:NIEAAO]2.0.CO;2.

Breure, A.M., Mulder, C., Rutgers, M., Schouten, T. and Zwart, D. (2004) A biological indicator for soil quality. In: *Agricultural Impacts on Soil Erosion and Soil Biodiversity: Developing indicators for Policy Analysis. Proceedings, OECD Expert Meeting*, Rome, pp. 485–494.

Butler, S.M., Melillo, J.M., Johnson, J.E., Mohan, J., Steudler, P.A. *et al.* (2012) Soil warming alters nitrogen cycling in a New England Forest: Implications for ecosystem function and structure. *Oecologia* 168(3), 819–828. DOI: 10.1007/s00442-011-2133-7.

Chen, B.-M., Wei, H.-J., Chen, W.-B., Zhu, Z.-C., Yuan, Y.-R. *et al.* (2018) Effects of plant invasion on soil nitrogen transformation processes and its associated microbial communities. *Chinese Journal of Plant Ecology* 42(11), 1071–1081. DOI: 10.17521/cjpe.2018.0138.

Cock, M.J., Biesmeijer, J.C., Cannon, R.J.C., Gerard, P.J., Gillespie, D. *et al.* (2012) The positive contribution of invertebrates to sustainable agriculture and food security. *CABI Reviews* 1–27. DOI: 10.1079/PAVSNNR20127043.

Coyle, D.R., Nagendra, U.J., Taylor, M.K., Campbell, J.H., Cunard, C.E. *et al.* (2017) Soil fauna responses to natural disturbances, invasive species, and global climate change: Current state of the science and a call to action. *Soil Biology and Biochemistry* 110, 116–133. DOI: 10.1016/j.soilbio.2017.03.008.

Culliney, T.W. (2013) Role of arthropods in maintaining soil fertility. *Agriculture* 3(4), 629–659. DOI: 10.3390/agriculture3040629.

Dangles, O., Gessner, M.O., Guerold, F. and Chauvet, E. (2004) Impacts of stream acidification on litter breakdown: Implications for assessing ecosystem functioning. *Journal of Applied Ecology* 41(2), 365–378. DOI: 10.1111/j.0021-8901.2004.00888.x.

Dassonville, N., Vanderhoeven, S., Vanparys, V., Hayez, M., Gruber, W. *et al.* (2008) Impacts of alien invasive plants on soil nutrients are correlated with initial site conditions in NW Europe. *Oecologia* 157(1), 131–140. DOI: 10.1007/s00442-008-1054-6.

Davis, M.B. (1987) Invasion of forest communities during the Holocene: Beech and hemlock in the Great Lakes region. In: Gray, A.J., Crawley, M.J. and Edwards, P.J. (eds) *Colonization, Succession and Stability. 26th Symposium of the British Ecological Society held jointly with the Linnean Society of London*, Blackwell, Oxford, pp. 373–393.

De Almeida, T., Forey, E. and Chauvat, M. (2022) Alien invasive plant effect on soil fauna is habitat dependent. *Diversity* 14(2), 61. DOI: 10.3390/d14020061.

de Souza Machado, A.A., Kloas, W., Zarfl, C., Hempel, S. and Rillig, M.C. (2018) Microplastics as an emerging threat to terrestrial ecosystems. *Global Change Biology* 24(4), 1405–1416. DOI: 10.1111/gcb.14020.

Didham, R.K., Tylianakis, J.M., Gemmell, N.J., Rand, T.A. and Ewers, R.M. (2007) Interactive effects of habitat modification and species invasion on native species decline. *Trends in Ecology & Evolution* 22(9), 489–496. DOI: 10.1016/j.tree.2007.07.001.

dos Santos Dias, D., Lima, R.W.S., Silva, C.A.R., Monique, É., Costa Santos, D.L.G. *et al.* (2022) Composition of the edaphic mesofauna in three environments: Conserved, in transition and altered, Atlantic Forest, Brazil. *Revista Brasileira de Geografia Física* 15(2), 736–749. DOI: 10.26848/rbgf.v15.2.p736-749.

Ehrenfeld, J.G. (2003) Effects of exotic plant invasions on soil nutrient cycling processes. *Ecosystems* 6(6), 503–523. DOI: 10.1007/s10021-002-0151-3.

Ehrenfeld, J.G. (2010) Ecosystem consequences of biological invasions. *Annual Review of Ecology, Evolution, and Systematics* 41(1), 59–80. DOI: 10.1146/annurev-ecolsys-102209-144650.

Fitter, A.H., Gilligan, C.A., Hollingworth, K., Kleczkowski, A., Twyman, R.M. *et al.* (2005) Biodiversity and ecosystem function in soil. *Functional Ecology* 19(3), 369–377. DOI: 10.1111/j.0269-8463.2005.00969.x.

Gerber, E., Krebs, C., Murrell, C., Moretti, M., Rocklin, R. *et al.* (2008) Exotic invasive knotweeds (*Fallopia* spp.) negatively affect native plant and invertebrate assemblages in European riparian habitats. *Biological Conservation* 141(3), 646–654. DOI: 10.1016/j.biocon.2007.12.009.

Gibbons, S.M., Lekberg, Y., Mummey, D.L., Sangwan, N., Ramsey, P.W. *et al.* (2017) Invasive plants rapidly reshape soil properties in a grassland ecosystem. *mSystems* 2(2), e00178–16. DOI: 10.1128/mSystems.00178-16.

Grant, T.A. and Paschke, M. (2012) Invasive plant impacts on soil properties, nutrient cycling, and microbial communities. In: Monaco, T.A. and Sheley, R.L. (eds) *Invasive Plant Ecology and Management: Linking Processes to Practice*. CABI Invasive Series, pp. 79–103. DOI: 10.1079/9781845938116.0079.

Gruss, I., Yin, R., Julia, S., Eisenhauer, N. and Schädler, M. (2023) The responses of Collembola biomass to climate and land-use changes vary with life form. *Soil and Tillage Research* 225, 105541. DOI: 10.1016/j.still.2022.105541.

Gruss, I., Lallaouna, R., Twardowski, J., Magiera-Dulewicz, J. and Twardowska, K. (2024) Collembola growth in heavy metal-contaminated soils. *Scientific Reports* 14, 27998. DOI: 10.1038/s41598-024-79766-5.

Hawkes, C.V., Wren, I.F., Herman, D.J. and Firestone, M.K. (2005) Plant invasion alters nitrogen cycling by modifying the soil nitrifying community. *Ecology Letters* 8(9), 976–985. DOI: 10.1111/j.1461-0248.2005.00802.x.

Hector, A., Joshi, J., Lawler, S., Spehn, E.M. and Wilby, A. (2001) Conservation implications of the link between biodiversity and ecosystem functioning. *Oecologia* 129(4), 624–628. DOI: 10.1007/s004420100759.

Heller, L. (2020) Effects of Invasive Species on Water Quality in Freshwater Ecosystems. Soka University of America. Available at: https://www.soka.edu/student-life/20th-anniversary-anthology/creative-coexistence-nature-humanity/effects-invasive (accessed 8 July 2025).

Hobbie, S.E. (2015) Plant species effects on nutrient cycling: Revisiting litter feedbacks. *Trends in Ecology & Evolution* 30(6), 357–363. DOI: 10.1016/j.tree.2015.03.015.

Hu, X., Arif, M., Ding, D., Li, J., He, X. *et al.* (2022) Invasive plants and species Richness impact litter decomposition in Riparian Zones. *Frontiers in Plant Science* 13, 955656. DOI: 10.3389/fpls.2022.955656.

Johnson, S.N. and Jones, T.H. (eds) (2017) *Global Climate Change and Terrestrial Invertebrates*. John Wiley & Sons, Oxford. DOI: 10.1002/9781119070894.

Jose, S., Cox, J., Miller, D.L., Shilling, D.G. and Merritt, S. (2002) Alien plant invasions: The story of cogongrass in Southeastern forests. *Journal of Forestry* 100(1), 41–44. DOI: 10.1093/jof/100.1.41.

Kammenga, J.E., Van Gestel, C.A.M. and Hornung, E. (2001) Switching life-history sensitivities to stress in soil invertebrates. *Ecological Applications* 11(1), 226–238. DOI: 10.1890/1051-0761(2001)011[0226:SLHSTS]2.0.CO;2.

Klimkowicz-Pawlas, A. (2021) Środowiskowe i zdrowotne skutki zanieczyszczenia gleb oraz metody przeciwdziałania [Environmental and health effects of soil contamination and methods of counteraction]. *Studia i Raporty IUNG-PIB* 66(20), 91–116.

Lachnicht, S.L., Hendrix, P.F., Potter, R.L., Coleman, D.C. and Crossley Jr, D.A. (2004) Winter decomposition of transgenic cotton residue in conventional-till and no-till systems. *Applied Soil Ecology* 27(2), 135–142. DOI: 10.1016/j.apsoil.2004.05.001.

Lavelle, P., Decaëns, T., Aubert, M., Barot, S., Blouin, M. *et al.* (2006) Soil invertebrates and ecosystem services. *European Journal of Soil Biology* 42, S3–S15. DOI: 10.1016/j.ejsobi.2006.10.002.

Liao, C., Peng, R., Luo, Y., Zhou, X., Wu, X. *et al.* (2008) Altered ecosystem carbon and nitrogen cycles by plant invasion: A meta-analysis. *New Phytologist* 177, 706–714. DOI: 10.1111/j.1469-8137.2007.02290.x.

Lindig-Cisneros, R. and Zedler, J.B. (2002) Relationships between canopy complexity and germination microsites for *Phalaris arundinacea* L. *Oecologia* 133, 159–167. DOI: 10.1007/s00442-002-1020-7.

Litt, A.R., Cord, E.E., Fulbright, T.E. and Schuster, G.L. (2014) Effects of invasive plants on arthropods. *Conservation Biology* 28, 1532–1549. DOI: 10.1111/cobi.12350.

Mahaney, W.M., Smemo, K.A. and Yavitt, J.B. (2006) Impacts of *Lythrum salicaria* invasion on plant community and soil properties in two wetlands in Central New York, USA. *Canadian Journal of Botany* 84(3), 477–484. DOI: 10.1139/b06-009.

Mantoani, M.C., Schmidt, O., Storey, S., Doyle, E., Vintro, L.L. *et al.* (2025) Changes in the soil detrital food chain associated with *Gunnera tinctoria* plant invasions. *Invasive Plant Science and Management* 18, e11. DOI: 10.1017/inp.2025.12.

Marjańska, E., Cwajna, A., Sztajnykier, I., Madej, D. and Moroń, D. (2023) *Inwazyjne obce gatunki nawłoci (Solidago spp.) zmieniają świat [Invasive alien species of goldenrod (Solidago spp.) are changing the world]*. Instytut Ochrony Przyrody PAN, Kraków.

McCary, M.A. and Wise, D.H. (2019) Plant invader alters soil food web via changes to fungal resources. *Oecologia* 191(3), 587–599. DOI: 10.1007/s00442-019-04510-0.

McGee, J.M. (2023) The effect of callery pear (*Pyrus calleryana*) on soil macrofauna (Diplopoda and Oligochaeta). Master's thesis, Georgia College & State University. Available at: https://kb.gcsu.edu/biology/35 (accessed 18 December 2025).

Mekonen, E.S. (2019) Soil fauna as webmasters, engineers and bioindicators in ecosystems: Implications for conservation ecology and sustainable agriculture. *American Journal of Life Sciences* 7(1), 17. DOI: 10.11648/j.ajls.20190701.14.

Menta, C. (2012) Soil fauna diversity-function, soil degradation, biological indices, soil restoration. In: Lameed, G.A. (ed.) *Biodiversity Conservation and Utilization in a Diverse World*. IntechOpen, pp. 60–84. DOI: 10.5772/51091.

Meyerson, L.A., Saltonstall, K., Windham, L., Kiviat, E. and Findlay, S. (2000) A comparison of *Phragmites australisin* freshwater and brackish marsh environments in North America. *Wetlands Ecology and Management* 8(2–3), 89–103. DOI: 10.1023/A:1008432200133.

Morales-Fonseca, D., Barantal, S., Buscot, F., Hättenschwiler, S., Milcu, A. *et al.* (2025) Functional diversity of soil macrofauna may contribute to microbial community stabilization under drought stress. *Frontiers in Microbiology* 16, 1597272. DOI: 10.3389/fmicb.2025.1597272.

National Park Service (2025) Species Spotlight – Japanese Barberry. U.S. National Park Service. Available at: https://www.nps.gov/articles/species-spotlight-japanese-barberry.htm (accessed 10 July 2025).

National Wildlife Federation (2025) Invasive Species. Available at: https://www.nwf.org/Educational-Resources/Wildlife-Guide/Threats-to-Wildlife/Invasive-Species (accessed 10 July 2025).

Natywa, M., Selwet, M. and Maciejewski, T. (2014) Wpływ wybranych czynników agrotechnicznych na liczebność i aktywność drobnoustrojów glebowych [Effect of some agrotechnical factors on the number and activity soil microorganisms]. *Fragmenta Agronomica* 31(2), 56–63.

Neira, C., Levin, L.A., Grosholz, E.D. and Mendoza, G. (eds) (2007) Influence of invasive *Spartina* growth stages on associated macrofaunal communities. *Biological Invasions* 9(8), 975–993. DOI: 10.1007/s10530-007-9097-x.

Parisi, V. and Menta, C. (2008) Microarthropods of the soil: Convergence phenomena and evaluation of soil quality using QBS-ar and QBS-C. *Fresenius Environmental Bulletin* 17, 1170–1174.

Pickett, B., Maltz, M. and Aronson, E. (2019) Impacts of invasive plants on soil fungi and implications for restoration. In: Sahu, S.C. and Kumar, S. (eds) *Diversity and Ecology of Invasive Plants. IntechOpen*. pp. 53–71. DOI: 10.5772/intechopen.82108.

Potapov, A., Bellini, B., Chown, S., Deharveng, L., Janssens, F. *et al.* (2020) Towards a global synthesis of Collembola knowledge – challenges and potential solutions. *Soil Organisms* 93(3), 161–188. DOI: 10.25674/SO92ISS3PP161.

Potapov, A.M., Beaulieu, F., Birkhofer, K., Bluhm, S.L., Degtyarev, M.I. *et al.* (2022) Feeding habits and multifunctional classification of soil-associated consumers from protists to vertebrates. *Biological Reviews of the Cambridge Philosophical Society* 97(3), 1057–1117. DOI: 10.1111/brv.12832.

Prabhakar, S.J. (2023) Influence of soil invertebrates on soil decomposition. *Journal of Survey in Fisheries Sciences* 10(3), 618–629.

Pyšek, P., Jarošík, V., Hulme, P.E., Pergl, J., Hejda, M. *et al.* (2012) A global assessment of invasive plant impacts on resident species, communities and ecosystems: The interaction of impact measures, invading species' traits and environment. *Global Change Biology* 18(5), 1725–1737. DOI: 10.1111/j.1365-2486.2011.02636.x.

Rai, P. and Singh, J.S. (2020) Invasive alien plant species: Their impact on environment, ecosystem services and human health. *Ecological Indicators* 111, 106020. DOI: 10.1016/j.ecolind.2019.106020.

Reinhart, K.O. and Callaway, R.M. (2006) Soil biota and invasive plants. *The New Phytologist* 170(3), 445–457. DOI: 10.1111/j.1469-8137.2006.01715.x.

Rejmanek, M. (2000) Invasive plants: Approaches and predictions. *Austral Ecology* 25, 497–506. DOI: 10.1111/j.1442-9993.

Reynolds, J.W. (1994) Earthworms of the world. *Global Biodiversity* 4(1), 11–16.

Ricciardi, A. and Cohen, J. (2007) The invasiveness of an introduced species does not predict its impact. *Biological Invasions* 9(3), 309–315. DOI: 10.1007/s10530-006-9034-4.

Rout, M.E. and Callaway, R.M. (2012) Interactions between exotic invasive plants and soil microbes in the rhizosphere suggest that 'everything is not everywhere. *Annals of Botany* 110(2), 213–222. DOI: 10.1093/aob/mcs06.

Roy, H., Peyton, J., Aldridge, D., Bantock, T., Blackburn, T. *et al.* (2014) Horizon scanning for invasive alien species with the potential to threaten biodiversity in Great Britain. *Global Change Biology* 20(12), 3859–3871. DOI: 10.1111/gcb.12603.

Ruwanza, S. and Shackleton, C.M. (2016) Effects of the invasive shrub, *Lantana camara*, on soil properties in the Eastern Cape, South Africa. *Weed Biology and Management* 16(2), 67–79. DOI: 10.1111/wbm.12094.

Scheu, S., Ruess, L. and Bonkowski, M. (2005) Interactions between microorganisms and soil micro- and mesofauna. In: Buscot, F. and Varma, A. (eds) *Microorganisms in Soils: Roles in Genesis and Functions*. Springer, Berlin, pp. 253–275. DOI: 10.1007/3-540-26609-7_12.

Schläpfer, F., Pfisterer, A.B. and Schmid, B. (2005) Non-random species extinction and plant production: Implications for ecosystem functioning. *Journal of Applied Ecology* 42(1), 13–24. DOI: 10.1111/j.1365-2664.2004.00987.x.

Schmitz, O.J., Grabowski, J.H., Peckarsky, B.L., Preisser, E.L., Trussell, G.C. *et al.* (2008) From individuals to ecosystem function: Toward an integration of evolutionary and ecosystem ecology. *Ecology* 89(9), 2436–2445. DOI: 10.1890/07-1030.1.

Schwartz, M.W., Brigham, C.A., Hoeksema, J.D., Lyons, K.G., Mills, M.H. *et al.* (2000) Linking biodiversity to ecosystem function: Implications for conservation ecology. *Oecologia* 122(3), 297–305. DOI: 10.1007/s004420050035.

Simberloff, D., Martin, J.-L., Genovesi, P., Maris, V., Wardle, D.A. *et al.* (2013) Impacts of biological invasions: What's what and the way forward. *Trends in Ecology & Evolution* 28(1), 58–66. DOI: 10.1016/j.tree.2012.07.013.

Sławska, M. (2004) Rola skoczogonków (*Collembola, Hexapoda*) w funkcjonowaniu ekosystemów leśnych [The role of springtails (*Collembola, Hexapoda*) in forest ecosystems processes]. *Sylwan* 148(7), 53–71.

Stefanowicz, A.M., Stanek, M., Majewska, M.L., Nobis, M. and Zubek, S. (2019) Invasive plant species identity affects soil microbial communities in a mesocosm experiment. *Applied Soil Ecology* 136, 168–177. DOI: 10.1016/j.apsoil.2019.01.004.

Sterzyńska, M., Shrubovych, J. and Nicia, P. (2017) Impact of plant invasion (*Solidago gigantea* L.) on soil mesofauna in a riparian wet meadows. *Pedobiologia* 64, 1–7. DOI: 10.1016/j.pedobi.2017.07.004.

Stinson, K.A., Campbell, S.A., Powell, J.R., Wolfe, B.E., Callaway, R.M. *et al.* (2006) Invasive plant suppresses the growth of native tree seedlings by disrupting belowground mutualisms. *PLoS Biology* 4(5), e140. DOI: 10.1371/journal.pbio.0040140.

Suding, K.N., LeJeune, K.D. and Seastedt, T.R. (2004) Competitive impacts and responses of an invasive weed: Dependencies on nitrogen and phosphorus availability. *Oecologia* 141(3), 526–535. DOI: 10.1007/s00442-004-1687-0.

Talley, T.S. and Levin, L.A. (2001) Modification of sediments and macrofauna by an invasive marsh plant. *Biological Invasions* 3(1), 51–68. DOI: 10.1023/A:1011453003168.

Tomimatsu, H., Sasaki, T., Kurokawa, H., Bridle, J.R., Fontaine, C. *et al.* (2013) FORUM: Sustaining ecosystem functions in a changing world: A call for an integrated approach. *Journal of Applied Ecology* 50(5), 1124–1130. DOI: 10.1111/1365-2664.12116.

Tonole, J. (2025) A Closer Look at Allelopathy: Better Living Through Chemicals. Finger Lakes Land Trust. Available at: https://www.fllt.org/a-closer-look-at-allelopathy-better-living-through-chemicals (accessed 8 July 2025).

Twardowski, J.P., Hurej, M. and Gruss, I. (2016) Diversity and abundance of springtails (Hexapoda: Collembola) in soil under 90-year potato monoculture in relation to crop rotation. *Archives of Agronomy and Soil Science* 62, 1–11. DOI: 10.1080/03650340.2015.1131270.

Vilà, M., Espinar, J.L., Hejda, M., Hulme, P.E., Jarošík, V. *et al.* (2011) Ecological impacts of invasive alien plants: A meta-analysis of their effects on species, communities and ecosystems. *Ecology Letters* 14(7), 702–708. DOI: 10.1111/j.1461-0248.2011.01628.x.

Vitousek, P.M. and Walker, L.R. (1989) Biological invasion by *Myrica faya* in Hawai'i: Plant demography, nitrogen fixation, ecosystem effects. *Ecological Monographs* 59(3), 247–265. DOI: 10.2307/1942601.

Walker, L.R. and Smith, S.D. (1997) Impacts of invasive plants on community and ecosystem properties. In: *Assessment and Management of Plant Invasions*. Springer, New York, pp. 69–86. DOI: 10.1007/978-1-4612-1926-2_7.

Wang, C., Xiao, H., Liu, J., Wang, L. and Du, D. (2015) Insights into ecological effects of invasive plants on soil nitrogen cycles. *American Journal of Plant Sciences* 06(1), 34–46. DOI: 10.4236/ajps.2015.61005.

Wardle, D.A. and Peltzer, D.A. (2017) Impacts of invasive biota in forest ecosystems in an aboveground–belowground context. *Biological Invasions* 19(11), 3301–3316. DOI: 10.1007/s10530-017-1372-x.

Weidenhamer, J.D. and Callaway, R.M. (2010) Direct and indirect effects of invasive plants on soil chemistry and ecosystem function. *Journal of Chemical Ecology* 36(1), 59–69. DOI: 10.1007/s10886-009-9735-0.

Wijesundara, S. (2010) Invasive alien plants in Sri Lanka. In: Wijesundara, S. (ed.) *Invasive Alien Plants in Sri Lanka*. Ministry of Environment, Sri Lanka, pp. 27–38.

Wilson, M., Freundlich, A. and Martine, C. (2017) Understory dominance and the new climax: Impacts of Japanese knotweed (*Fallopia japonica*) invasion on native plant diversity and recruitment in a riparian woodland. *Biodiversity Data Journal* 5, e20577. DOI: 10.3897/BDJ.5.e20577.

Wolfe, B.E. and Klironomos, J.N. (2005) Breaking new ground: Soil communities and exotic plant invasion. *BioScience* 55(6), 477. DOI: 10.1641/0006-3568(2005)055[0477:BNGSCA]2.0.CO;2.

Wu, D., Du, E., Eisenhauer, N., Mathieu, J. and Chu, C. (2025) Global engineering effects of soil invertebrates on ecosystem functions. *Nature* 640(8057), 120–129. DOI: 10.1038/s41586-025-08594-y.

Xu, H., Liu, Q., Wang, S., Yang, G. and Xue, S. (2022) A global meta-analysis of the impacts of exotic plant species invasion on plant diversity and soil properties. *Science of The Total Environment* 810, 152286. DOI: 10.1016/j.scitotenv.2021.152286.

Xu, Y., Chen, X., Ding, L. and Kong, C.H. (2023) Allelopathy and allelochemicals in grasslands and forests. *Forests* 14(3), 562. DOI: 10.3390/f14030562.

Zhang, Z., Liu, J., Wang, L.C., Wang, Y. and Deng, L.L. (2020) Impact of the invasive plant *Alternanthera philoxeroides* (Mart.) Griseb on soil meso- and microfaunal community. *Applied Ecology and Environmental Research* 18(2), 3073–3085. DOI: 10.15666/aeer/1802_30733085.

Zhukov, O., Kunakh, O., Yorkina, N. and Tutova, A. (2023) Response of soil macrofauna to urban park reconstruction. *Soil Ecology Letters* 5(2), 220156. DOI: 10.1007/s42832-022-0156-0.

Zizolfi, M., Maisto, G., Santini, G., Memoli, V., Trifuoggi, M. *et al.* (2025) Invasive plants do not exert univocal responses on soil abiotic and biotic characteristics. *Forests* 16(2), 209. DOI: 10.3390/f16020209.

8 Structural Alterations Induced by Invasive Vegetation: Consequences for Microbial Communities

Mohammad Danish[1]*, Mohammad Shahid[2] and S. Maqbool Ahmed[1]
[1]Section of Botany, School of Sciences, Maulana Azad National Urdu University, Hyderabad, India; [2]Marwadi University Research Centre, Department of Agriculture, Faculty of Science, Marwadi University, Rajkot, Gujrat, India

Introduction

Exotic plant invasion is generally blamed on globalization, agriculture and human activities (Linders *et al.*, 2019; Jin *et al.*, 2022). The reduction of native biodiversity as a result of plant invasion is the main cause to the significant negative impacts on ecosystem processes and services (Wang *et al.*, 2020; Zhang and Suseela, 2021). Microorganisms may even become key factors in plant invasion, although they normally react to an invasion (Zhang *et al.*, 2022). Invasive species are major factors that can alter ecosystems by influencing both the above- and belowground environment (Wardle and Peltzer, 2017).

In addition, aboveground invasive species allow the development of dense canopies and suppress native plants by modifying light and litter dynamics. Simultaneously, belowground they alter root morphology, rhizosphere chemistry and organic matter deposition, which initiates a chain reaction of changes in soil properties (Hejda *et al.*, 2021). Such changes move beyond direct plant–plant competition and thus affect microbial communities that are crucial for ecosystem processes. Soil microbial communities are the 'hidden drivers' of ecosystem functioning. They are nutrient cycling agents, decomposers, soil aggregators, carbon sequestration agents and plant–microbe symbionts (Daba and Dejene, 2018).

When invasive vegetation alters soil conditions, it can lead to major changes in the types of microbes and their activity in the soil. For example, invasive species sometimes release allelopathic compounds or change the C:N ratio of the soil, thus microbial groups that are best adapted to consume such inputs become dominant (Wang *et al.*, 2022). There are cases in which invasive plants recruit microbes that become beneficial to their growth; on the other hand, these plants suppress native plant–microbe mutualisms. Understanding these interactions is crucial because microbial changes that are closely involved with invasion success also have an effect on ecosystem stability or collapse over time (Zhang *et al.*, 2023).

Besides forming ecological clusters and living very close to each other, soil microorganisms have various relationships, which may be competitive or cooperative, for example (Berry and Widder, 2014). Cooperation

*Corresponding author: danish.botanica@gmail.com

DOI: 10.1079/9781836991298.0008

among microbial species can enhance both the efficiency of resource transmission and ecological processes (Lin *et al.*, 2021). Recent work has been done in the context of ecological co-occurrence networks and their topological features. This has been done to understand the intricacy and the degree of interconnection between the taxa. The topological features together reflect the structural complexity, connectivity patterns and potential functional resilience of the microbial network (Shi *et al.*, 2020; Xu *et al.*, 2022).

Co-occurrence patterns have been demonstrated to be a powerful tool for investigating the organization of microbial communities at different taxonomic levels, thereby unravelling interaction networks and identifying keystone taxa under the challenging conditions of the soil (Fan *et al.*, 2021). The key to understanding microbial community composition, stability and their contributions to ecosystem processes is through the recognition of keystone taxa; that is, strongly connected taxa which, whether singly or in a guild, possess the highest explanatory power of network structure and function independent of their abundance (Chen *et al.*, 2019).

This chapter provides an analysis of changes in microbial community structure resulting from invasive plants. It also emphasizes the importance of integrating the knowledge of plant invasions with that of soil microbes to understand, predict and control the cascading impacts of invasive species on ecosystem services.

Structural Alterations Induced by Invasive Vegetation

Invasive species very often have aboveground parts that are extremely visible and can thus help them to dominate and outgrow native plants significantly. Features of these parts include rapid growth, higher photosynthetic efficiency, greater biomass accumulation, higher seed production and the development of denser canopy structures. With such features, invasive species may become monopolists, thereby preventing the regeneration and growth of native plants (Peter *et al.*, 2021). Apart from the drastic reduction of light accessibility, a thick canopy will also alter the local temperature and humidity regimes, thereby modifying the microclimate experienced by soil organisms (Zhang *et al.*, 2020). In addition, invasive plants usually create more litter than native ones, and this litter is often different in quality – it may have more lignin or secondary metabolite content – and thus the decomposition rate will be slower and nutrient cycling will be altered (Zhang *et al.*, 2020). Changes in aboveground inputs resulting from the loss of native plants eventually affect the soil, creating long-term effects on microbial communities that regulate decomposition, carbon sequestration and nutrient availability (Daba and Dejene, 2018).

The effect of invasive vegetation is not be only seen by canopy dominance: belowground structural changes can have deeper and longer-lasting impacts. Invasive species usually allocate a large proportion of their energy to roots that are longer, more branched or more proliferative compared to those of native plants. These root structures thus heavily penetrate the soil, take more water and absorb more nutrients, enabling invasive species to persist even in resource-limited environments. However, invasive roots also secrete more complex exudates, including sugars, amino acids, organic acids and secondary metabolites, into the rhizosphere than native plant roots (Yu *et al.*, 2022). These compounds become a selective resource that the soil microbial populations must utilize, and hence the microbial taxa supporting the success of the invasive plants dominate, while those beneficial to native species are suppressed (Núñez-Mir and McCary, 2024). In this way root exudation is the main factor by which invasive species manipulate soil microbial assemblages to their benefit.

Consequently, plant invasion mechanisms that involve soil microbial communities have been elucidated. To illustrate, the novel weapon hypothesis, inter alia, suggests that imported plants use allelopathy (Bais *et al.*, 2003) through root exudates, weakening or disrupting symbiotic relationships between native plants and microorganisms to help the colonization of exotic plants (Callaway and Ridenour, 2004). The enemy release hypothesis suggests that newly introduced plants lack local pathogens, reducing 'defence costs' and giving them a 'head start' in competition with native plants (Dai *et al.*, 2015). In addition, the allelopathic compounds released by root exudates intensify these changes by inhibiting neighbouring plant germination and growth while at the same time modifying microbial community composition. As an example, some allelochemicals suppress the beneficial symbionts such as mycorrhizal fungi or nitrogen-fixing bacteria that native plants depend on, while promoting the microbial groups that are resistant to or can metabolize these compounds (Yu *et al.*, 2022).

Such dual effects, suppressing plant competitors and changing the soil microbial communities, inevitably result in positive feedback loops that maintain the dominance of invasive vegetation. The feedback loops gradually create new soil environments resistant to recolonization by native species, contributing to long-term transformation of an ecosystem (Zhang *et al.*, 2023).

In summary, invasive vegetation is the main driver of the structural changes both above and belowground, which eventually reshapes entire ecosystems. By changing canopy structures, altering litter dynamics and restructuring root traits and rhizosphere chemistry, invasive species not only suppress native flora but also restructure soil microbial communities. These changes underscore the tight linkages existing between plant traits and microbial ecology, whereby changes in plant structure – both obvious and hidden – have a trickle-down effect on belowground biodiversity and ecosystem processes (Xu *et al.*, 2022). Understanding these interactions is thus essential for predicting invasion pathways and designing management strategies that address both an invasive's visible dominance as well as the invisible restructuring of microbial communities. These multiple changes in above- and belowground plant structures are outlined in Table 8.1, which highlights the major components altered by invasive vegetation and the ecological consequences.

Soil Physical and Chemical Modifications

The root systems of invasive plants that are dense and fibrous are the main factors that change soil porosity, aggregation and compaction, which are ultimately responsible for water infiltration, retention and aeration (Unger *et al.*, 2022). These alterations in the soil lead to the

Table 8.1. Structural alterations induced by invasive vegetation and the ecological consequences.

Plant component	Alteration by invasive vegetation	Ecological consequence	Reference
Aboveground biomass	Increased litter deposition, altered canopy structure	Changes light availability, affects native plant growth	Zhao *et al.* (2024)
Root architecture	Denser or deeper root systems, altered root exudates	Modifies nutrient and water uptake, influences soil microbes	Núñez-Mir and McCary (2024)
Soil physical properties	Compaction, reduced porosity, altered soil moisture	Impacts aeration, water infiltration and microbial habitats	Jamin *et al.* (2022)
Soil chemical properties	Changes in pH, nutrient availability, organic matter content	Favours invasive species, suppresses natives	Afzal *et al.* (2023)
Microbial community	Shifts in composition, abundance and diversity	Enhances invader-associated microbes, creates positive feedbacks	Zhao *et al.* (2024)
Plant soil feedback	Strengthened positive feedback loops for invaders	Increased competitive advantage, reduced native plant fitness	Ali *et al.* (2024)

formation of different subterranean biomes, which can have either positive or negative effects on microbial activity. For instance, increased root penetration and turnover might help soil aggregation, whereas compaction of the upper soil layers due to aggressive colonization by roots may cause decreased oxygen availability (Zhao *et al.*, 2024). In the same way, the increased litter deposition by invasive plants changes the dynamics of soil organic matter. Sometimes it leads microbial decomposition along different pathways (Souza-Alonso *et al.*, 2024).

Invasive species, when looked at from a chemical standpoint, cause almost equal changes in soil nutrient dynamics and biogeochemical cycles. A good number of invasive plants speed up nutrient cycling, in particular nitrogen and phosphorus availability, by changing microbial activity and enzyme production. Li *et al.* (2024) found that the microbial biomass was higher in soils under invasive vegetation and that enzyme activity related to carbon and nitrogen turnover was increased. The authors explained that such processes not only increase nutrient fluxes but also change the ratio of fast- and slow-cycling nutrient pools, with the risk of long-term soil fertility becoming instable. It is worth mentioning that invasive plants frequently release allelopathic secondary metabolites that can suppress native microbial taxa while favouring those microbes that are tolerant or metabolically compatible (Yu *et al.*, 2022). The aforementioned compounds, in combination with changes in litter chemistry, have the potential to alter soil pH, salinity and moisture levels, thus changing the abiotic environment that controls microbial community assembly. Invasive plants are a strong force of nature, as they change the soil's physical and chemical environment. The alterations brought about by these changes restructure the ecological filters that determine microbial community composition and function, thus creating soil conditions that gradually become more and more suitable for invasive dominance while the resilience of native ecosystems is lowered (Unger *et al.*, 2022). Hence, it is very important to comprehend these transformations at the soil level in order to be able to link plant invasions with microbial community dynamics and their cascading ecological consequences.

Consequences for Microbial Communities

Invasive vegetation-induced changes to the structure and chemical composition of the soil have a wide-ranging influence on the composition, diversity and functioning of microbial communities. Microbes, being extremely sensitive to changes in soil organic matter and nutrient availability, can undergo significant restructuring even due to subtle alterations by invasive plants (Canini *et al.*, 2024). For instance, an invasive species that significantly increases litter inputs with a chemically distinct profile may result in the dominance of microbial groups specializing in the degradation of complex compounds, and at the same time the population of microbes adapted to the native litter quality may decrease (Souza-Alonso *et al.*, 2024). These changes in microbial composition have the effect not only of changing decomposition rates but also of altering nutrient cycling and ecosystem productivity on a larger scale.

Bacterial communities are highly responsive when plants invade because bacteria have short generation times and are metabolically versatile. Usually, invasive plants increase the number of copiotroph bacteria like Proteobacteria and Bacteroidetes which live in nutrient-rich environments that are created by the increase of root exudation and accelerated nutrient turnover (Zhang *et al.*, 2023). On the other hand, oligotrophic bacterial groups like Acidobacteria, which are more typical of resource-limited soils, decrease in invasive-dominated areas (Canini *et al.*, 2024). In the same way, archaeal populations that are involved in nitrogen cycling, for instance, ammonia-oxidizing archaea, frequently show changed abundance and activity in ecosystems that have been invaded, thus leading to altered nitrogen transformation pathways (Yu *et al.*, 2022). These variations demonstrate how invasive vegetation acts as a main driver in reassembling microbial functional guilds. This, in turn, renews nutrient dynamics at the ecosystem level in a way that is beneficial to invasive species. Moreover, changes in soil microbial community structure are frequently accompanied by shifts in soil enzymatic profiles. Elevated activities of urease, phosphatase and dehydrogenase are examples of enzymatic changes that support fast nutrient turnover in such soils. The continuous operation of this feedback system may thus result in the formation of self-stabilizing microbial consortia that keep soil conditions stable for the further spread of invasive vegetation. Consequently, the functional reorganization of bacterial and archaeal communities reduces ecosystem resistance to the invasive species, making it harder for native microbial assemblages to recover, even after management interventions.

While the impacts on fungal communities have been shown to be comparable, the effects are usually more complicated due to the strong functional specialization of fungi and their complex relations with plants. Mycorrhizal fungi, in particular arbuscular mycorrhizal fungi (AMF), are understandably the most affected in a scenario of invasive vegetation (Singh *et al.*, 2025b). Several experiments have illustrated that an invasive plant can either lessen AMF colonization of native plants or change fungal community composition in a way that supports the organisms that are less beneficial to native vegetation and more compatible with the invasive species (Singh *et al.*, 2025b). In this way, disruptions to native plants can go quite deep as nutrient acquisition is severely compromised due to the indispensable role of AMF in phosphorus uptake and soil structure stabilization. The dwindling of mutualistic fungi, therefore, becomes a negative feedback loop that weakens the resistance of native plant communities and pollutes the natural recovery process of ecosystems even after invasive removal. On the other hand, invasive species may manipulate fungal communities to favour the fungi that are capable of breaking down their own hard-to-degrade litter, thereby releasing nutrients more rapidly and supporting the invasives' dominance (Magar, 2024). For example, one study has documented those certain taxa of Ascomycota and Basidiomycota greatly increase their abundance under dense invasive stands, thus contributing more to litter decomposition and soil organic carbon turnover (Zhao *et al.*, 2024). In addition, introduction of invasive plants may change afungal community's balance between mutualistic and pathogenic species. Pathogenic fungi may dominate soils that have been invaded, leading to high disease pressure on native species that eventually causes their competitive exclusion (Kowalski and Bilański, 2024). So, changes in fungal community composition and functionality that accompany alterations in belowground biodiversity have an influence on the dynamics of aboveground vegetation as well. The loss of mycorrhizal mutualisms alongside the increased saprotrophic and pathogenic activities that synergize the ecological displacement caused by invasive plants is the main theme of this paper. Such a chain effect shows the role of fungi as the main mediators of the invasion-driven ecosystem change and at the same time points to the necessity of fungal indicators being included in future invasion ecology assessments. The intricate webs of microbial communities that are altered by invasive vegetation along with the natural processes that are affected are shown in Fig. 8.1.

The functional consequences of microbial restructuring are substantial. Enhanced microbial enzyme activity, accelerated nutrient mineralization and altered carbon and nitrogen cycling create positive feedback loops that support invasive plant dominance (Li *et al.*, 2024). For example, microbes associated with invasive plants can increase nitrogen availability, directly benefiting invasive plants. Simultaneously, disruption of symbiotic networks results in the decrease of native plant resilience. These microbial changes are not only a passive reaction to invasion but a key factor of invasion success. Moreover, these modifications often continue for a long time after the removal of invasives and thus they cause long-lasting legacy effects in soil microbial communities (Shen *et al.*, 2022). Table 8.2 summarizes the main characteristic responses of different microbial groups to invasive vegetation, the changes that were observed and their functional implications.

Impacts on Microbial Community Composition and Functions

One of the most significant consequences of restructuring microbial communities by invasive plants is the change of decomposition and nutrient mineralization rates through the shift in decomposer microbial taxa and their enzyme activities to reshape organic matter turnover (Torres *et al.*, 2021). In some ecosystems, such

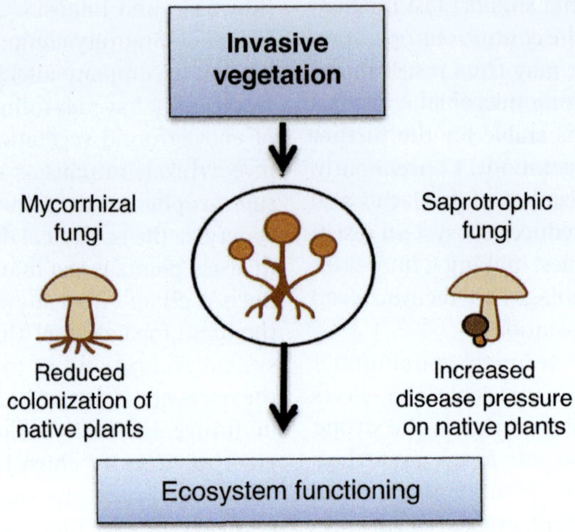

Fig. 8.1. Schematic representation of invasive plants alter microbial community composition and ecosystem functions.

Table 8.2. Effects of invasive vegetation on microbial communities.

Microbial group	Observed changes under invasive vegetation	Functional consequences	Reference
Bacteria: copiotrophs	Increased abundance due to high root exudates and nutrient inputs	Accelerated decomposition and nutrient cycling	Zhang *et al.* (2023)
Bacteria: oligotrophs	Decline in resource-limited invaded soils	Reduced stability of slow nutrient cycles	Yu *et al.* (2022)
Archaea: ammonia-oxidizing	Altered abundance and activity	Modified nitrification and denitrification pathways	Li *et al.* (2024)
Mycorrhizal fungi: AMF	Decline in colonization of native plants; shift toward invasive-compatible taxa	Reduced native plant nutrient uptake	Singh *et al.* (2025b)
Saprotrophic fungi	Increased abundance due to litter from persistent invasive plants	Enhanced decomposition of complex organic matter	Magar (2024)
Pathogenic fungi	Often enriched in invaded soils	Increased disease pressure on native plants	Shen *et al.* (2022)

microbial changes lead to an acceleration of nutrient release, thus increasing nitrogen, phosphorus and other essential element availability, which in turn nourishes invasive plant growth and increases their competitive dominance. This kind of fast nutrient cycling usually results in a positive feedback loop, which allows invaders to keep their biomass production and restrain native flora (Lee *et al.*, 2012). Alternatively, invasive species may deposit more persistent litter that slows microbial activity, thus further biasing decomposition dynamics in their favour by lowering the availability of resources for native competitors (Canini *et al.*, 2024).

Moreover, symbiotic interactions with microbial partners are one of the key factors that determines invasion dynamics for non-native species. Quite a few invasive organisms have the capability to either take advantage of or change the nature of helpful microorganisms like mycorrhizal fungi and nitrogen-fixing rhizobia, which in turn improves nutrient acquisition and adaptability to new environments. Jesus *et al.* (2025) provided evidence that the microbiomes of *Acacia longifolia* seeds and root nodules played a major role in its invasive success by providing functional advantages in nutrient uptake and stress tolerance. Such results show that microbial partnerships can be the main facilitators of invasion rather than being only passive processes. In the same way, the invaders may strengthen their mycorrhizal associations selectively, which in turn changes soil mutualistic networks that support native plant diversity (Magar, 2024). Invasive plant-driven microbial restructuring that goes beyond nutrient cycling and mutualistic symbioses can also impact ecosystem services. Changes in microbial communities have an impact on systems that rely on soil, such as soil stabilization, water infiltration and the suppression of plant pathogens. Zhao *et al.* (2024) observed that in a large number of invaded soils, diseases caused by pathogenic microbes decreased in their relative abundance, thus reducing pathogen pressure on invaders. As a result, the natural checks and balances that would otherwise limit their proliferation are weakened. These cascading effects show that microbial dynamics not only dominate plant competition but also become the main factors in ecosystem service delivery, thus leading to long-term functional changes. In other words, the ability of invasive vegetation to change microbial communities emphasizes the key role of belowground processes in driving aboveground ecological changes (Wardle and Peltzer, 2017; Wang *et al.*, 2022). The responses of major fungal groups to invasive vegetation and their resulting ecological consequences are depicted in Fig. 8.2.

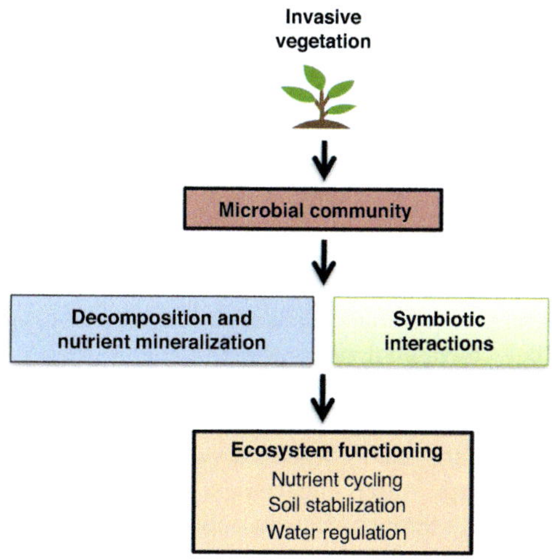

Fig. 8.2. Effect of invasive plant species on soil microbial communities.

Case Studies and Examples

A number of recent studies have showcased the varied and sometimes complicated effects that invasive plants have on their interactions with microbes, ranging from global syntheses to localized ecosystem analyses. At the global level, Zhao *et al.* (2024) carried out a field study over 8 years and came up with robust empirical evidence showing how invasive alien plants change the structure of soil microbial communities so that these changes favour their dominance in an unbalanced way. The study found that the invading species not only alter microbial biomass but also selectively accelerate nutrient cycling by stimulating the activity of those microbial groups that eventually benefit the invaders themselves. Significantly, changes in microbes that accompanied this transition were not evenly distributed but showed a strong directional bias which frequently resulted in less resistance from indigenous microbial networks. The findings thus underscore that invasive species manipulate microbial communities to sustain their dominance instead of being merely passive recipients of already existing soil conditions.

Nunez-Mir *et al.* (2019) conducted a worldwide meta-analysis that associated root traits of invasive plant species with microbial assemblages. Their research brought to light that invasive plants commonly manifest unique belowground characteristics such as fine root density, high exudation rates and elevated production of allelochemicals, which by restructuring microbial community assembly modify their surroundings. These alterations were independent of geography and ecosystems, thus making root traits general predictors of microbial changes resulting from invasion. By uncovering this relationship, they offered a mechanistic model for understanding invasion as a consequence of plant–microbe co-adaptation rather than competition alone.

Many case studies provide more detailed understanding at regional and ecosystem-specific levels. For instance, Torres *et al.* (2021) aggregated data from 46 empirical studies conducted in temperate, tropical and Mediterranean ecosystems. Their meta-analysis provided initial evidence that plant invasions have the overall tendency to reduce microbial alpha-diversity and hence increase the relative dominance of specific microbial groups. These changes were demonstrated to propagate through shifts in nutrient cycling, soil respiration and enzymatic activity thus, providing strong evidence that invasive plants affect both community composition and microbial function at the same time. Interestingly, it was also highlighted that the size of these effects depended on the biome and the invader's identity, stressing that the outcomes of the invasion should not be generalized across different contexts.

Environmental gradients are essential factors that determine how invasive plants interact with microbial communities. Canini *et al.* (2024) studied *Carpobrotus edulis* invasions across coastal gradients in Mediterranean ecosystems. Their work demonstrated that the microbial consequences of the invasion were not consistent but changed with the abiotic factors such as soil salinity, moisture and organic matter content. To illustrate, in sandy soils with lower organic matter, the invasion of *C. edulis* greatly increased fungal-to-bacterial ratios, while in nutrient-rich soils bacterial dominance was elevated. This example illustrates that the effect of invasive vegetation is dependent on local environmental filters, which implies that the same species can lead to different microbial results in different ecosystems.

An interesting case of plant–microbe co-adaptation is from research on *A. longifolia*. Jesus *et al.* (2025) revealed how microbial partners associated with this invasive legume contribute to its ecological strategy in novel ecosystems. By studying the microbiomes of seeds and root nodules, the research showed that *A. longifolia* harbours specific nitrogen-fixing rhizobia and beneficial mycorrhizal fungi that together enable the plant to grow and survive under environmental stress. These symbiotic partnerships not only help the plant to get more nutrients but also protect it from native soil pathogens; thus the plant is surrounded by a microbial shield that strengthens its competitive advantage. The results obtained strongly support the idea that invasions are not only plant-driven processes but also microbial-mediated phenomena.

These studies, in unison, delineate how complicated and changeable the effects of invasion to microbial communities are. Some processes, like increased microbial activity that helps invaders to become successful, remain consistent, but local environmental filters, species-specific traits and the partner microbial community all determine the results. Such variation emphasizes the necessity of integrative, multi-scale strategies combining global meta-analyses with local-scale investigations. Thus, by these contrasts, scientists can obtain better capabilities to forecast invasion patterns and craft management strategies.

Management and Mitigation Strategies

The control of invasive plants that threaten the environment should follow a strategy that takes into account the aspects of both the plants and the microbes. Among these strategies, microbe-based approaches are considered as a potential means, for example, for the employment of microbial inoculants to re-establish native plant–microbe interaction or to disrupt those of invasive species (Chen *et al.*, 2020). Shahrtash and Brown (2021) emphasized the use of microbial consortia and the targeting of plant–microbe interactions as new technologies for controlling invasives. The issues of invasive species have been taken care of by ecological restoration practices. The use of organic amendments and modifying soil characteristics like pH can facilitate the transition of soil microbial communities from those of the invaded to the native ones and thus, restore the native plant species (Weidlich *et al.*, 2020). Such strategies check the spread of invasive species and, at the same time, enhance the resilience of native vegetation. Biocontrol agents that specifically target invasive species may be used together with microbial interventions to make the ecosystem more stable (Shahrtash and Brown, 2021). Furthermore, long-term monitoring of invasive species over a long period of time and the use of microbial indicators in management plans are two very important things that lead to sustainability at the policy and conservation levels. The use of microbial strategies together with the conventional management methods provides a comprehensive model to lessen the ecological and functional influences of invasive vegetation.

Future Perspectives

To advance our understanding of invasive vegetation–microbe interactions, mechanistic and integrative research must be conducted. A myriad of possibilities is presented by various -omics techniques, such as metagenomics, transcriptomics and metabolomics, to discover how microbial community functions and metabolic pathways are altered during plant invasions. These methods can identify the molecular mechanisms that cause microbial changes and the ecological impact that follow. Long-term field experiments, such as those conducted by Zhao *et al.* (2024), have the potential to reveal the temporal dynamics of microbial succession and ecosystem changes over time. By extending such research across different ecosystems, we will be able to generalize patterns and understand context-dependent effects. The use of predictive modelling and machine learning approaches can be very helpful in anticipating the effects of invasions under different environmental conditions and thus help in the planning of management strategies that are ahead of their time (Singh *et al.*, 2025a). Combining different scales – from tightly controlled microcosm studies to vast landscape-level observations – will help understand how microbial changes at a very local level affect ecosystem processes at a larger scale. There are still significant knowledge gaps in understanding, which include the ecological roles of rare microbial taxa, interactions between fungi and bacteria, and the combined effects of climate change stressors on invasive species and their associated microbes. Filling in these gaps will increase the mechanistic understanding and also be a source of fresh, ground-breaking ideas for managing invasions in such a way that the resilience of ecosystems is preserved.

Conclusion

Invasive plants can, literally, change ecosystems both above and below ground in such a way that eventually soil physical and chemical properties are affected. Often, these changes remodel soil microbial communities to the extent that they become dominated by those microorganisms associated with the invaders, thus creating positive feedback loops that increase the latter's competitive advantage. Nevertheless, effects of invasions on microbial communities and other components of ecosystems can vary greatly, according to research, and they depend on species, environmental conditions and time. Effective management of invasive plants requires methods that combine microbial considerations with traditional restoration methods. It is possible to restore ecological resilience through native microbial communities and to disrupt invader-associated microbes, guaranteeing that native plants can regain space and resources. A more holistic and sustainable pathway to ecosystem recovery emerges when microbial approaches are combined with conventional interventions like the reintroduction of native species or the adjustment of soil properties. Future studies should employ more integrated and mechanistic strategies: tools such as metagenomics, transcriptomics and metabolomics can uncover the molecular and functional aspects of microbial communities that change during invasions. Long-term experiments and predictive modelling can also reveal temporal dynamics and forecast invasion impact under changing environmental conditions. Focusing on rare microbial taxa, fungal–bacterial interactions and the effect of climate change will be crucial for closing current knowledge gaps. Understanding and managing the complex of interactions between invasive vegetation and soil microbes are essential not only for conserving biodiversity but also for maintaining ecosystem functions. Science-driven interventions together with long-term monitoring and adaptive management will help mitigate invasions impacts and maintain the resilience of native ecosystems.

References

Afzal, M.R., Naz, M., Ashraf, W. and Du, D. (2023) The legacy of plant invasion: Impacts on soil nitrification and management implications. *Plants (Basel, Switzerland)* 12(16), 2980. DOI: 10.3390/plants12162980.

Ali, H.E., Al-Wahaibi, A.M. and Shahid, M.S. (2024) Plant-soil feedback and plant invasion: Effect of soil conditioning on native and invasive prosopis species using the plant functional trait approach. *Frontiers in Plant Science* 15, 1321950. DOI: 10.3389/fpls.2024.1321950.

Bais, H.P., Vepachedu, R., Gilroy, S., Callaway, R.M. and Vivanco, J.M. (2003) Allelopathy and exotic plant invasion: From molecules and genes to species interactions. *Science* 301(5638), 1377–1380. DOI: 10.1126/science.1083245.

Berry, D. and Widder, S. (2014) Deciphering microbial interactions and detecting keystone species with co-occurrence networks. *Frontiers in Microbiology* 5, 219. DOI: 10.3389/fmicb.2014.00219.

Callaway, R.M. and Ridenour, W.M. (2004) Novel weapons: Invasive success and the evolution of increased competitive ability. *Frontiers in Ecology and the Environment* 2(8), 436–443. DOI: 10.1890/1540-9295(2004)002[0436:NWISAT]2.0.CO;2.

Canini, F., Borruso, L., Magrini, S., D'Acqui, L.P., Buzzini, P. *et al.* (2024) Soil microbial communities dynamics in response to invasive groundcover Carpobrotus: Insights into native species interactions and edaphic influence. *Applied Soil Ecology* 202, 105537. DOI: 10.1016/j.apsoil.2024.105537.

Chen, L., Jiang, Y., Liang, C., Luo, Y., Xu, Q. *et al.* (2019) Competitive interaction with keystone taxa induced negative priming under biochar amendments. *Microbiome* 7(1), 77. DOI: 10.1186/s40168-019-0693-7.

Chen, L., Shi, Y.T. and Zhang, H.B. (2020) Effects of modifying soil microbes on restoring vegetation degraded by an invasive plant crofton weed (*Ageratina adenophora*). *Applied Ecology and Environmental Research* 18(2), 3087–3100. DOI: 10.15666/aeer/1802_30873100.

Daba, M.H. and Dejene, S.W. (2018) The role of biodiversity and ecosystem services in carbon sequestration and its implication for climate change mitigation. *Environmental Sciences and Natural Resources* 11(2), 1–10. DOI: 10.19080/IJESNR.2018.11.555810.

Dai, Z.-C., Qi, S.-S., Miao, S.-L., Liu, Y.-T., Tian, Y.-F. *et al.* (2015) Isolation of NBS-LRR rgas from invasive *Wedelia trilobata* and the calculation of evolutionary rates to understand bioinvasion from a molecular evolution perspective. *Biochemical Systematics and Ecology* 61, 19–27. DOI: 10.1016/j.bse.2015.05.004.

Fan, K., Delgado-Baquerizo, M., Guo, X., Wang, D., Zhu, Y. *et al.* (2021) Biodiversity of key-stone phylotypes determines crop production in a 4-decade fertilization experiment. *The ISME Journal* 15(2), 550–561. DOI: 10.1038/s41396-020-00796-8.

Hejda, M., Sádlo, J., Kutlvašr, J., Petřík, P., Vítková, M. *et al.* (2021) Impact of invasive and native dominants on species richness and diversity of plant communities. *Preslia* 93(3), 181–201. DOI: 10.23855/preslia.2021.181.

Jamin, J., Diehl, D., Meyer, M., David, J., Schaumann, G.E. *et al.* (2022) Physico-chemical soil properties affected by invasive plants in Southwest Germany (Rhineland-Palatinate)—a case study. *Soil Systems* 6(4), 93. DOI: 10.3390/soilsystems6040093.

Jesus, J.G., Anjos, A., Máguas, C. and Trindade, H. (2025) Unraveling the microbiome dynamics of the invasive *Acacia longifolia*: A closer look at seeds and nodules. *NeoBiota* 99, 93–108. DOI: 10.3897/neobiota.99.144628.

Jin, H., Chang, L., van Kleunen, M. and Liu, Y. (2022) Soil mesofauna may buffer the negative effects of drought on alien plant invasion. *Journal of Ecology* 110(10), 2332–2342. DOI: 10.1111/1365-2745.13950.

Kowalski, T. and Bilański, P. (2024) Recognition of *Davidsoniella virescens* on *Fagus sylvatica* wood in Poland and assessment of its pathogenicity. *Journal of Fungi* 10(7), 465. DOI: 10.3390/jof10070465.

Lee, M.R., Flory, S.L. and Phillips, R.P. (2012) Positive feedbacks to growth of an invasive grass through alteration of nitrogen cycling. *Oecologia* 170(2), 457–465. DOI: 10.1007/s00442-012-2309-9.

Li, J., He, J.-Z., Liu, M., Yan, Z.-Q., Xu, X.-L. *et al.* (2024) Invasive plant competitivity is mediated by nitrogen use strategies and rhizosphere microbiome. *Soil Biology and Biochemistry* 192, 109361. DOI: 10.1016/j.soilbio.2024.109361.

Lin, W., Lu, J., Yao, H., Lu, Z., He, Y. *et al.* (2021) Elevated pCO_2 alters the interaction patterns and functional potentials of rearing seawater microbiota. *Environmental Pollution* 287, 117615. DOI: 10.1016/j.envpol.2021.117615.

Linders, T.E.W., Schaffner, U., Eschen, R., Abebe, A., Choge, S.K. *et al.* (2019) Direct and indirect effects of invasive species: Biodiversity loss is a major mechanism by which an invasive tree affects ecosystem functioning. *Journal of Ecology* 107(6), 2660–2672. DOI: 10.1111/1365-2745.13268.

Magar, D.G. (2024) Effect of *Ageratina adenophora* invasion on litter decomposition, fine root production and mycelial growth of selected saprophytic fungi. Doctoral dissertation, Tribhuvan University Kathmandu, Nepal. Available at: https://hdl.handle.net/20.500.14540/23825

Nunez-Mir, G.C., Guo, Q., Rejmánek, M., Iannone, III, B.V. and Fei, S. (2019) Predicting invasiveness of exotic woody species using a traits-based framework. *Ecology* 100(10), e02797. DOI: 10.1002/ecy.2797.

Núñez-Mir, G.C. and McCary, M.A. (2024) Invasive plants and their root traits are linked to the homogenization of soil microbial communities across the United States. *Proceedings of the National Academy of Sciences USA* 121(44), e2418632121. DOI: 10.1073/pnas.2418632121.

Peter, A., Žlabur, J.Š., Šurić, J., Voća, S., Purgar, D.D. *et al.* (2021) Invasive plant species biomass-evaluation of functional value. *Molecules (Basel, Switzerland)* 26(13), 3814. DOI: 10.3390/molecules26133814.

Shahrtash, M. and Brown, S.P. (2021) A path forward: Promoting microbial-based methods in the management of invasive plants. *Plants* 10(5), 943. DOI: 10.3390/plants10050943.

Shen, M.-C., Zhang, Y.-Z., Bo, G.-D., Yang, B., Wang, P. *et al.* (2022) Microbial responses to the reduction of chemical fertilizers in the rhizosphere soil of flue-cured tobacco. *Frontiers in Bioengineering and Biotechnology* 9, 812316. DOI: 10.3389/fbioe.2021.812316.

Shi, Y., Delgado-Baquerizo, M., Li, Y., Yang, Y., Zhu, Y.-G. *et al.* (2020) Abundance of kinless hubs within soil microbial networks are associated with high functional potential in agricultural ecosystems. *Environment International* 142, 105869. DOI: 10.1016/j.envint.2020.105869.

Singh, G., Rosman, B., Byrne, M.J. and Reynolds, C. (2025a) An earth observation and explainable machine learning approach for determining the drivers of invasive species - a water hyacinth case study. *Environmental Monitoring and Assessment* 197(11), 1172. DOI: 10.1007/s10661-025-14517-1.

Singh, M., Chauhan, A. and Singh, P.K. (2025b) Impacts of invasive plant species on arbuscular mycorrhizal fungi: Consequences for ecosystem functions and plant communities. *International Journal of Environmental Studies* 82(4), 1526–1542. DOI: 10.1080/00207233.2025.2506345.

Souza-Alonso, P., Guisande-Collazo, A., Lechuga-Lago, Y. and González, L. (2024) Changes in decomposition dynamics, soil community function and the growth of native seedlings under the leaf litter of two invasive plants. *Biological Invasions* 26(11), 3695–3714. DOI: 10.1007/s10530-024-03405-3.

Torres, N., Herrera, I., Fajardo, L. and Bustamante, R.O. (2021) Meta-analysis of the impact of plant invasions on soil microbial communities. *BMC Ecology and Evolution* 21(1), 172. DOI: 10.1186/s12862-021-01899-2.

Unger, I.M., Kremer, R.J., Veum, K.S. and Goyne, K.W. (2022) Immediate and long-term effects of invasive plant species on soil characteristics. *Soil Ecology Letters* 4(3), 276–288. DOI: 10.1007/s42832-021-0104-4.

Wang, C., Wang, W.Q., Sardans, J., Ouyang, L.M., Tong, C. *et al.* (2020) Higher fluxes of C, N and P in plant/soil cycles associated with plant invasion in a subtropical estuarine wetland in China. *Science of the Total Environment* 730, 139124. DOI: 10.1016/j.scitotenv.2020.139124.

Wang, X., Wang, X., Wang, W., Wang, J. and Yu, F. (2022) Effects of invasive plant diversity on soil microbial communities. *Diversity* 14(11), 992. DOI: 10.3390/d14110992.

Wardle, D.A. and Peltzer, D.A. (2017) Impacts of invasive biota in forest ecosystems in an aboveground–belowground context. *Biological Invasions* 19(11), 3301–3316. DOI: 10.1007/s10530-017-1372-x.

Weidlich, E.W.A., Flórido, F.G., Sorrini, T.B. and Brancalion, P.H.S. (2020) Controlling invasive plant species in ecological restoration: A global review. *Journal of Applied Ecology* 57(9), 1806–1817. DOI: 10.1111/1365-2664.13656.

Xu, H.W., Liu, Q., Wang, S.Y., Yang, G.S. and Xue, S. (2022) A global meta-analysis of the impacts of exotic plant species invasion on plant diversity and soil properties. *Science of The Total Environment* 810, 152286. DOI: 10.1016/j.scitotenv.2021.152286.

Yu, H., He, Y., Zhang, W., Chen, L., Zhang, J. *et al.* (2022) Greater chemical signaling in root exudates enhances soil mutualistic associations in invasive plants compared to natives. *The New Phytologist* 236(3), 1140–1153. DOI: 10.1111/nph.18289.

Zhang, Z.L. and Suseela, V. (2021) Nitrogen availability modulates the impacts of plant invasion on the chemical composition of soil organic matter. *Soil Biology and Biochemistry* 156, 108195. DOI: 10.1016/j.soilbio.2021.108195.

Zhang, G., Bai, J., Tebbe, C.C., Huang, L., Jia, J. *et al.* (2022) Plant invasion reconstructs soil microbial assembly and functionality in coastal salt marshes. *Molecular Ecology* 31(17), 4478–4494. DOI: 10.1111/mec.16600.

Zhang, H.-Y., Goncalves, P., Copeland, E., Qi, S.-S., Dai, Z.-C. *et al.* (2020) Invasion by the weed *Conyza canadensis* alters soil nutrient supply and shifts microbiota structure. *Soil Biology and Biochemistry* 143, 107739. DOI: 10.1016/j.soilbio.2020.107739.

Zhang, M., Shi, C., Li, X., Wang, K., Qiu, Z. *et al.* (2023) Changes in the structure and function of rhizosphere soil microbial communities induced by *Amaranthus palmeri* invasion. *Frontiers in Microbiology* 14, 1114388. DOI: 10.3389/fmicb.2023.1114388.

Zhao, M., Yang, Y., Zhang, H., Li, Q., Zhao, X. *et al.* (2024) Asymmetric succession in soil microbial communities enhances the competitive advantage of invasive alien plants. *Microbiome* 12(1), 265. DOI: 10.1186/s40168-024-01989-5.

9 Interactions of Soil Hydrobionts (Nematoda) with Invasive Plants in their Introducer Ranges

Marek Renčo*
Laboratory of Plant Nematology, Institute of Parasitology, Slovak Academy of Sciences, Košice, Slovakia

Introduction

The study of patterns in distribution, diversity and abundance of species in relation to environmental variables in nature and faunal and floral species interactions has had a long tradition in community ecology (Brussaard, 1998; Crowther *et al.*, 2019). As reported by MacDougall and Turkington (2005) the current rate of non-native species invasions to new localities across and between continents is the highest in history. Research has revealed that 37% of all first records of invasive species were reported within the period 1970–2014 (Seebens *et al.*, 2017). Of these organisms, an important role is played by invasive plants, which can wreak havoc in an ecosystem. According to Pyšek *et al.* (2020) the numbers of invasive alien species – the subset of alien species that spread widely in areas where they are not native, affecting the environment or human livelihood – are constantly increasing. Synergies with other global changes (global heating) are exacerbating current invasions and facilitating new ones, thereby escalating the extent and impacts of invaders (Pyšek *et al.*, 2020).

The generally accepted definition of an invasive plant is as an 'introduced species' that must be able to establish self-sustaining populations and produce significant changes in the composition, structure or ecosystem processes of natural or semi-natural vegetation (Hughes, 1994; Cronk and Fuller, 1995). Alien plants have several biological traits that allow them to outcompete with native species: (1) the rapid germination of seedlings and quick growth into reproductive adults shortens generation time and allows rapid production of the next generation (Baker, 1974); (2) yearly high seed production allows the survival and establishment of sufficient numbers of new individuals on a site (Anderson *et al.*, 1996); (3) the ability to alter light availability and other resources needed for native species growth (Reinhart *et al.*, 2006); (4) vegetative (clonal) growth allows many invasive species to reproduce and survive in the absence of suitable pollinators and thus cover a site efficiently and quickly (Pyšek, 1997); and (5) alien species can grow and form monospecific communities under a variety of biotic conditions due to high plasticity (Droste *et al.*, 2010).

Without a doubt, invasive alien species are considered to be one of the leading threats to the ecological integrity of native flora and fauna by

*Corresponding author: renco@saske.sk

DOI: 10.1079/9781836991298.0009

establishment of monospecific stands (Jose *et al.*, 2013). Some of these species become extremely abundant in their new range, especially in habitats that have ideal conditions for their establishment (Richards *et al.*, 2006). Many invasive alien species can change ecosystem functioning and the delivery of ecosystem services by altering nutrient and contaminant cycling, hydrology, habitat structure and disturbance regimes. These biodiversity and ecosystem impacts are accelerating and will increase further in the future (Pyšek *et al.*, 2020).

The Role of Plants in the Ecosystem

Soil ecosystem functions depend on the development, activity and maintenance of soil biology which in turn depend on soils dwelling biota, small invertebrates, insects, macro- or microorganisms and plants (Chen *et al.*, 2007; Blankinship *et al.*, 2011; Van der Putten *et al.*, 2013; Ponge, 2015; Dervash *et al.*, 2018; Vezzani *et al.*, 2018). However, plants are clearly considered the primary source of organic matter, supporting soil microbial activity and thus maintaining many important functions of the soil ecosystem (Ponge, 2015; Vezzani *et al.*, 2018). During growth, plants draw nutrients from the soil and return them to it through litter fall, root exudates, mucilage and root remnants, thus maintaining the cycle of nitrogen, phosphorus, carbon and other elements. Through photosynthesis, plants convert CO_2 into sugars and other organic substances, which become humus after parts of the plants (leaves, roots) die off through the process of decomposition (Van der Putten *et al.*, 2013).

Some substantial soil physical and chemical properties directly and indirectly influence each other and can be affected by plants, such as soil moisture, temperature, soil acidity and nutrient stock. Conversely these factors have an impact on plant growth, survival, diversity and distribution in space. Soil moisture directly affects plant establishment and growth, leaf phenology and competitive interactions and succession processes (Asbjornsen *et al.*, 2011). Plants play an important role in the water cycle since they influence soil moisture heterogeneity and complexity over time and space, which alters the hydrological cycle in many aspects (Wang *et al.*, 2019). Breshears *et al.* (1998) quantified the effects of woody plants on the microclimate with respect to soil temperature and evaporation. Results showed higher temperatures, higher evaporation rates and lower soil moisture in inter-canopy than canopy patches during warmer months, all of which in turn negatively affect the germination and plant-available water. In addition, soil temperature is one of the pivotal factors regulating nutrient availability and uptake by plants. Changes in soil temperature are unavoidably linked to changes in soil moisture at the ecosystem level. Changes in plant and microbial activities are very tightly linked to changes in soil temperature and moisture and these micro-ecological interactions play crucial roles in nutrient availability and net ecosystem productivity (Mondal *et al.*, 2016).

The Role of Soil Organisms in the Ecosystem

Soil, a non-renewable resource critical for human well-being (Delgado-Baquerizo *et al.*, 2025), supports a vast array of organisms, called the edaphon. The edaphon comprises all organisms living in the soil, which can be divided based on body diameter into microflora (bacteria, archaea, fungi and algae), microfauna (e.g. protists and Nematoda), mesofauna and macrofauna (Paul, 2015). Microflora and microfauna form the community of microorganisms in soils whose life cycle stages all take place in the soil, and are therefore permanent soil organisms (Blume *et al.*, 2016). They are generally well adapted to occasional desiccation and/or shortage of food in their micro-environment. The spatial scale at which they operate is that of a few millimetres and their generation times are of the order of a few hours to a few days.

Mesofauna include Microarthropoda and Oligochaeta (Enchytraeidae), with bodies 0.1–2 mm wide, allowing them to move freely in the extensive network of pores in the surface litter and soil (Lavelle and Spain, 2001). They operate over a range of space in the order of centimetres and over weeks to months. Macrofauna, larger invertebrates, are much wider than most soil pores; they live in surface litter (e.g. most arthropods), move the soil around and create tunnels that alter soil in a beneficial way (most earthworms, termites or endogenic larvae of Coleoptera). Average time and space scales covered are months to years and decimetres to metres (Lavelle and Spain, 2001). Together these organisms play important roles in nutrient cycling, pollination and organic matter decomposition, making soil a living part of our planet.

However, some taxa are known to play a more important role than others in maintaining soil multifunctionality. Larger organisms (such as annelids, tardigrades, arthropods and flatworms) are responsible for processing large amounts of plant and animal detritus and might ultimately determine the availability of fresh resources in the soil food web (Delgado-Baquerizo *et al.*, 2020). These organisms act as a manufacturing line that processes detritus and infuses the soil with physically smaller and chemically decomposed resources. It is therefore assumed that the diversity of these soil invertebrates might play critical roles in supporting multiple functions (that is, rates and availabilities) operating at high levels of functioning (Byrnes *et al.*, 2014). On the other hand, the biodiversity of soil microorganisms (such as protists, bacteria and fungi) can be essential for the maintenance of multiple functions and energy flow in the soil but they are still dependent on the activities of larger organisms. It is therefore assumed that the smallest soil organisms are responsible for bottom-up (producers) and top-down (consumers) energy through activation of nutrients from the soil and through predation; the recirculation of energy from larger organisms to smaller ones through microbial loops (Bonkowski, 2004; Geisen, 2016). In other words, these soil organisms recirculate available resources in the soil, thereby ensuring the functioning of terrestrial ecosystems.

Nematodes

The hydrobionts are aquatic organisms which live and feed in the free water and soil (Lavelle and Spain, 2001). Nematodes, perfectly adapted morphologically to living in edaphic and aquatic habitats, are considered the most numerous and diverse hydrobionts in the soils globally, with densities up to millions of individuals per square metre (van den Hoogen *et al.*, 2019).

The species found in soil are usually small, live in soil capillaries and detritus particles, and feed on a wide range of substrata, including other nematodes preyed upon by many species of the same group. Their size rarely exceeds 5 mm but most are only a few millimetres long (0.2–2 mm). It is estimated that approximately one in every five animals in the world is a nematode (Orgiazzi *et al.*, 2016). As a species-rich and abundant component of the belowground system, soil nematodes occupy a wide range of trophic positions and contribute significantly to soil food-web interactions (Ferris *et al.*, 2001).

Most nematodes are free-living, encompassing bacterivores, fungivores, algal feeders, predators and omnivores (Yeates *et al.*, 1993), but a special group comprises obligate and facultative plant parasitic nematodes (Yeates *et al.*, 1993; Perry and Moens, 2006) that altogether co-occur in soil. In natural ecosystems, the nematodes contribute to spatial and temporal diversity in plant communities

Bacterivores

Bacteria-feeding nematodes have an oral cavity opening into a muscular, sucking pharynx; they suck in their food by contracting the pharynx (Fig. 9.1). They are able to migrate to places of high bacterial and protozoan activity (Griffiths and Caul, 1993) where they aggregate and are important grazers of both bacteria and amoebae and eventually become the dominant predators (Elliott *et al.*, 1980; Rønn *et al.*, 1996).

Bacterivorous nematodes are known to contribute significantly to mobilize nutrients and accelerate mineralization processes in the soil. For example, bacterivorous nematodes may affect nitrogen mineralization, an important ecosystem process, directly by the excretion of nitrogen and indirectly by grazing on bacteria and protozoa, thereby stimulating bacterial turnover and production (Freckman, 1988; Ferris *et al.*, 1998; Postma-Blaauw *et al.*, 2005). Thus, their indirect contributions to nutrient cycling may be even more important than their direct effects because grazing stimulates microbial mineralization processes.

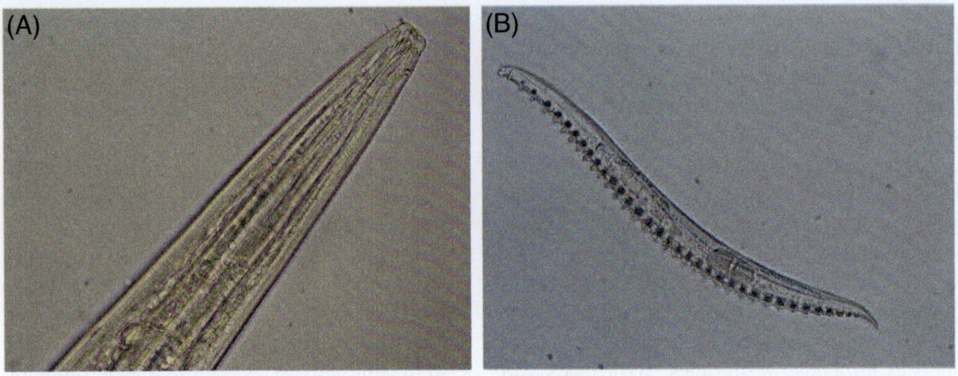

Fig. 9.1. Bacterivorous nematodes of the genus *Plectus* (A) and *Bunonema* (B).

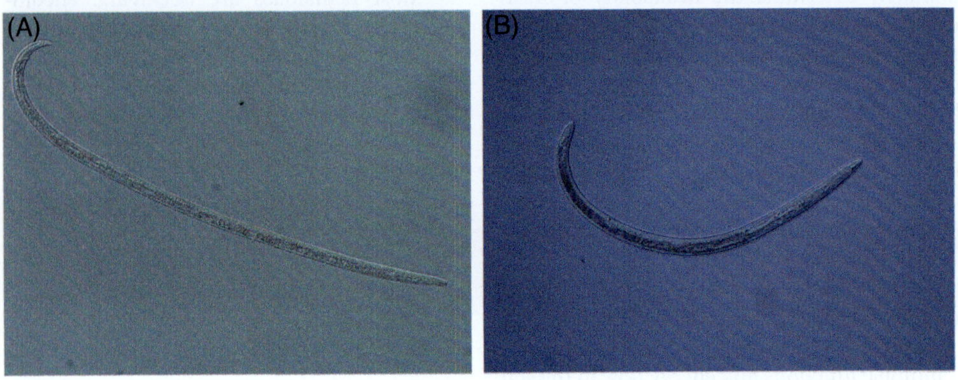

Fig. 9.2. Fungivorous nematodes of the genus *Aphelenchoides* (A) and *Tylencholaimus* (B).

Fungivores

Fungivores use their stylets to pierce fungal hyphae (Fig. 9.2). Generally, fungivores are polyphagous species, feeding on a wide range of species of soil fungi including saprophytic, pathogenic, beneficial and mycorrhizal fungi (Ruess and Dighton, 1996). By feeding on fungi, they regulate fungal populations and contribute to nutrient cycling (Okada *et al.*, 2005).

By consuming fungal hyphae and spores, they help maintain a balance between beneficial and pathogenic fungi in the soil. As they digest fungi, fungivorous nematodes release nutrients such as nitrogen and phosphorus in plant-available forms, enhancing soil fertility (McGonigle, 1995). They indirectly affect the decomposition of organic matter by modifying fungal community structure, which can accelerate or alter decomposition processes. By feeding on pathogenic fungi, they can reduce the severity of certain plant diseases, acting as a natural form of biocontrol (Okada, 2006; Lagerlöf *et al.*, 2011)

Omnivores

Omnivorous nematodes occupy mid- to upper-level positions in the soil food web, and have a stylet- or odontostyle-based feeding apparatus (Fig. 9.3) but with modifications that allow them to feed on both plant material and small invertebrates including other nematodes, fungi and bacteria (Potapov *et al.*, 2022). Their feeding structures are more complex than those of strictly bacterivorous or plant parasitic

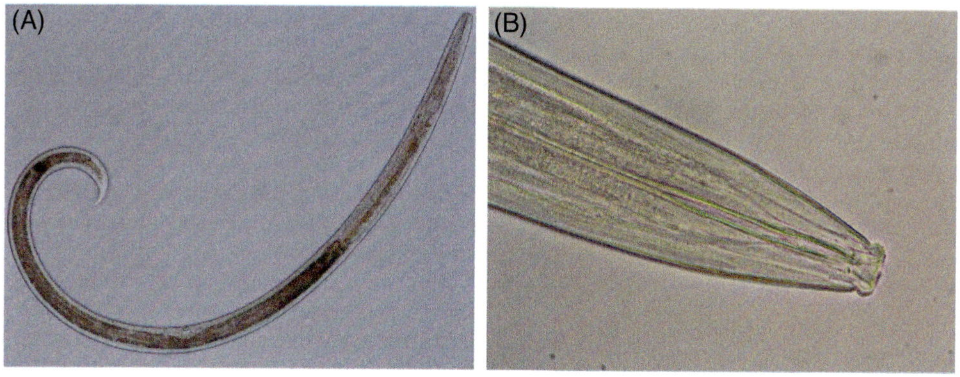

Fig. 9.3. Omnivorous nematodes of the genus *Eudorylaimus* (A) and *Aporcelaimus* (B).

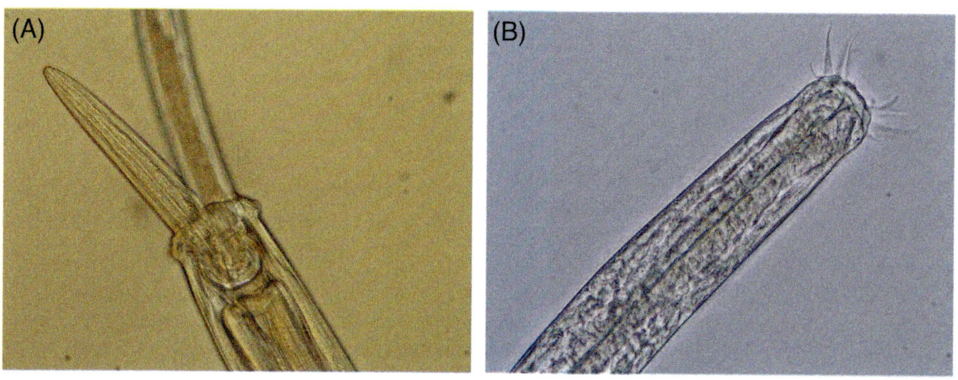

Fig. 9.4. Predatory nematodes of the genus *Anatochnus* (A) and *Tripylina* (B).

nematodes. Their diverse diet allows them to buffer fluctuations in the food web by switching food sources depending on availability, as well as control populations of other soil organisms, including plant parasitic nematodes, helping to protect plant roots (Perry and Moens, 2006). Due to their sensitivity to changes in soil conditions and pollutants, the presence and abundance of omnivorous nematodes can serve as a bioindicator of soil quality and ecosystem stability (Bongers and Bongers, 1998; Ferris *et al.*, 2001)

Predators

Predatory nematodes belong to one of four orders of the Nematoda; Mononchida (Ahmad and Jairajpuri, 2010), Dorylaimida (Jairajpuri and Ahmad, 1992); Diplogasterida (Bilgrami, 1990) and Aphelenchida (Kim, 2015). They are equipped with different types of feeding apparatus, with distinct feeding mechanisms and food sources (Kanwar *et al.*, 2021) (Fig. 9.4). Their main prey are nematodes, although they also feed upon nematode eggs, rotifers, oligochaetes, enchytraeids, tardigrades and arthropods. Therefore, predators are frequently studied as biocontrol agents in the management of plant pests, including plant parasitic nematodes (Khan and Kim, 2007). They reduce populations of plant parasitic nematodes in virtually all soils because of their constant association with plant parasitic nematodes in the rhizosphere, and also release nutrients in plant-available forms, which may enable plants to better withstand nematode burden on their roots (Stirling, 2014). They are usually among the least represented trophic groups within soil nematode communities, in terms of both abundance and species diversity. Similar as omnivores, with the higher sensitivity of predators to changes in soil conditions and pollutants, their presence and abundance can serve as bioindicators of soil quality and ecosystem stability (Bongers and Bongers, 1998; Ferris *et al.*, 2001).

Plant parasites

The feeding strategy of plant parasitic nematodes disrupts plant function and facilitates disease (Perry and Moens, 2006). We recognize two main groups of plant parasites. Ectoparasitic nematodes remain outside the roots to feed externally, puncturing individual cells with their stylet, extracting food and nutrients for short periods and then withdrawing the stylet to move on to another cell to repeat the process (Fig. 9.5). Endoparasitic nematodes invade the host plant and migrate through host tissues, spending most of their life inside bulbs, corms, roots, tubers, galls and cysts; hence they can be unknowingly spread and escape detection (Nicol *et al.*, 2011). They cause crop damage leading to stunted growth, wilting, yellowing of leaves, root galls and poor yield. Damage by nematodes can make plants more vulnerable to bacterial and fungal pathogens, acting as entry points for secondary infections. The estimated global crop losses caused by plant parasitic nematodes amount to over US$100 billion annually (Nicol *et al.*, 2011). Actual losses may be even higher because there are no data from many countries where nematological expertise is lacking. Moreover, yield loss data are also difficult to obtain because of the complex interactions of plants, nematodes, other soil organisms and soils (Singh *et al.*, 2013).

While plant parasitic nematodes have mainly economic importance, they are often dominant in the nematode community (Valocká *et al.*, 2001; Renčo, 2003; Viketoft *et al.*, 2005). Like other groups of soil nematodes, parasitic nematodes are an important component of the biomass of soil fauna, a food source for other organisms including nematodes (predators) and a source of nutrients (Khan and Kim, 2007; Kanwar *et al.*, 2021).

Nematode Communities as Bioindicators and the Indices used for Measurement

The scientific discipline of indicator (or bioindicator) development is clearly becoming a more important part of environmental study (Burger, 2006). Since the early 1970s, when only one or two papers a year self-identified as indicator papers, the number has grown steadily. Partly this is a function of the need to assess environmental changes in a manner

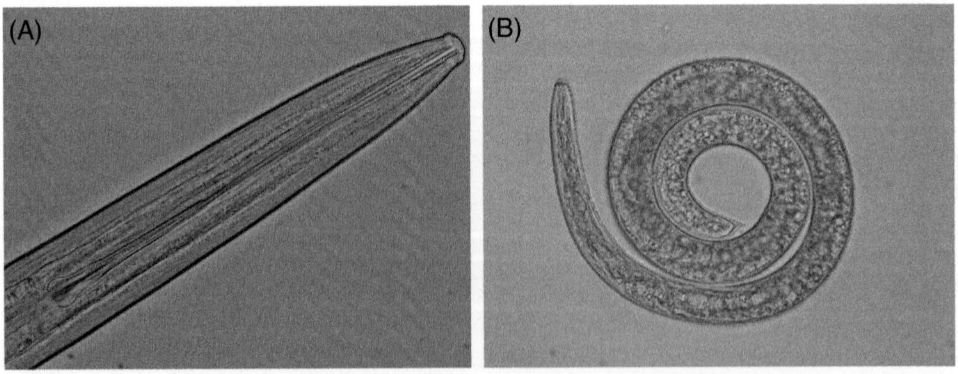

Fig. 9.5. Plant parasitic nematodes of the genus *Xiphinema* (A) and *Helicotylenchus* (B).

that can be repeated regularly and will provide early warning of future risk (Burger, 2006). In a general ecological sense bioindicators, according to Fränzle (2006), are organelles, organisms or groups of organisms suited to determine qualitatively or quantitatively the state of the environment. However, the organisms to be used as bioindicators must be widespread and abundant, easily collected and identified with known biology, ecology, specificity and ecological relevance.

Soil nematodes meet the following criteria. Nematodes are ubiquitous, extremely abundant and diverse and are therefore important biological components in the soil ecosystem. Soil nematodes have transparent bodies, so their diagnostic internal features can be seen without dissection, and their life course is short. In addition, transparent bodies allow individuals to be identified microscopically to the genus and species level. Based on mouth and body morphology nematodes can be divided to various trophic groups. These features enhance their bioindicator potential and provide a cost-effective, ecologically meaningful way for assessment of environmental health (Bongers and Bongers, 1998; Wilson and Kakouli-Duarte, 2009).

Various descriptive and evaluative indicators of soil nematode communities are used to determine the state of the environment, qualitatively or quantitatively. Of the descriptive indicators, number of species/genera, total nematode abundance and total/relative abundance of nematodes per trophic group are widely used. Of the evaluative indicators, some ecological indices developed specifically for soil nematodes have been proven to be useful tools for evaluating soil conditions (Zhang *et al.*, 2012).

As the basic nematode ecological index the maturity index for free-living taxa (MI) was designed by Bongers (1990) to monitor colonization and subsequent development of the nematode fauna. The maturity index is calculated according to the formula:

$$\mathrm{MI} = \frac{\sum(\nu_i \times f_i)}{n}$$

where ν_i is the colonizer-persister (c-p) value of taxon i, f_i is the frequency of taxon i in a sample and n is the total number of individuals in a sample.

In addition, the plant parasite index (PPI) for plant parasitic taxa (Bongers, 1990) is used to monitor pests' pressure and prevalence of obligate and facultative plant parasites under various conditions; the maturity index MI(2-5) (Korthals *et al.*, 1996) indicates disturbances resulting from chronic soil pollution, such as heavy metals, but excludes the enrichment (see EI, below) opportunistic nematodes (c-p1) in the calculation; and the sigma maturity index (ΣMI), includes all nematode taxa (Yeates, 1994). All these indices are calculated similarly to basic MI, using c-p values for nematode taxa that represent the life history characteristics associated with *r* and *K* selection. Species with c-p values of 1 or 2 are *r*-selected; that is, colonizers. These species are very tolerant to disturbances due to their short generation times, large population fluctuations and high fecundities. Species with a c-p value of 5 are *K*-selected, or persisters, with long life cycles, low reproductive rates, low metabolic activities and slow movement; they are thus very sensitive to disturbances.

Ferris *et al.* (2001) expanded Bongers' concept of the maturity index to a weighted faunal analysis to allow an extended assessment of soil food webs through the analysis of nematode communities. Nematode taxa are assigned to functional guilds, where all nematodes in a functional guild have the same feeding habit (Yeates *et al.*, 1993) and the same c-p values (e.g. the functional guild Ba_3 includes all bacterivores with a c-p value of 3).

These functional guilds are indicators of *basal* (b), *structured* (s) and *enriched* (e) soil food webs. An enrichment index (EI) and structure index (SI) can be calculated using an indicator weighting system based on the importance of the functional guilds along hypothesized trajectories of enrichment and structure (Ferris *et al.*, 2001). The EI provides location of the food web along the enrichment trajectory and is calculated as:

$$\mathrm{EL} = 100 \times \frac{e}{(e + b)}$$

Similarly, the SI provides location of the food web along the structure trajectory and is calculated as:

$$\mathrm{SI} = 100 \times \frac{s}{(s + b)}$$

For example, *b* component is calculated as $\Sigma k_b n_b$, where k_b are the weightings assigned to guilds that indicate basal characteristics of the food web (Ba_2, Fu_2) and n_b are the abundances of nematodes in sample. The *e* and *s* components are calculated similarly, using those guilds indicating enrichment (Ba_1, Fu_2) and structure (Ba_3–Ba_5, Fu_3–Fu_5, Om_3–Om_5, Ca_2–Ca_5), respectively. For more details see the original work (Ferris *et al.*, 2001).

EI is based on the expected responsiveness of the opportunistic guilds (bacterivorous nematodes with a c-p value of 1) to organic-resource enrichment. EI thus indicates if the soil environment is nutrient-enriched (high EI) or -depleted (low EI). SI represents an aggregation of functional guilds with c-p values of 3–5 and indicates if the soil ecosystem is structured with more trophic links (high SI) or degraded with fewer trophic links (low SI). Low to moderate disturbances tend to have soils with high EIs and SIs, a high level of disturbance tends to have soils with a high EI and a lower SI, undisturbed soil with moderate levels of nutrient enrichment tends to have soils with a low EI and a higher SI and stressed systems with relatively depleted nutrients tend to have soils with both a low EI and a low SI (Ferris *et al.*, 2001).

Additionally, the channel index (CI) indicates the predominant decomposition channel in the soil food web, calculated as:

$$CI = \frac{0.8\ Fu_2}{(3.2\ Ba_1 + 0.8\ Fu_2)}$$

where Ba_1 = abundance of nematode families Rhabditidae, Panagrolaimidae and Diplogasteridae in a sample and Fu_2 = abundance of nematode families Aphelenchidae, Aphelenchoididae and Anguinidae. A high CI (>50%) indicates a higher proportion of fungal decomposition (fungal decomposition channels), a high relative abundance of c-p2 fungal feeders (Aphelenchoididae and Aphelenchidae) and a correspondingly low abundance of c-p1 bacterial feeders (e.g. Rhabditidae and Panagrolaimidae). A low CI (< 50 %) suggests bacterial decomposition channels (Ferris *et al.*, 2001).

Analysis of nematode functional metabolic footprints provides another metric for the magnitude of ecosystem functions and services provided by component organisms of the soil food web (Ferris, 2010). The footprint has a production (*P*) and a respiration (*R*) component. *P* is the lifetime amount of carbon partitioned into growth and egg production and *R* assesses carbon use in metabolic activity. The footprint has various forms: (1) enrichment footprint, the metabolic footprint of those nematodes most rapidly responsive to resource enrichment, (2) structure footprint, the metabolic footprint of higher trophic levels that may have regulatory functions in the food web and that are indicative of the abundance of organisms of similar non-nematode taxa, (3) functional footprint, total area of the two functional footprints (enrichment and structure) and (4) composite footprint, the metabolic footprint of the complete nematode assemblage, regardless of trophic role or ecosystem function. The expanded equation for the metabolic footprint of nematodes (*F*), as an index of the carbon use of the component taxa, is a sum of the production and respiration components:

$$F = P + R$$

The metabolic footprint is depicted graphically. The point in the centre of a rhombus represents the intersection of EI and SI, and the lengths of the vertical and horizontal axes of the rhombus correspond to the footprints of the enrichment and structure components, respectively (Ferris *et al.*, 2001). Dotted lines on both sides of the rhombus indicate the standard deviation, and solid lines represent the mean values for a treatment. The metabolic footprint is maximal when the rhomboid becomes a square, indicating that the productivity and turnover rates of the enrichment indicators, representative of the prey, are sufficient to maintain the needs of the predators (the structure indicators) so that the system is metabolically balanced.

Besides the above-mentioned nematode community indicators, several other nematode specific indices and ratios have been proposed and described. The Wasilewska index ($WI = F + B/PP$) is the ratio of fungivores + bacterivores to plant parasites which describes the relative balance of positive-to-negative impacts of nematodes on primary productivity or stage of decomposition (Wasilewska, 1994). A ratio greater than 1 suggests that the positive impact of nematodes outweighs the negative impact of plant parasitises on plant productivity. The nematode channel ratio (NCR = $B/B + F$), where B and F are the numbers of bacterivores and fungivores in the total soil nematode population (Yeates, 2003), assesses the relative activity of the bacterium-based energy channel (pathway) and slower fungus-based channel in decomposition processes in soil. This index is constrained to have values between 1 (totally bacterium-mediated decomposition) and 0 (totally fungus-mediated decomposition).

Finally, various indices and mathematical expressions are used to quantify the biodiversity and species richness of the nematode community. The Shannon–Weaver index of diversity ($H' = -\sum p_i \ln p_i$) where p_i is the proportion of individuals in the ith taxon, is a species diversity measure that gives more weight to rare species (Shannon and Weaver, 1949). The Simpson diversity index ($D = 1/\sum p_{i^2}$) gives more weight to common species, in which p_i is the proportion of each taxon in the total population (Simpson, 1949). The Pielou evenness index ($J' = H'/\log(S)$) where H' is the Shannon–Weaver diversity index and S is the number of species in the community, estimates evenness of species distribution (Pielou, 1966). The Margalef index (SR = $(S - 1) / \ln N$), where N is total nematode abundance, estimates species richness (Margalef, 1958).

Factors Modifying Nematode Community Composition

Correlations between aboveground and belowground organisms can be positive, negative or non-significant. Plant composition, identity and diversity (Viketoft *et al.*, 2005; Eisenhauer *et al.*, 2011; Dietrich *et al.*, 2021), as well as various soil physical, chemical and microbial properties (Griffiths *et al.*, 2003; Briar *et al.*, 2007; Nisa *et al.*, 2021), all shape the soil nematode communities in natural and semi-natural ecosystems. On the other hand, soil nematode communities can negatively reflect the disruption of their living space from top to bottom or vice versa by various natural and human-induced effects. Extensive research has indicated that wildfires (Bastow, 2020; Gongalsky *et al.*, 2021; Renčo *et al.*, 2022a), wind and storms (Čerevková and Renčo, 2009; Renčo *et al.*, 2019a), clearcutting and harvesting (Panesar *et al.*, 2000; Háněl, 2004), soil degradation by heavy metal pollutions (Renčo *et al.*, 2022b), agricultural practices (tillage) and chemical fertilizers (Okada and Harada, 2007; Hu and Cao, 2008; Pothula *et al.*, 2022) have, in most cases, a negative impact on nematode community structure.

One factor, however, has a partly natural as well as a human cause: the spread of non-native plant species into new regions. The widespread introduction of species into ecosystems has prompted important research on their ecology, dispersal mechanisms and management. The list of ecological (environmental) impacts of invasive plant establishment is long and relatively well recorded, and is considered as evidence for ranking invasive species as one of the 'big five' environmental issues of the 21st century (Sala *et al.*, 2000). However, it is the ecological impacts that are the most difficult to quantify, integrate and rank (Vilà *et al.*, 2011; Barney *et al.*, 2013), but naturally occurring bioindicators can be used to assess the health of the environment, detecting changes in the environment, either positive or negative (Holt and Miller, 2011).

Soil Nematode Communities under Invasive Plant Stress

The main threat to global biodiversity is direct anthropogenic degradation of natural habitats due to inappropriate resource use or pollution. However, another serious and often unappreciated problem is the widespread and often permanent invasion of natural and semi-natural habitats by alien organisms (Cronk and Fuller, 1995). According to van Kleunen *et al.* (2015), around 13,000 plant species are growing outside their native ranges, creating their own populations, but only some of this number are considered to be invasive organisms. The list of invasive plant species differs from region to region and from country to country; however, some are considered highly widespread nowadays. Among them, *Heracleum sosnowskyi* Manden, *Heracleum mantegazzianum* Sommier and Levier, *Fallopia (Reynoutria) japonica* (Houtt.) Ronse Decr, *Solidago gigantea* Ait., *Solidago canadensis* L., *Asclepias syriaca* L., *Impatiens*

Fig. 9.6. The most serious invasive weeds in Slovakia: (A) *Heracleum mantegazzianum*, (B) *Asclepias syriaca*, (C) *Solidago gigantea* and (D) *Fallopia japonica*.

parviflora DC. and *Impatiens glandulifera* Royle belong to the most invasive and spreading herbs in Europe (DAISIE, 2020), while dozens of others are invasive around the globe (Fig. 9.6).

The interest of scientists in their impacts on native vegetation and above- and belowground biota, including soil nematodes, has therefore increased in recent decades. Our research group has also investigated the impact of the most widespread but hitherto little-studied invasive herbaceous species on soil nematofauna, as well as the physical and microbial properties of soil and native vegetation in recent years.

For example, invasion impacts of *H. sosnowskyi* in various habitats, namely abandoned land, grassland on a roadside slope, the edges of afforested land in Lithuania (Renčo and Baležentienė, 2015), a wet dump ground depression, a non-cultivated line between two agricultural fields, a road edge leading to farm, an occasionally inundated abandoned alluvial meadow in the Poland (Renčo *et al.*, 2019b) and native forest or grassland in Russia (Čerevková *et al.*, 2020a) on soil nematode communities have all been investigated. The data showed that the responses of nematode communities to aboveground vegetation changes caused by *H. sosnowskyi* varied slightly, but in general nematode abundance was negatively affected by *H. sosnowskyi* in almost all habitats where the invasive plant had become dominant in the vegetation cover. In contrast, in the habitats where a solitary plant of *H. sosnowskyi* grew no significant changes in nematode abundance were observed. Considering diversity, the studies revealed that *H. sosnowskyi* decreased nematode taxa diversity in Lithuania (Renčo and Baležentienė, 2015), but no significant changes between control and invaded stands in the habitats investigated in Poland and Russia have been observed, disregarding habitat type (Renčo *et al.*, 2019b; Čerevková *et al.*, 2020a). Considering trophic structure, plant parasitic nematodes were the trophic group that differed the most in all habitats, while bacterivorous and fungivorous nematode populations were significantly higher in invaded habitats in Lithuania (Renčo and Baležentienė, 2015). Similarly, plant parasitic nematodes, together with stress-sensitive omnivores and fungivores, best reflected changes in the nematode community trophic structure under the influence of *H. sosnowskyi* invasion in Polish habitats investigated. The negative effect of *H. sosnowskyi* was most evident on *Aphelenchus*, *Tylencholaimus*, *Geocenamus*, *Helicotylenchus*, *Pratylenchus*, *Tylenchorhynchus* and *Aporcelaimellus* (Renčo *et al.*, 2019b). In contrast, however, the numbers of bacterivorous, fungivorous and predaceous nematodes were not affected by *H. sosnowskyi* invasion in the Russian forest and grassland investigated (Čerevková *et al.*, 2020a).

Many years ago, grassland on the riverbanks of the Uh River in Slovakia was occupied by a second highly invasive species of the genus *Heracleum*, namely *H. mantegazzianum*, and its spread is ongoing. Our study (Renčo *et al.*, 2021a) detected significantly fewer nematode species in *H. mantegazzianum*-invaded soil than in plots with natural grassland vegetation, but nematode species diversity was not affected by invasion throughout the two vegetation seasons investigated. Total nematode abundance slightly increased under giant hogweed, while total nematode biomass did not differ between the invaded and uninvaded plots. The higher abundance of bacterivores and fungivores but lower number of omnivorous nematodes illustrated the negative impact of giant hogweed invasion on soil food webs, supported by low values of all maturity indices and the channel index. Surprisingly, the hogweed-invaded plots showed a higher abundance of plant parasitic nematodes, mainly *Paratylenchus microdorus*.

Overall, the negative impact on the structure of nematode communities was also found in the case of Japanese knotweed (*F. japonica*) (Čerevková *et al.*, 2019; Renčo *et al.*, 2021b). This globally successful invasive plant spreads rapidly after introduction, mainly along riparian or road corridors where dense stands often extend over several hundred square metres.

In three natural habitats (forest, grassland and wetland) invaded by *F. japonica* the total number of identified nematode species and their abundance were higher in the uninvaded compared to the invaded plots. A negative effect of *F. japonica* on omnivores, plant parasites and nematodes feeding on root fungi was confirmed by their lower abundance in the invaded compared to the uninvaded plots. In the invaded plots, we also confirmed lower maturity and channel indices, but a higher enrichment index (Čerevková *et al.*, 2019).

Similarly, an unfavourable effect of *F. japonica* expansion on nematode communities was also observed in ruderal forest habitats in Tatra National Park, Slovakia (Renčo *et al.*, 2021b). The study showed a reduction in the number and coverage of native plant species and modification of nematode communities and their structures. Total nematode abundance, species number and nematode biomass were significantly lower in invaded than uninvaded plots, but species diversity remained unaffected throughout the study. The overall abundance of all nematode trophic groups illustrated well the negative impact of *F. japonica* invasion on soil food webs, supported by low values of all maturity indices, a structural index and the Jaccard index of faunal similarity.

In contrast, our research showed that some invasive species had no, or only minimal, impact on the composition of nematode communities and food web conditions in the soil. Moreover, their impact on several measured community parameters was positive and not negative. For example, stands with invasive *S. gigantea*, a widespread and important invasive species in most European countries (Weber, 1998), showed higher soil nematode abundance than uninvaded sites, whereas the number of identified nematode species was lower, indicated by a lower species diversity index (Čerevková *et al.*, 2020b). Herbivorous nematodes were the most affected trophic group. Herbivore abundance was higher in invaded than uninvaded sites and in grassland than forest. The herbivorous species *Boleodorus thylactus*, *Geocenamus* sp., *Helicotylenchus* spp., *Paratylenchus bukowinensis*, *Pratylenchoides crenicauda* and *Rotylenchus robustus* were more abundant at the invaded sites. Abundances of nematodes in the other tropic groups were limited or not affected. The *S. gigantea* invasion did not significantly affect ecological and functional indices, except for the channel index (Čerevková *et al.*, 2020b).

Similarly, research indicates that invasive common milkweed (*A. syriaca*) and small balsam (*I. parviflora*) did not negatively affect soil nematode community structure over a period of two vegetation seasons (Jurová *et al.*, 2020; Renčo *et al.*, 2024). Common milkweed did not decrease the abundance of plant parasites, bacterivores or fungivores and did not affect species diversity. Moreover, basic physicochemical soil parameters were similar in both milkweed-invaded and non-invaded sites, although there is evidence that invasive milkweed considerably decreases native plant species cover. In the case of *I. parviflora* in Carpathian beech forests, overall, nematode abundance and species diversity, relative abundances of fungivores, omnivores and predators as well as nematode biomass did not differ between invaded and non-invaded plots. Invaded plots had a higher abundance of bacterivores and relatively low values of CI and *F*(*F* + *B*) ratio across all seasons, suggesting that *I. parviflora* invasion can have a lasting impact on the decomposition pathway. The lower abundance of plant parasitic nematodes under small balsam in some seasons was probably more a consequence of changes in soil moisture than the presence *I. parviflora* with modest root systems. Moreover, the impact of small balsam on nematode communities did not differ between sampling years and across seasons, indicating a stable and unchanging response to invasive plant growth (Renčo *et al.*, 2024).

Van der Putten *et al.* (2005) tested the hypotheses that in its natural range, the invasive foredune grass *Ammophila arenaria* is associated with more taxa of root-feeding nematodes and more specialist feeders (i.e. sedentary, endoparasite nematodes) than in its introduced range (Mitchell and Power, 2003; the taxon diversity hypothesis) and that in its introduced range, *A. arenaria* has fewer root-feeding nematode taxa than do similar native species (Mitchell and Power, 2003; the native species hypothesis). The results revealed that in most introduced regions *A. arenaria* did not have fewer root-feeding nematode taxa than the average number in its natural range, and native plant species did not have more nematode taxa than the introduced species. This contradicts the findings by Zhang *et al.* (2018), who reported that in the rhizosphere of invasive grass *Spartina alterniflora* plant-feeding nematodes were markedly less abundant compared than in that of native grass *Phragmites australis*, as well as the findings of García-De la Cruz *et al.* (2019) from sites invaded by spotted knapweed (*Centaurea stoebe*) in its invaded range in western North America. Moreover, *S. alterniflora* simplified the soil food web with fewer omnivorous nematodes and lower diversity compared to native *P. australis* (Zhang *et al.*, 2018). The invasion of grass *Bromus tectorum* into stands of the native perennial grass *Hilaria jamesii* resulted in a decrease in abundance and richness and caused a species shift in nematode communities (Belnap *et al.*, 2005). The invasive plant *Heterotheca subaxillaris* had a considerable effect on free-living nematodes, resulting in a decrease in abundance and diversity, and affecting stability, which is mirrored by low c-p values and the disappearance of c-p4 and c-p5, mostly represented by omnivores and predators (Fitoussi *et al.*, 2016).

In addition to invasive alien herbaceous species, many tree species have been designated invasive status. Among them, *Ailanthus altissima* (Mill.) Swingle, *Acer negundo* L. and *Robinia pseudoacacia* L. are considered some of the worst invasive woody plant species in Europe (DAISIE, 2020). Little data are available on their effects on soil fauna, particularly nematodes.

A study by Lazzaro *et al.* (2018) showed that *R. pseudocacia* poses a severe threat to ecosystems, decreasing local biodiversity of nematodes, microarthropods and plant communities, especially in areas where it becomes a dominant species. Nematode richness was significantly lower in invaded plots, mainly plant parasitic nematodes. Indeed, fewer taxa of root-feeding specialist nematodes, such as those of the families Longidoridae and Trichodoridae, were found associated with *R. pseudoacacia* in comparison to native oak stands. However, the proportion of nematodes in the feeding groups was similar for bacterial feeders (the major group), fungal feeders, omnivores and predators in both types of stand (Lazzaro *et al.*, 2018).

Renčo *et al.* (2025) carried out a pilot study on stands dominated by *A. negundo* (an invasive, non-native species), *Fraxinus excelsior* L. and *Alnus glutinosa* Gaertn. (both as native representatives) in flood plain forest habitats of the Morava River. The invasive ash-leaved maple *A. negundo* did not modify overall structure of soil nematodes when compared to the nematode fauna under the two above-mentioned native tree species at the locality. The total nematode abundance, genus richness, composition of nematode trophic groups and functional guilds were nearly identical in the cluster of ash, alder and ash-leaved maple. A cumulative total of 96 nematode genera (the highest was 87 genera for *A. glutinosa*), belonging to 52 nematode families, were recorded in the investigated flood plain forest sites, including the genus *Aprutides* not previously reported in Slovakia. Most nematode genera occurred in the rhizosphere of all tree species, with few exceptions; for example, the genera *Monhystrella* and *Prodorylaimus* were only found in clusters of ash-leaved maple, while the genus *Hexatylus* was only observed in alder clumps. A relatively high total nematode abundance and genus diversity across all tree species with no significant differences suggests that nematode communities were supported by all three species.

Conclusion

It would be interesting to know the composition of the community of organisms that we want to follow before a disturbance and subsequently to evaluate the changes after the disturbance, in our case ecosystem colonization by an invasive plant species. Unfortunately, we cannot predict when and where invasions will take place, thus, we can only compare the community structure of nematodes inhabiting invaded plots with communities in nearby uninvaded plots, assuming both plots had similar nematofauna structure before the invasion.

Every day new reports about the occurrence and spread of various invasive plants species arise, which vary by region, continent and the significance of the alien species in question. Extensive research and observations have revealed which alien plant species are generally more invasive with potential negatively effects on multiple ecological, biological or abiotic parameters in the introduced range. But in many cases, such impacts become apparent or problematic only when invaders are well established and have large ranges. Nevertheless, in many countries, invasions receive little attention. As Pyšek *et al.* (2020) invasive alien species break down biogeographic realms, affect native species richness and abundance, increase the risk of native species extinction, affect the genetic composition of native populations, change native animal behaviour, alter phylogenetic diversity across communities and modify trophic networks.

Although nematode communities are considered a reliable tool for assessing the condition of the soil environment, there are relatively few studies on the effects of non-native plant on this group of soil fauna. These studies have suggested that some invasive plant species may have a long-term negative impact on soil nematode fauna. On the other hand, they reveal that some invasive plant species do not disrupt the structure of soil nematode communities compared to the original vegetation in the introduced range.

References

Ahmad, W. and Jairajpuri, M.S. (2010) Mononchida: The predaceous nematodes. In: *Nematology Monographs and Perspectives*, Vol. 7. Brill, Leiden, p. 298.

Anderson, R.C., Dhillion, S.S. and Kelley, T.M. (1996) Aspects of the ecology of an invasive plant, garlic mustard (*Alliaria petiolata*), in central Illinois. *Restoration Ecology* 4(2), 181–191. DOI: 10.1111/j.1526-100X.1996.tb00118.x.

Asbjornsen, H., Goldsmith, G.R., Alvarado-Barrientos, M.S., Rebel, K., Osch, F.P. *et al.* (2011) Ecohydrological advances and applications in plant-water relations research: A review. *Journal of Plant Ecology* 4(1–2), 3–22. DOI: 10.1093/jpe/rtr005.

Baker, H.G. (ed.) (1974) The evolution of weeds. *Annual Review of Ecology and Systematics* 5(1), 1–24. DOI: 10.1146/annurev.es.05.110174.000245.

Barney, J.N., Tekiela, D.R., Dollete, E.S. and Tomasek, B.J. (2013) What is the “real” impact of invasive plant species? *Frontiers in Ecology and the Environment* 11(6), 322–329. DOI: 10.1890/120120.

Bastow, J. (2020) The impacts of a wildfire in a semiarid grassland on soil nematode abundances over 4 years. *Biology and Fertility of Soils* 56(5), 675–685. DOI: 10.1007/s00374-020-01441-4.

Belnap, J., Phillips, S.L., Sherrod, S.K. and Moldenke, A. (2005) Soil biota can change after exotic plant invasion: Does this affect ecosystem processes? *Ecology* 86(11), 3007–3017. DOI: 10.1890/05-0333.

Bilgrami, A.L. (1990) Diplogasterid predators. In: Jairajpuri, M.S., Alam, M.M. and Ahmad, I. (eds) *Nematode Biocontrol: Aspects and Prospects*. CBS Publisher & Distributor Pvt. Ltd, New Delhi, pp. 136–142.

Blankinship, J.C., Niklaus, P.A. and Hungate, B.A. (2011) A meta-analysis of responses of soil biota to global change. *Oecologia* 165(3), 553–565. DOI: 10.1007/s00442-011-1909-0.

Blume, H.P., Brümmer, G.W., Fleige, H., Horn, R., Kandeler, E. *et al*. (2016) Soil organisms and their habitat. *Scheffer/Schachtschabel Soil Science* 87–122. DOI: 10.1007/978-3-642-30942-7_4.

Bongers, T. (1990) The maturity index: An ecological measure of environmental disturbance based on nematode species composition. *Oecologia* 83(1), 14–19. DOI: 10.1007/BF003.

Bongers, T. and Bongers, M. (1998) Functional diversity of nematodes. *Applied Soil Ecology* 10(3), 239–251. DOI: 10.1016/S0929-1393(98)00123-1.

Bonkowski, M. (2004) Protozoa and plant growth: The microbial loop in soil revisited. *New Phytologist* 162(3), 617–631. DOI: 10.1111/j.1469-8137.2004.01066.x.

Breshears, D.D., Nyhan, J.W., Heil, C.E. and Wilcox, B.P. (1998) Effects of woody plants on microclimate in a semiarid woodland: Soil temperature and evaporation in canopy and intercanopy patches. *International Journal of Plant Sciences* 159(6), 1010–1017. DOI: 10.1086/314083.

Briar, S.S., Grewal, P.S., Somasekhar, N., Stinner, D. and Miller, S.A. (2007) Soil nematode community, organic matter, microbial biomass and nitrogen dynamics in field plots transitioning from conventional to organic management. *Applied Soil Ecology* 37(3), 256–266. DOI: 10.1016/j.apsoil.2007.08.004.

Brussaard, L. (1998) Soil fauna, guilds, functional groups and ecosystem processes. *Applied Soil Ecology* 9(1–3), 123–135. DOI: 10.1016/S0929-1393(98)00066-3.

Burger, J. (2006) Bioindicators: A review of their use in the environmental literature 1970–2005. *Environmental Bioindicators* 1(2), 136–144. DOI: 10.1080/15555270600701540.

Byrnes, J.E.K., Gamfeldt, L., Isbell, F., Lefcheck, J.S., Griffin, J.N. *et al*. (2014) Investigating the relationship between biodiversity and ecosystem multifunctionality: Challenges and solutions. *Methods in Ecology and Evolution* 5(2), 111–124. DOI: 10.1111/2041-210X.12143.

Čerevková, A. and Renčo, M. (2009) Soil nematode community changes associated with windfall and wildfire in forest soil at the high Tatras national park, Slovak Republic. *Helminthologia* 46(2), 123–130. DOI: 10.2478/s11687-009-0024-9.

Čerevková, A., Bobuľská, L., Miklisová, D. and Renčo, M. (2019) A case study of soil food web components affected by *Fallopia japonica* (Polygonaceae) in three natural habitats in Central Europe. *Journal of Nematology* 51, 1–16. DOI: 10.21307/jofnem-2019-042.

Čerevková, A., Ivashchenko, K., Miklisová, D., Ananyeva, N. and Renčo, M. (2020a) Influence of invasion by Sosnowsky's hogweed on nematode communities and microbial activity in forest and grassland ecosystems. *Global Ecology and Conservation* 21, e00851. DOI: 10.1016/j.gecco.2019.e00851.

Čerevková, A., Miklisová, D., Bobuľská, L. and Renčo, M. (2020b) Impact of the invasive plant *Solidago gigantea* on soil nematodes in a semi-natural grassland and a temperate broadleaved mixed forest. *Journal of Helminthology* 94, e51. DOI: 10.1017/S0022149X19000324.

Chen, X., Liu, M., Hu, F., Mao, X. and Li, H. (2007) Contributions of soil micro-fauna (protozoa and nematodes) to rhizosphere ecological functions. *Acta Ecologica Sinica* 27(8), 3132–3143. DOI: 10.1016/S1872-2032(07)60068-7.

Cronk, Q.C. and Fuller, J.L. (1995) *Plant Invaders: The Threat to Natural Ecosystems*. Chapman & Hall, London.

Crowther, T.W., van der Hoogen, J., Wan, J., Mayes, M.A., Keiser, A.D. *et al*. (2019) The global soil community and its influence on biogeochemistry. *Science* 365(6455), eaav0550. DOI: 10.1126/science.aav0550.

DAISIE (2020) Delivering alien invasive species inventories for Europe (DAISIE). Available at: https://www.gbif.org/dataset/39f36f10-559b-427f-8c86-2d28afff68ca

Delgado-Baquerizo, M., Reich, P.B., Trivedi, C., Eldridge, D.J. and Abades, S. (2020) Multiple elements of soil biodiversity drive ecosystem functions across biomes. *Nature Ecology & Evolution* 4(2), 210–220. DOI: 10.1038/s41559-019-1084-y.

Delgado-Baquerizo, M., Eldridge, D.J., Liu, Y.-R., Liu, Z.-W., Coleine, C. *et al*. (2025) Soil biodiversity and function under global change. *PLoS Biology* 23(3), e3003093. DOI: 10.1371/journal.pbio.3003093.

Dervash, M., Bhat, R., Mushtaq, N. and Singh, V. (2018) Dynamics and importance of soil mesofauna. *International Journal of Advance Research in Science and Engineering* 7, 2010–2019.

Dietrich, P., Cesarz, S., Liu, T., Roscher, C. and Eisenhauer, N. (2021) Effects of plant species diversity on nematode community composition and diversity in a long-term biodiversity experiment. *Oecologia* 197(2), 297–311. DOI: 10.1007/s00442-021-04956-1.

Droste, T., Flory, S.L. and Clay, K. (2010) Variation for phenotypic plasticity among populations of an invasive exotic grass. *Plant Ecology* 207(2), 297–306. DOI: 10.1007/s11258-009-9673-5.

Eisenhauer, N., Migunova, V.D., Ackermann, M., Ruess, L. and Scheu, S. (2011) Changes in plant species richness induce functional shifts in soil nematode communities in experimental grassland. *PloS ONE* 6(9), e24087. DOI: 10.1371/journal.pone.0024087.

Elliott, E.T., Anderson, R.V., Coleman, D.C. and Cole, C.V. (1980) Habitable pore space and microbial trophic interactions. *Oikos* 35(3), 327. DOI: 10.2307/3544648.

Ferris, H. (2010) Form and function: Metabolic footprints of nematodes in the soil food web. *European Journal of Soil Biology* 46(2), 97–104. DOI: 10.1016/j.ejsobi.2010.01.003.

Ferris, H., Venette, R.C., van der Meulen, H.R. and Lau, S.S. (1998) Nitrogen mineralization by bacterial-feeding nematodes: Verification and measurement. *Plant and Soil* 203(2), 159–171. DOI: 10.1023/A:1004318318307.

Ferris, H., Bongers, T. and de Goede, R.G.M. (2001) A framework for soil food web diagnostics: Extension of the nematode faunal analysis concept. *Applied Soil Ecology* 18(1), 13–29. DOI: 10.1016/S0929-1393(01)00152-4.

Fitoussi, N., Pen-Mouratov, S. and Steinberger, Y. (2016) Soil free-living nematodes as bio-indicators for assaying the invasive effect of the alien plant *Heterotheca subaxillaris* in a coastal dune ecosystem. *Applied Soil Ecology* 102, 1–9. DOI: 10.1016/j.apsoil.2016.02.005.

Fränzle, O. (2006) Complex bioindication and environmental stress assessment. *Ecological Indicators* 6(1), 114–136. DOI: 10.1016/j.ecolind.2005.08.015.

Freckman, D.W. (1988) Bacterivorous nematodes and organic-matter decomposition. *Agriculture, Ecosystems & Environment* 24(1–3), 195–217. DOI: 10.1016/0167-8809(88)90066-7.

García-De la Cruz, R., Knudsen, G.R., Dandurand, L.M., Carta, L.K. and Newcombe, G. (2019) Nematodes associated with invasive spotted knapweed. *Nematropica* 49, 200–207.

Geisen, S. (2016) The bacterial-fungal energy channel concept challenged by enormous functional versatility of soil protists. *Soil Biology and Biochemistry* 102, 22–25. DOI: 10.1006/1.soilbio.2016.06.013.

Gongalsky, K.B., Zaitsev, A.S., Korobushkin, D.I., Saifutdinov, R.A., Butenko, K.O. *et al*. (2021) Forest fire induces short-term shifts in soil food webs with consequences for carbon cycling. *Ecology Letters* 24(3), 438–450. DOI: 10.1111/ele.13657.

Griffiths, B.S. and Caul, S. (1993) Migration of bacterial-feeding nematodes, but not protozoa, to decomposing grass residues. *Biology and Fertility of Soils* 15(3), 201–207. DOI: 10.1007/BF00361612.

Griffiths, B.S., Neilson, R. and Bengough, A.G. (2003) Soil factors determined nematode community composition in a two year pot experiment. *Nematology* 5(6), 889–897. DOI: 10.1163/156854103773040808.

Háněl, L. (2004) Response of soil nematodes inhabiting spruce forests in the šumava mountains to disturbance by bark beetles and clear-cutting. *Forest Ecology and Management* 202(1–3), 209–225. DOI: 10.1016/j.foreco.2004.07.027.

Holt, E.A. and Miller, S.W. (2011) Bioindicators: Using organisms to measure. *Nature Education Knowledge* 3, 8–13.

Hu, C. and Cao, Z.P. (2008) Nematode community structure under compost and chemical fertilizer management practice, in the North China plain. *Experimental Agriculture* 44(4), 485–496. DOI: 10.1017/S0014479708006716.

Hughes, C.E. (1994) Risks of species introductions in tropical forestry. *The Commonwealth Forestry Review* 74, 243–252. DOI: 10.2307/42607066.

Jairajpuri, M.S. and Ahmad, W. (1992) *Dorylaimida: Free-Living, Predaceous and Plant-Parasitic Nematodes*. Brill, Leiden. DOI: 10.1163/9789004630475.

Jose, S., Singh, H.P., Batish, D.R. and Kohli, R.K. (2013) *Invasive Plant Ecology*. CRC Press, Boca Raton, FL. DOI: 10.1201/b13865.

Jurová, J., Renčo, M., Gömöryová, E. and Čerevková, A. (2020) Effects of the invasive common milkweed (*Asclepias syriaca*) on nematode communities in natural grasslands. *Nematology* 22(4), 423–438. DOI: 10.1163/15685411-00003314.

Kanwar, R.S., Patil, J.A. and Yadav, S. (2021) Prospects of using predatory nematodes in biological control for plant parasitic nematodes – A review. *Biological Control* 160, 104668. DOI: 10.1016/j.biocontrol.2021.104668.

Khan, Z. and Kim, Y.H. (2007) A review on the role of predatory soil nematodes in the biological control of plant parasitic nematodes. *Applied Soil Ecology* 35(2), 370–379. DOI: 10.1016/j.apsoil.2006.07.007.

Kim, Y.H. (2015) Predatory nematodes as biocontrol agents of phytonematodes. In: Askary, T.H. and Martinelli, P.R.P. (eds) *Biocontrol Agents of Phytonematodes*. CAB International, Wallingford, pp. 393–420. DOI: 10.1079/9781780643755.0393.

Korthals, G.W., Goede, R.D., Kammenga, J.E. and Bongers, T. (1996) The maturity index as an instrument for risk assessment of soil pollution. In: *Proceedings of the NATO Advanced Research Workshop on New Approaches to the Development of Bioindicator Systems for Soil Pollution*, pp. 85–93.

Lavelle, P. and Spain, A.V. (2001) *Soil Ecology*. Kluwer Academic Publishers, Dordrecht, The Netherlands.

Lagerlöf, J., Insunza, V., Lundegårdh, B. and Rämert, B. (2011) Interaction between a fungal plant disease, fungivorous nematodes and compost suppressiveness. *Acta Agriculturae Scandinavica, Section B-Soil & Plant Science* 61, 377–372. DOI: 10.1007/0-306-48162-6.

Lazzaro, L., Mazza, G., d'Errico, G., Fabiani, A., Giuliani, C. *et al.* (2018) How ecosystems change following invasion by *Robinia pseudoacacia*: Insights from soil chemical properties and soil microbial, nematode, microarthropod and plant communities. *The Science of the Total Environment* 622–623, 1509–1518. DOI: 10.1016/j.scitotenv.2017.10.017.

MacDougall, A.S. and Turkington, R. (2005) Are invasive species the drivers or passengers of change in degraded ecosystems? *Ecology* 86(1), 42–55. DOI: 10.1890/04-0669.

Margalef, R. (1958) Information theory in ecology. *General Systems* 3, 36–71.

McGonigle, T.P. (1995) The significance of grazing on fungi in nutrient cycling. *Canadian Journal of Botany* 73(S1), 1370–1376. DOI: 10.1139/b95-399.

Mitchell, C.E. and Power, A.G. (2003) Release of invasive plants from fungal and viral pathogens. *Nature* 421(6923), 625–627. DOI: 10.1038/nature01317.

Mondal, S., Ghosal, S. and Barua, R. (2016) Impact of elevated soil and air temperature on plants growth, yield and physiological interaction: A critical review. *Scientia Agriculturae* 14(3), 293–305. DOI: 10.15192/PSCP.SA.2016.14.3.293305.

Nicol, J.M., Turner, S.J., Coyne, D.L., Nijs, L.D., Hockland, S. *et al.* (2011) Current nematode threats to world agriculture. In: Jones, J., Gheysen, G. and Fenoll, C. (eds) *Genomics and Molecular Genetics of Plant-Nematode Interactions*. Springer, Dordrecht, pp. 21–43. DOI: 10.1007/978-94-007-0434-3.

Nisa, R.U., Tantray, A.Y., Kouser, N., Allie, K.A., Wani, S.M. *et al.* (2021) Influence of ecological and edaphic factors on biodiversity of soil nematodes. *Saudi Journal of Biological Sciences* 28(5), 3049–3059. DOI: 10.1016/j.sjbs.2021.02.046.

Okada, H.H. (2006) Ecology of fungivorous nematodes and their use for suppression of plant diseases. *Bulletin of the National Agricultural Research Centre for Tohoku* 105, 155–197.

Okada, H. and Harada, H. (2007) Effects of tillage and fertilizer on nematode communities in a Japanese soybean field. *Applied Soil Ecology* 35(3), 582–598. DOI: 10.1016/j.apsoil.2006.09.008.

Okada, H., Harada, H. and Kadota, I. (2005) Fungal-feeding habits of six nematode isolates in the genus *Filenchus*. *Soil Biology and Biochemistry* 37(6), 1113–1120. DOI: 10.1016/j.soilbio.2004.11.010.

Orgiazzi, A., Bardgett, R.D. and Barrios, E. (2016) Global soil biodiversity atlas. *European Commission, Luxembourg* 167. DOI: 10.1093/oso/9780199668564.003.0007.

Panesar, T.S., Marshall, V.G. and Barclay, H.J. (2000) The impact of clearcutting and partial harvesting systems on population dynamics of soil nematodes in coastal douglas-fir forests. *Pedobiologia* 44(6), 641–665. DOI: 10.1078/S0031-4056(04)70079-7.

Paul, E.A. (2015) Soil microbiology, ecology, and biochemistry: An exciting present and great future built on basic knowledge and unifying concepts. *Soil Microbiology, Ecology and Biochemistry* 1–15. DOI: 10.1016/B978-0-12-415955-6.00001-3.

Perry, R.N. and Moens, M. (2006) *Plant Nematology*. CAB International, Wallingford.

Pielou, E.C. (1966) Species-diversity and pattern-diversity in the study of ecological succession. *Journal of Theoretical Biology* 10(2), 370–383. DOI: 10.1016/0022-5193(66)90133-0.

Ponge, J.F. (2015) The soil as an ecosystem. *Biology and Fertility of Soils* 51(6), 645–648. DOI: 10.1007/s00374-015-1016-1.

Postma-Blaauw, M.B., de Vries, F.T., de Goede, R.G.M., Bloem, J., Faber, J.H. *et al.* (2005) Within-trophic group interactions of bacterivorous nematode species and their effects on the bacterial community and nitrogen mineralization. *Oecologia* 142(3), 428–439. DOI: 10.1007/s00442-004-1741-x.

Potapov, A.M., Beaulieu, F., Birkhofer, K., Bluhm, S.L., Degtyarev, M.I. *et al.* (2022) Feeding habits and multifunctional classification of soil-associated consumers from protists to vertebrates. *Biological Reviews of the Cambridge Philosophical Society* 97(3), 1057–1117. DOI: 10.1111/brv.12832.

Pothula, S.K., Phillips, G. and Bernard, E.C. (2022) Increasing levels of physical disturbance affect soil nematode community composition in a previously undisturbed ecosystem. *Journal of Nematology* 54(1), 20220022. DOI: 10.2478/jofnem-2022-0022.

Pyšek, P. (1997) Clonality and plant invasions: Can a trait make a difference. In: de Kroon, H. and Van Groenendael, J.M. (eds) *The Ecology and Evolution of Clonal Plants*. Backhuys Publishers, Leiden, pp. 405–427.

Pyšek, P., Hulme, P.E., Simberloff, D., Bacher, S., Blackburn, T.M. *et al.* (2020) Scientists' warning on invasive alien species. *Biological Reviews* 95(6), 1511–1534. DOI: 10.1111/brv.12627.

Reinhart, K.O., Gurnee, J., Tirado, R. and Callaway, R.M. (2006) Invasion through quantitative effects: Intense shade drives native decline and invasive success. *Ecological Applications* 16(5), 1821–1831. DOI: 10.1890/1051-0761(2006)016[1821:ITQEIS]2.0.CO;2.

Renčo, M. (2003) The communities of nematodes in agroecosystems of sugar beet, cereals and lucerne in Eastern Slovakia. *Helminthologia* 40, 55–58.

Renčo, M. and Baležentiené, L. (2015) An analysis of soil free-living and plant-parasitic nematode communities in three habitats invaded by *Heracleum sosnowskyi* in central Lithuania. *Biological Invasions* 17(4), 1025–1039. DOI: 10.1007/s10530-014-0773-3.

Renčo, M., Čerevková, A. and Gömöryová, E. (2019a) Soil nematode fauna and microbial characteristics in an early-successional forest ecosystem. *Forests* 10(10), 888. DOI: 10.3390/f10100888.

Renčo, M., Kornobis, F.W., Domaradzki, K., Jakubska-Busse, A., Jurová, J. *et al.* (2019b) How does an invasive *Heracleum sosnowskyi* affect soil nematode communities in natural conditions? *Nematology* 21(1), 71–89. DOI: 10.1163/15685411-00003196.

Renčo, M., Jurová, J., Gömöryová, E. and Čerevková, A. (2021a) Long-term giant hogweed invasion contributes to the structural changes of soil nematofauna. *Plants* 10(10), 2103. DOI: 10.3390/plants10102103.

Renčo, M., Čerevková, A. and Homolová, Z. (2021b) Nematode communities indicate the negative impact of *Reynoutria japonica* invasion on soil fauna in ruderal habitats of Tatra national park in Slovakia. *Global Ecology and Conservation* 26, e01470. DOI: 10.1016/j.gecco.2021.e01470.

Renčo, M., Adámek, M., Jílková, V. and Devetter, M. (2022a) Post-fire recovery of soil nematode communities depends on fire severity. *Diversity* 14(12), 1116. DOI: 10.3390/d14121116.

Renčo, M., Čerevková, A. and Hlava, J. (2022b) Life in a contaminated environment: How soil nematodes can indicate long-term heavy-metal pollution. *Journal of Nematology* 54(1), 20220053. DOI: 10.2478/jofnem-2022-0053.

Renčo, M., Jurová, J. and Čerevková, A. (2024) Invasiveness of impatiens Parviflora in carpathian beech forests: Insights from soil nematode communities. *Diversity* 16(7), 393. DOI: 10.3390/d16070393.

Renčo, M., Čerevková, A. and Gömöryová, E. (2025) Floodplain forest soil nematode communities as influenced by non-native *Acer negundo* L. invasion. *Diversity (Basel)* 17(6). DOI: 10.3390/d17060376.

Richards, C.L., Bossdorf, O., Muth, N.Z., Gurevitch, J. and Pigliucci, M. (2006) Jack of all trades, master of some? On the role of phenotypic plasticity in plant invasions. *Ecology Letters* 9(8), 981–993. DOI: 10.1111/j.1461-0248.2006.00950.x.

Rønn, R., Griffiths, B.S., Ekelund, F. and Christensen, S. (1996) Spatial distribution and successional pattern of microbial activity and micro-faunal populations on decomposing barley roots. *Journal of Applied Ecology* 33(4), 662. DOI: 10.2307/2404938.

Ruess, L. and Dighton, J. (1996) Cultural studies on soil nematodes and their fungal hosts. *Nematologica* 42(3), 330–346. DOI: 10.1163/004425996X00065.

Sala, O.E., Chapin, F.S., Armesto, J.J., Berlow, E. and Bloomfield, J. (2000) Global biodiversity scenarios for the year 2100. *Science* 287(5459), 1770–1774. DOI: 10.1126/science.287.5459.1770.

Seebens, H., Blackburn, T.M., Dyer, E.E., Genovesi, P., Hulme, P.E. *et al.* (2017) No saturation in the accumulation of alien species worldwide. *Nature Communications* 8, 14435. DOI: 10.1038/ncomms14435.

Shannon, C.E. and Weaver, W. (1949) *A Mathematical Model of Communication*, Vol. 11. University of Illinois Press, Urbana, IL, pp. 11–20.

Simpson, E.H. (1949) Measurement of diversity. *Nature* 163(4148), 688–688. DOI: 10.1038/163688a0.

Singh, S.K., Hodda, M. and Ash, G.J. (2013) Plant-parasitic nematodes of potential phytosanitary importance, their main hosts and reported yield losses. *EPPO Bulletin* 43(2), 334–374. DOI: 10.1111/epp.12050.

Stirling, G.R. (2014) *Biological Control of Plant-parasitic Nematodes: Soil Ecosystem Management in Sustainable Agriculture*. CAB International, Wallingford.

Valocká, B., Sabová, M. and Renčo, M. (2001) Soil and plant nematode communities of two types of ecosystems. *Helminthologia* 38, 105–110.

van den Hoogen, J., Geisen, S., Routh, D., Ferris, H., Traunspurger, W. *et al.* (2019) Soil nematode abundance and functional group composition at a global scale. *Nature* 572(7768), 194–198. DOI: 10.1038/s41586-019-1418-6.

Van der Putten, W.H., van der Yeates, G.W., Duyts, H., Reis, C. and Karssen, G. (2005) Invasive plants and their escape from root herbivory: A worldwide comparison of the root-feeding nematode communities of the dune grass *Ammophila arenaria* in natural and introduced ranges. *Biological Invasions* 7(4), 733–746. DOI: 10.1007/s10530-004-1196-3.

Van der Putten, W.H., Bardgett, R.D., Bever, J.D., Bezemer, T.M., Casper, B.B. *et al.* (2013) Plant–soil feedbacks: The past, the present and future challenges. *Journal of Ecology* 101(2), 265–276. DOI: 10.1111/1365-2745.12054.

van Kleunen, M., Dawson, W., Essl, F., Pergl, J., Winter, M. *et al.* (2015) Global exchange and accumulation of non-native plants. *Nature* 525(7567), 100–103. DOI: 10.1038/nature14910.

Vezzani, F.M., Anderson, C., Meenken, E., Gillespie, R., Peterson, M. *et al.* (2018) The importance of plants to development and maintenance of soil structure, microbial communities and ecosystem functions. *Soil and Tillage Research* 175, 139–149. DOI: 10.1016/j.still.2017.09.002.

Viketoft, M., Palmborg, C., Sohlenius, B., Huss-Danell, K. and Bengtsson, J. (2005) Plant species effects on soil nematode communities in experimental grasslands. *Applied Soil Ecology* 30(2), 90–103. DOI: 10.1016/j.apsoil.2005.02.007.

Vilà, M., Espinar, J.L., Hejda, M., Hulme, P.E., Jarošík, V. *et al.* (2011) Ecological impacts of invasive alien plants: A meta-analysis of their effects on species, communities and ecosystems. *Ecology Letters* 14(7), 702–708. DOI: 10.1111/j.1461-0248.2011.01628.x.

Wang, C., Fu, B., Zhang, L. and Xu, Z. (2019) Soil moisture–plant interactions: An ecohydrological review. *Journal of Soils and Sediments* 19(1), 1–9. DOI: 10.1007/s11368-018-2167-0.

Wasilewska, L. (1994) The effect of age of meadows on succession and diversity in soil nematode communities. *Pedobiologia* 38(1), 1–11. DOI: 10.1016/S0031-4056(24)01407-0.

Weber, E. (1998) The dynamics of plant invasions: A case study of three exotic goldenrod species (*Solidago* L.) in Europe. *Journal of Biogeography* 25(1), 147–154. DOI: 10.1046/j.1365-2699.1998.251119.x.

Wilson, M.J. and Kakouli-Duarte, T. (2009) *Nematodes as Environmental Indicators*. CAB International, Wallingford. DOI: 10.1079/9781845933852.0000.

Yeates, G.W. (1994) Modification and qualification of the nematode maturity index. *Pedobiologia* 38(2), 97–101. DOI: 10.1016/S0031-4056(24)00113-6.

Yeates, G.W. (2003) Nematodes as soil indicators: Functional and biodiversity aspects. *Biology and Fertility of Soils* 37(4), 199–210. DOI: 10.1007/s00374-003-0586-5.

Yeates, G.W., Bongers, T., De Goede, R.G., Freckman, D.W. and Georgieva, S.S. (1993) Feeding habits in soil nematode families and genera-an outline for soil ecologists. *Journal of Nematology* 25(3), 315–331.

Zhang, P., Neher, D.A., Li, B. and Wu, J. (2018) The impacts of above- and belowground plant input on soil microbiota: Invasive *Spartina alterniflora* versus native *Phragmites australis*. *Ecosystems* 21(3), 469–481. DOI: 10.1007/s10021-017-0162-8.

Zhang, X., Li, Q., Zhu, A., Liang, W., Zhang, J. *et al.* (2012) Effects of tillage and residue management on soil nematode communities in North China. *Ecological Indicators* 13(1), 75–81. DOI: 10.1016/j.ecolind.2011.05.009.

10 Invasive Plant Impacts on Soil Macrofauna through Allelopathy and Environmental Modification

Fredrick Ojija[1]*, Giovanni Bacaro[2], Matthew Chidozie Ogwu[3], Gbenga Festus Akomolafe[4], Hudson Laizer[5], Philipina Shayo[1] and Gisandu Malunguja[1]
[1]Department of Earth Sciences, College of Science and Technical Education, Mbeya University of Science and Technology, Mbeya, Tanzania; [2]Department of Life Sciences, University of Trieste, Trieste, Italy; [3]Goodnight Family Department of Sustainable Development, Appalachian State University, Boone, NC, USA; [4]Department of Plant Science and Biotechnology, Federal University of Lafia, Lafia, Nasarawa State, Nigeria; [5]School of Life Sciences and Bioengineering, Nelson Mandela African Institution of Science and Technology, Arusha, Tanzania

Introduction

Soil macrofauna taxonomic groups such as annelids (earthworms), arthropods (insects and crustaceans) and molluscs (gastropods/snails) play an essential role in soil ecosystems (van Hengstum *et al.*, 2014; Zulu *et al.*, 2022; Mamabolo *et al.*, 2024). These organisms help convert organic matter into forms that can be utilized by plants, ultimately supporting soil fertility (Sofo *et al.*, 2020; Mamabolo *et al.*, 2024). They influence soil structure, organic matter decomposition, nutrient cycling, soil aeration and ecosystem functioning (van Hengstum *et al.*, 2014; Ibrahima *et al.*, 2017). Thus, their presence and diversity are integral to maintaining soil fertility and productivity, making them important bioindicators for assessing soil health (Rousseau *et al.*, 2013; Sofo *et al.*, 2020). Soil macrofauna physically and biologically alter the soil structure, enhancing the permeability of the soil and improving water infiltration (Zhou *et al.*, 2022). Besides, they contribute to the creation of microhabitats for other organisms, promoting biodiversity conservation (Huerta and van der Wal, 2012). For instance, earthworms, through their burrowing and feeding activities, not only recycle nutrients but also create channels in the soil that facilitate the movement of water and air (Ibrahima *et al.*, 2017; Yang *et al.*, 2024).

Soil macrofauna – ants, termites and beetles – are involved in the breakdown of organic matter, including plant litter and animal remains, thereby playing a role in nutrient cycling (Jouquet *et al.*, 2011; Ibrahima *et al.*, 2017; Mamabolo *et al.*, 2024). The presence and activity of these macrofauna can also influence the abundance and diversity of soil microorganisms, further enhancing nutrient cycling (Zhou *et al.*, 2022). Research conducted by Liu *et al.* (2020) and Singh *et al.* (2019) reveals

*Corresponding author: fredrick.ojija@must.ac.tz

DOI: 10.1079/9781836991298.0010

Table 10.1. Some common examples of soil macrofauna.

Common name	Scientific name	Family name
Common earthworm	*Lumbricus terrestris*	Lumbricidae
Termites or white ant	*Macrotermes bellicosus*	Termitidae
Formosan termite	*Coptotermes formosanus*	Rhinotermitidae
Red wood ant	*Formica rufa*	Formicidae
Garden snail	*Cornu aspersum*	Helicidae
African millipede	*Archispirostreptus gigas*	Spirostreptidae
Sacred scarab/beetle	*Scarabaeus sacer*	Scarabaeidae
Woodlouse/sow bug	*Armadillidium vulgare*	Armadillidiidae
Springtail	*Folsomia candida*	Hypogastruridae
African centipede	*Ethmostigmus trigonopodus*	Scolopendridae

that earthworms are particularly effective at enhancing soil aggregation, which improves the structural stability of the soil. This effect is also seen in termite species, which build intricate tunnel systems that help aerate the soil and facilitate nutrient movement (Black and Okwakol, 1997; Sileshi *et al.*, 2010; Jouquet *et al.*, 2011, 2018; Ali *et al.*, 2013). Table 10.1 provides some common examples of soil macrofauna species.

In addition to earthworms (Fig. 10.1) and termites, other arthropods like beetles and ants contribute to soil structure by burrowing and mixing soil layers, leading to improved soil aeration and reduced compaction. These soil macrofauna also contribute to the formation of soil aggregates, which are crucial for maintaining soil porosity and preventing erosion (Sofo *et al.*, 2020; Mamabolo *et al.*, 2024). Moreover, ants and termites play a role in regulating pest populations by preying on insects and other invertebrates that could harm crops and other beneficial plants (Bohlen *et al.*, 2004).

Despite their importance in ecosystems, soil macrofauna are negatively impacted by invasive plant species in addition to other numerous threats (Blakeslee *et al.*, 2011). Native vegetation cover is considered to be among the factors influencing soil macrofauna (Wiatrowska *et al.*, 2024). But, the invasive plants alter native vegetation communities (i.e. composition and structure), transforming the habitat quality because they modify soil physicochemical properties, most often exerting a directional effect on soil macrofauna (Jesse *et al.*, 2020; Wiatrowska *et al.*, 2024). Thus, invasive plants may pose a significant ecological threat, not only through their disruption of native plant communities but also via their profound impacts on soil health and ecosystem functioning (Stiers *et al.*, 2011; Wiatrowska *et al.*, 2024). They impact soil macrofauna by altering soil chemistry, nutrient availability (or nutrient cycling), microbial communities (Gwate *et al.*, 2021) and organic matter dynamics (Table 10.2). For instance, some invasive plants often produce allelopathic compounds or allelochemicals that increase bacterial diversity but simultaneously disrupt the balance of soil macrofauna vital for nutrient cycling and plant health (Ojija and Manyanza, 2021; Torres *et al.*, 2021; Ojija *et al.*, 2024). Garlic mustard (*Alliaria petiolata*), for example, was claimed to suppress beneficial mycorrhizal fungi, undermining native plant resilience and soil macrofauna (Flaherty *et al.*, 2024; Nunez-Mir and McCary, 2024).

Additionally, invasive plant litter and root exudates modify decomposition and nutrient mineralization processes, often slowing the release of essential nutrients like nitrogen and phosphorus (Ojija and Manyanza, 2021). This can lead to a reduction in soil nitrification rates, further constraining native species and delaying ecosystem recovery (Ehrenfeld, 2010; Afzal *et al.*, 2023). Physical soil characteristics are also affected, with species such as black wattle (*Acacia mearnsii*) degrading soil structure and reducing

Fig. 10.1. Some examples of soil macrofauna: (A) earthworm (*Lumbricus terrestris*), (B) millipede (*Eumillipes* sp.), (C) centipede (*Scolopendra* sp.), (D) beetle larva (*Scarabaeidae* sp.) and (E) sowbug or woodlouse (*Porcellio scaber*).

organic matter, which diminishes soil fertility and agricultural productivity (Gwate *et al.*, 2021; Lusizi *et al.*, 2024). These biological and physical transformations generate feedback loops that favour the continued dominance of invasive plants, reduce soil macrofauna biodiversity and compromise ecosystem services (Vilà *et al.*, 2011; Aloo *et al.*, 2013; Coakley and Petti, 2021). The disruption and decrease of soil macrofauna populations can lead to long-term declines in soil health and ecosystem stability (Ibrahima *et al.*, 2017; Yang *et al.*, 2024). Thus, as invasive plants continue to expand their range due to climate change and human activity (Müller-Schärer *et al.*, 2004; O'Reilly-Nugent *et al.*, 2016; Mavimbela *et al.*, 2018), further studies are needed to understand long-term soil macrofauna responses and develop sustainable conservation strategies (Blakeslee *et al.*, 2011). Monitoring soil health and soil macrofauna biodiversity in invaded ecosystems can provide valuable insights into ecosystem resilience and inform future restoration efforts. By addressing the impact of invasive plants on soil macrofauna, conservationists can work towards maintaining soil fertility, enhancing ecosystem stability and preserving biodiversity in affected regions (Fig. 10.2).

This chapter delves into the effects of invasive plants on the populations of soil macrofauna, with a particular focus on earthworms, arthropods and gastropod communities. It highlights the direct and indirect impacts of invasive plants, including disruptions to food

Table 10.2. A comparison of soil properties in native and invasive plant-dominated ecosystems, illustrating the impact of invasive plants on soil conditions that affect soil macrofauna.

Soil property	Native vegetation	Invaded vegetation	Source
pH	6.5–7.2	5.5–6.0	Bohlen *et al.* (2004)
	6.5 ± 6.6	6.4 ± 6.9	Ojija and Manyanza (2021)
Organic matter (%)	4.5–6.0	2.5–4.0	Ibrahima *et al.* (2017), Yang *et al.* (2024)
Soil moisture (%)	20–35	10–25	Hendrix *et al.* (2008)
Electrical conductivity (mS/cm)	0.6 ± 0.7	0.4 ± 0.5	Ojija and Manyanza (2021)
Earthworm density (individuals/m^2)	150–300	50–120	Hale *et al.* (2005)
Phosphorus (mg/kg)	68.6 ± 73.3	48.1 ± 51.8	Ojija and Manyanza (2021)

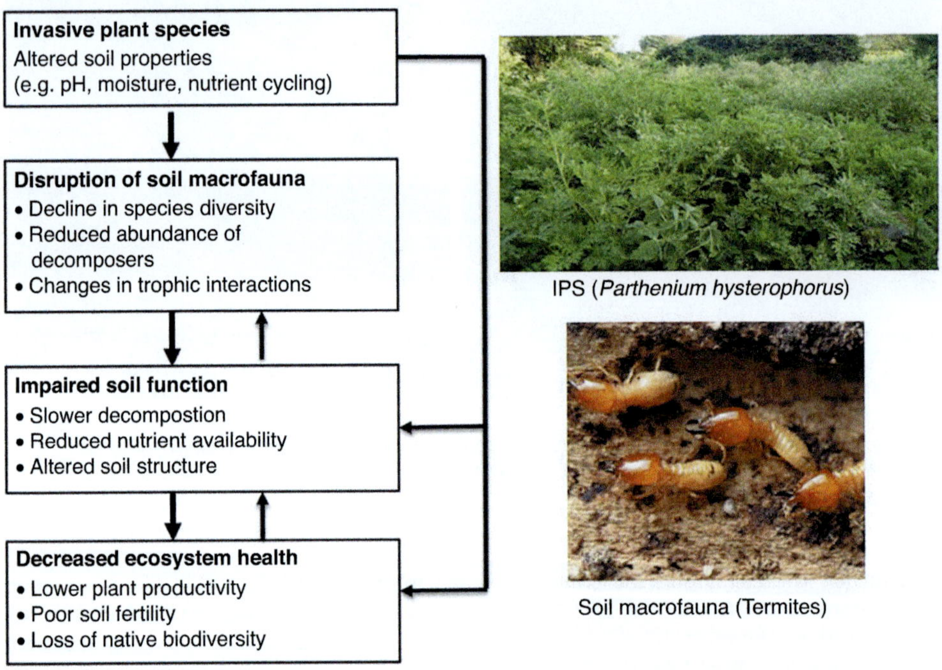

Fig. 10.2. The relationship between an invasive plant species (IPS) and its effects on soil macrofauna and ecosystem health.

webs, altered litter quality and quantity, and the release of allelopathic compounds that modify soil chemistry. The chapter elucidates various strategies for safeguarding soil macrofauna from the negative effects of invasive plants, such as implementing targeted management interventions, enhancing soil biodiversity monitoring and adopting ecosystem-based conservation practices. Furthermore, it draws on a rich body of evidence from previous studies to demonstrate the global implications of invasive plants on soil health and biodiversity conservation. By synthesizing insights from diverse ecosystems, the chapter underscores the importance of integrating soil biodiversity considerations into invasive species management to foster long-term ecological resilience and sustainability.

Impacts of Invasive Plants on Earthworm Populations

Earthworms play a vital role in maintaining soil health by contributing to soil structure, organic matter decomposition and nutrient cycling (Singh *et al.*, 2019; Liu *et al.*, 2020). Through their burrowing activities, they improve soil aeration, enhance water infiltration and stimulate microbial activity, facilitating nutrient release essential for plant growth, yet they are vulnerable to environmental changes induced by invasive plants. The introduction of invasive plants has been shown to disrupt earthworm populations by altering key soil properties such as pH, moisture levels, organic matter availability and microbial communities (Hendrix *et al.*, 2008). These disruptions can have cascading effects on soil ecosystem functioning, influencing nutrient availability, plant growth and broader biodiversity.

One of the primary ways invasive plants impact earthworm communities is through allelopathy, the release of biochemical compounds that negatively affect other organisms (Ojija *et al.*, 2019, 2024; Pyšek *et al.*, 2020). Some invasive plant species produce allelopathic chemicals that may be toxic to earthworms (Table 10.3), reducing their survival and reproductive rates (Ehrenfeld and Scott, 2001; Rodgers *et al.*, 2008; Nunez-Mir and McCary, 2024). For instance, *A. petiolata*, a widespread invasive plant in North America, produces glycosylates that inhibit the growth of mycorrhizal fungi, an essential component of soil microbial communities (Stinson *et al.*, 2006). Since earthworms rely on decomposed organic material rich in microbial activity, such disruptions lead to reduced food availability, ultimately decreasing earthworm abundance and diversity (Table 10.3). Similarly, *Acacia dealbata* (silver wattle), an invasive tree species in Mediterranean ecosystems, releases phenolic compounds that suppress soil microbial activity, leading to a decline in earthworm populations (Lazzaro *et al.*, 2014).

Invasive plants have also been reported to modify soil chemistry (Table 10.2), creating unfavourable conditions for native earthworm species (Simberloff *et al.*, 2013; Ojija and Manyanza, 2021). For instance, an invasive plant *Lonicera maackii* (Amur honeysuckle) alters soil pH and increases carbon and nitrogen levels, disrupting microbial community structures and making the soil less suitable for some earthworm species (Hale *et al.*, 2005). *Phragmites australis* (common reed), a dominant invasive plant in wetland ecosystems, tends to raise soil salinity and slows organic matter decomposition thereby influencing the decrease of earthworm abundance (Srivastava *et al.*, 2014).

Table 10.3. Some case studies on the impacts of invasive plants on earthworm communities.

Invasive plant	Effects on earthworms	Study location	Source
Alliaria petiolata	Releases glucosinolates that suppress microbial activity, reducing food availability for earthworms	North America	Stinson *et al.* (2006)
Lonicera maackii	Alters soil pH and nutrient balance, leading to declines in native earthworm species	North America	Hale *et al.* (2005)
Phragmites australis	Increases soil salinity, reducing earthworm abundance and biomass	Africa, South America, Australia	Srivastava *et al.* (2014)
Acacia dealbata	Produces phenolic compounds that inhibit microbial activity and reduce earthworm population	Mediterranean ecosystem	Lazzaro *et al.* (2014)
Fallopia japonica	Reduces fungal biomass, affecting decomposition process and earthworm food availability	Europe	Ehrenfeld (2010)

These chemical alterations not only impact earthworms directly but also influence soil food web dynamics, affecting other organisms dependent on decomposed organic material (Gwate *et al.*, 2021; Lusizi *et al.*, 2024).

A study by Liu *et al.* (2020) assessed the allelopathic effects of invasive asterid species (*Ageratina adenophora, Bidens pilosa* and *Erigeron annuus*) on the common composting worm *Eisenia fetida*. The results showed significant physiological stress; that is, reduced body mass and increased oxidative damage in earthworms exposed to these invasive plants compared to native plants (Liu *et al.*, 2020). This supports the novel weapons hypothesis, which suggests that invasive plants produce unique chemical compounds unfamiliar to native soil macrofauna, leading to harmful effects (Liu *et al.*, 2020). On the other hand, Bugiel *et al.* (2018) investigated the effects of *Vincetoxicum rossicum* (dog-strangling vine) in Rouge National Urban Park, Canada. The presence of this invasive species significantly altered soil bacterial communities, which indirectly affected earthworm populations by disrupting their food base and microhabitat conditions (Bugiel *et al.*, 2018). Similarly, *Fallopia japonica* (Japanese knotweed) was reported to reduce soil invertebrate diversity by changing litter quality and soil chemistry, leading to less favourable conditions for detritivores (Ehrenfeld and Scott, 2001; Ehrenfeld, 2010).

In wetlands, *P. australis* was reported to alter soil properties – namely oxygen levels and pH – which can suppress native earthworm species (Srivastava *et al.*, 2014). In addition, *Lythrum salicaria* (purple loosestrife) invasions can negatively affect soil invertebrates including earthworms by shifting soil moisture and litter dynamics according to Blossey *et al.* (2001). Another invasive plant, *Melaleuca quinquenervia*, a problematic tree in Florida's wetlands, was asserted to modify soil hydrology and chemistry to the detriment of native earthworm populations (Martin *et al.*, 2011). Its thick leaf litter and canopy reduce sunlight penetration and water infiltration, which compromises the soil conditions earthworms rely on (Martin *et al.*, 2011).

Another mechanism through which invasive plants impact earthworm populations is by altering soil moisture and organic matter availability (Table 10.2). Many invasive plants have aggressive growth patterns that lead to changes in litter decomposition rates (Jordan *et al.*, 2008). In Europe, the spread of *F. japonica* has been linked to decline in the availability of organic matter for earthworms and other decomposers because this invasive plant produces dense litter with high lignin content that decomposes slowly (Ehrenfeld, 2010). As a result, earthworm populations decline due to a lack of accessible nutrients. Similarly, research in North American forests has shown a significant reduction in earthworm diversity in areas invaded by *A. petiolata*, affecting forest regeneration by altering soil nutrient cycling (Wolfe *et al.*, 2008; O'Sullivan *et al.*, 2019).

Implications for soil health due to loss of earthworms

The decline in earthworm populations due to invasive plants can have severe ecological consequences and cascading effects on soil health and ecosystem functioning. For instance, a reduction in earthworm density leads to soil compaction, decreased aeration and slower organic matter decomposition, all of which negatively affect soil fertility and plant growth (Hendrix *et al.*, 2008). This in turn can alter entire ecosystem dynamics, as soil organisms play a foundational role in nutrient cycling. The loss of earthworms may also impact higher trophic levels, such as birds and mammals that rely on them as a food source. Moreover, the decline in microbial diversity associated with invasive species-induced soil changes can further disrupt soil stability, making it more prone to erosion and degradation. Besides, decreased earthworm activity can negatively impact plant growth, as reduced burrowing limits nutrient mixing and root penetration (Singh *et al.*, 2019; Liu *et al.*, 2020). These implications highlight the importance of understanding and mitigating the effects of invasive plants on soil macrofauna biodiversity. Understanding the mechanisms through which invasive plants influence earthworms can help develop conservation strategies that mitigate their negative effects. Hence, future studies may focus on identifying soil management practices that enhance earthworm resilience in invasive plant-affected habitats. Additionally, restoration efforts may prioritize the removal of invasive plants and the reintroduction of native vegetation to support earthworm biodiversity.

Impacts of invasive plants on arthropod populations

Like annelids and earthworms, soil arthropods – such as ants, mites, springtails and beetles – play a vital role in ecosystem functioning through nutrient cycling, soil formation, decomposition, food web dynamics and soil aeration (Rousseau *et al.*, 2013; Sofo *et al.*, 2020). However, invasive plants can profoundly affect these ecosystem functional roles of soil arthropods (van Hengstum *et al.*, 2014). Invasive plants use various mechanisms to impact soil arthropods, for instance, changes in litter quality and quantity, alteration of soil chemistry, structure and microbial communities, and microhabitat modification (Bohlen *et al.*, 2004; Hale *et al.*, 2005; Hendrix *et al.*, 2008; Ibrahima *et al.*, 2017; Ojija and Manyanza, 2021). Often, invasive plants produce litter with different chemical and physical properties compared to native vegetation. This influences the composition of detritivores, arthropods that rely on organic matter for food and habitat (Wiatrowska *et al.*, 2024). Some invasive plants modify soil pH, nitrogen content and microbial activity, indirectly impacting soil arthropod populations that depend on specific soil conditions or microbe-arthropod interactions. Additionally, dense invasive plant cover may affect soil temperature, moisture and aeration, all of which are important for soil arthropod survival and activity (Herrera and Dudley, 2003; Ogura-Yamada and Krushelnycky, 2020; Wekhanya *et al.*, 2020).

Invasive plants can affect soil arthropod communities (i.e. abundance, diversity and composition) because they alter composition of native flora and fauna (Braschi *et al.*, 2021; Pearson *et al.*, 2024), disrupting soil nutrient cycles and trophic interactions (Ehrenfeld, 2010; Ojija, 2024). Also, by altering habitat complexity, food resource availability and microclimate conditions, invasive plants exert varying effects on arthropod taxa (van Hengstum *et al.*, 2014; Bobuľská *et al.*, 2025). Earlier studies have shown that an invasive plant *Lantana camara* (Lantana) decreased Collembola and Acari diversity (Wekhanya *et al.*, 2020; Tlou and Duffy, 2022), *Mikania micrantha* (bitter vine) shifted Formicidae and Araneae composition (Ehrenfeld and Scott, 2001) and *Carpobrotus* species negatively affected spider assemblage dynamics (Braschi *et al.*, 2021).

Moreover, some invasive plants have been reported to change the physical environment by increasing soil organic matter due to rapid litter accumulation, altering soil pH and moisture levels, and modifying shading, which influences soil temperature and microbial communities (Ojija and Manyanza, 2021; Torres *et al.*, 2021). It was also reported that invasive plants can disrupt soil trophic interactions by reducing detritivore abundance, affecting decomposition rates, altering predator–prey relationships and changing herbivore populations (Aloo *et al.*, 2013; Shuvar *et al.*, 2021; Nunez-Mir and McCary, 2024). A study from East Africa on *L. camara* demonstrated its negative effects on soil arthropod communities by increasing soil organic matter, which altered microbial activity and shifted fungal-feeding arthropod populations (Sharma *et al.*, 2005; Wekhanya *et al.*, 2020; Tlou and Duffy, 2022).

Given these profound ecosystem-level changes, management strategies can focus on restoring native vegetation, implementing biological control (Banks-Leite *et al.*, 2020; Ojija *et al.*, 2024) and enhancing soil health through conservation agriculture (Huerta and van der Wal, 2012; Zulu *et al.*, 2022). Understanding these changes is critical for ecosystem conservation, and future research should integrate long-term monitoring and experimental approaches to assess arthropod responses in various ecosystems.

Case studies of invasive plant impacts on arthropod communities

A study by Flaherty *et al.* (2024) showed that *A. petiolata* invasion led to reduced abundance and diversity of Collembola and oribatid mites due to altered soil pH and microbial shifts. This is beause the *A. petiolata* releases allelochemicals that disrupt mycorrhizal fungi and reduce the quality of leaf litter (Flaherty *et al.*, 2024). In riparian and urban ecosystems, *F. japonica* was

Table 10.4. Some examples of invasive plants and their impact on soil arthropod communities.

Invasive plant	Affected arthropod group	Change in abundance and diversity	Source
Lantana camara	Collembola, Acari	Decline in diversity	Wekhanya *et al.* (2020), Tlou and Duffy (2022)
Eichhornia crassipes	Coleoptera, Diptera	Increase in detritivores	Getahun and Kefale (2023)
Mikania micrantha	Araneae, Formicidae	Shift in species composition	Ehrenfeld and Scott (2001)
Euphorbia esula and *Cirsium arvense*	Ground-dwelling invertebrates (arthropods)	Decrease in abundance and species richness	Pritekel *et al.* (2006)
Phytolacca americana	Ground-dwelling forest arthropods (Carabidae and Formicidae)	Changes in species composition and abundance	Schirmel (2020)
Arundo donax	Ground-dwelling insects	Changes in habitat quality and vegetation structure in riparian systems	Herrera and Dudley (2003), Maceda-Veiga *et al.* (2016)
Spiraea tomentosa	Uropodina mites	Altered community composition	Wiatrowska *et al.* (2024)
Frangula alnus	Hemiptera, Diptera	Increase in abundance and diversity of certain arthropods	Greenleaf *et al.* (2023)
Carpobrotus spp.	Spiders	Changes in assemblage dynamics	Braschi *et al.* (2021)
Asclepias syriaca	Ground-dwelling arthropods (diplopods, spiders and ants)	Altered community structure and reduced diversity	Gallé *et al.* (2015)

reported to significantly change litter decomposition rates and soil microhabitats (Ehrenfeld and Scott, 2001; Ehrenfeld, 2010). Gerber *et al.* (2008) found a decrease in abundance and richness of soil macroarthropods, including Carabidae and Staphylinidae beetles, due to the low palatability and slow-decomposing litter of the invader. Furthermore, Herrera and Dudley (2003) reported significant reductions in soil arthropod abundance and diversity in areas invaded by *Arundo donax* (the giant reed), attributing these declines to the plant's dense canopy cover and alterations in litter decomposition (Table 10.4). Similarly, another study observed that arthropods in *A. donax*-invaded soils were smaller, and the absence of specific taxa within trophic guilds – or even entire trophic groups such as parasitoids – indicates that changes in the food web structure in these invaded areas are likely (Maceda-Veiga *et al.*, 2016).

Moreover, invasive *Spartina alterniflora* was reported to alter the soil structure and salinity in Chinese estuaries, thereby causing the decline in abundance and diversity of soil arthropods (Huang *et al.*, 2022). Huang *et al.* (2022) documented changes in soil macrofauna including declines in native detritivore and predator arthropods, attributed to reduced oxygen levels and altered soil pH in *S. alterniflora*-invaded habitats (Table 10.4). Additionally, modest but statistically non-significant increases in the abundance and species richness of forest floor-dwelling invertebrate communities, particularly arthropods, were reported following the suppression of the invasive species *Solenopsis papuana* (Ogura-Yamada and Krushelnycky, 2020). In addition, in East African wetlands, accumulation of decomposing *Eichhornia crassipes* (water hyacinth) was claimed to reduce oxygen availability and cause the soil to become compacted (Table 10.4), leading to decreases in detritivorous arthropods like springtails and woodlice (Aloo *et al.*, 2013).

Despite the overall negative impact of invasive plants on soil arthropod communities, some studies highlight that not all invasive plants have adverse effects on these macrofauna (Srivastava *et al.*, 2014; Gallé *et al.*, 2015; Mantoani *et al.*, 2022; Getahun and Kefale, 2023). In fact, some invasive plants can promote or increase the abundance of certain arthropod groups (Srivastava *et al.*, 2014). For instance, it was reported that *E. crassipes* contributed to an increase in detritivore populations, particularly among groups such as Diptera and Coleoptera (Getahun and Kefale, 2023). This increase is likely driven by the enhanced availability of organic matter provided by the plant's decaying biomass, supporting higher detritivore abundance (Getahun and Kefale, 2023). Furthermore, Gallé *et al.* (2015) observed that the invasive plant *Asclepias syriaca* increased the abundance of ant species, likely due to indirect food web interactions where ants are attracted to aphids that inhabit the plant. Additionally, it was reported that *Gunnera tinctoria* (giant rhubarb), a nitrogen-fixing invasive plant, increased earthworm abundance and diversity in some ecosystems (Mantoani *et al.*, 2022).

Impacts of invasive plants on gastropod communities

Gastropods, including snails and slugs, play essential roles in ecosystems as decomposers and grazers, contributing to nutrient cycling and soil structure maintenance (Ovando *et al.*, 2019). Nevertheless, their populations can be significantly impacted by environmental changes (Strong *et al.*, 2008), particularly those induced by invasive plants. Invasive plants can alter habitat structure, microclimatic conditions and food availability, leading to shifts in gastropod communities (Ojija *et al.*, 2024). One of the primary ways invasive plant species impact gastropod communities is through habitat modification (Strong *et al.*, 2008; Ovando *et al.*, 2019). Many invasive plants alter the physical structure of ecosystems, affecting soil moisture, temperature and leaf litter accumulation, all of which are critical for gastropod survival. Previous studies assert that dense stands of *Impatiens glandulifera* reduced ground-level light penetration and created humid microhabitats that may have favoured certain slug species while negatively affecting others that preferred drier conditions (Ruckli *et al.*, 2013; Kiełtyk and Delimat, 2019; Seeney *et al.*, 2019; Coakley and Petti, 2021). Similarly, it was asserted by Vacek *et al.* (2020) and Casati *et al.* (2023) that *Rhododendron ponticum*, an aggressive invader in European woodlands, altered leaf litter composition and reduced understorey plant diversity, leading to a decline in native gastropod populations.

Since gastropods rely heavily on leaf litter, fungi and decaying organic matter as their primary food sources, the introduction of invasive plants might affect their diet quality and availability (Sharma *et al.*, 2005; Ojija *et al.*, 2024). This is because invasive plants tend to alter the nutritional composition and decomposition rates of leaf litter. Previous studies have shown that the leaf litter of some invasive plants, such as *Ailanthus altissima* (tree of heaven), has a lower nutrient content and decomposes at a slower rate, reducing the availability of high-quality organic matter for gastropods (Schweizer *et al.*, 2019). Additionally, invasive grasses like *Microstegium vimineum* outcompete native plant species and produce litter with high silica content, which is less palatable to decomposer gastropods (Strickland *et al.*, 2011). Besides, invasive plants can indirectly impact gastropods by facilitating the spread of non-native gastropod competitors or predators (Watz and Nyqvist, 2022). For instance, invasive plants that create dense vegetation cover and abundant leaf litter may provide ideal habitats for invasive gastropods; one example is *Arion vulgaris* (Spanish slug), which has been shown to outcompete native species (Watz and Nyqvist, 2022). Furthermore, invasive plants that attract non-native predators, like *Platydemus manokwari* (predatory flatworms), can increase predation pressure on native gastropod populations, further threatening their survival (Doherty *et al.*, 2016).

Changes in microclimatic conditions due to invasive plants can further create physiological stress for gastropods. Invasive species like *L. camara* form thick, impenetrable thickets that alter soil moisture levels and reduce temperature fluctuations, creating conditions that may not be suitable for native gastropods (Sharma *et al.*, 2005). In contrast, invasive grasses that dry out quickly can lead to desiccation stress in moisture-dependent gastropods, reducing their population densities in affected areas (Keller *et al.*, 2011). The decline of gastropod populations due to invasive plant impacts can have broader ecological consequences. Gastropods contribute to the breakdown of organic matter and facilitate nutrient cycling in forest and grassland ecosystems (Wronski and Hausdorf, 2010). A decline in gastropod diversity can lead to slower decomposition rates, altered microbial communities, and changes in soil structure (Vilà *et al.*, 2011; Zulu *et al.*, 2022). Furthermore, many gastropods serve as prey for birds, amphibians and small mammals, meaning their population declines can disrupt food webs and impact higher trophic levels (Dhiman *et al.*, 2020; Mohammadi and Ahmadzadeh, 2024).

In general, invasive plants exert significant effects on gastropod communities through habitat modification, changes in food availability, increased competition and predation, and shifts in microclimatic conditions. These impacts contribute significantly to population declines and biodiversity loss, with cascading effects on ecosystem functions such as nutrient cycling and decomposition (Dhiman *et al.*, 2020). Thus, future research may focus on long-term monitoring of gastropod responses to invasive plant management strategies, as well as conservation efforts aimed at preserving native plant communities to maintain gastropod diversity.

Strategies for protecting soil macrofauna populations from invasive plants

Effectively mitigating the impact of invasive plants on soil macrofauna – including earthworms, arthropods and gastropods – requires an integrated approach that combines ecological restoration, biological control, soil enhancement techniques and proactive monitoring (Moran *et al.*, 2005). These strategies help restore soil conditions, promote soil macrofauna biodiversity and maintain overall ecosystem stability.

Restoration of native vegetation

Replacing invasive plants with native vegetation is a fundamental step in restoring soil ecosystems. Native plants contribute to soil health by supporting diverse soil macrofauna communities, including earthworms, arthropods and gastropods. They provide suitable microhabitats, increase organic matter input and restore natural nutrient cycling, creating a balanced environment for invertebrate populations. Invasive species removal should be conducted using environmentally friendly methods such as manual removal or targeted herbicide application to minimize harm to soil organisms. In cases where invasive plants have significantly altered soil conditions, gradual restoration strategies should be implemented to ensure the successful re-establishment of native vegetation without further disruption to soil biodiversity.

Biological control methods

Biological control offers a sustainable approach to managing invasive plant species without the use of chemical herbicides that could negatively affect soil invertebrates (Ojija and Ngimba, 2021; Ojija *et al.*, 2024). The introduction of natural enemies, such as herbivorous insects, fungal pathogens or grazing animals, can help reduce the spread and dominance of invasive plants (Gaskin *et al.*, 2021). This, in turn, benefits soil macrofauna by restoring habitat conditions (Moran *et al.*, 2005). For instance, introducing specialist insect herbivores has been effective in controlling invasive plants like *F. japonica* and *P. australis*, helping to improve soil structure and nutrient availability. However, biological control programmes must be carefully monitored to avoid unintended ecological consequences, such as the introduction of new invasive species or disruption of native arthropod populations.

Soil amendments and habitat restoration

Enhancing soil quality through the addition of organic matter can help counteract the negative effects of invasive plants on soil macrofauna. The application of compost, mulch or biochar improves soil structure, increases moisture retention and promotes microbial activity, creating a more hospitable environment for earthworms, arthropods and gastropods. Organic soil amendments also enhance decomposition processes, restoring food availability for detritivores and decomposer organisms. Additionally, habitat restoration strategies like controlled grazing, conservation tillage and agroforestry practices can support soil macrofauna biodiversity by maintaining suitable habitats for soil invertebrates.

Early detection and rapid response

Monitoring soil communities in areas at risk of invasion is essential for detecting early changes in soil macrofauna populations and implementing timely interventions (Mgobozi *et al.*, 2008). Regular assessments of soil biodiversity, including earthworm abundance, arthropod diversity and gastropod distribution, can provide valuable insights into the effects of invasive plants on soil ecosystems. Rapid response strategies such as targeted invasive plant removal and habitat restoration can help prevent further degradation of soil macrofauna communities (Mgobozi *et al.*, 2008). Collaborative efforts between researchers, conservation practitioners and local communities can enhance monitoring efforts and facilitate early management actions before invasive plants establish dominance.

Integrated management approaches

A combination of restoration, biological control, soil amendments and monitoring is the most effective way to protect arthropod, earthworm and gastropod populations from invasive plants. Integrating these strategies into broader land management and conservation policies can help maintain soil macrofauna biodiversity and ecosystem resilience. Public awareness campaigns and stakeholder engagement in invasive species management can also promote sustainable land-use practices that support soil health and macrofauna biodiversity conservation. As climate change and human activity continue to drive the spread of invasive plants, adaptive management strategies will be vital in safeguarding soil macrofauna populations and preserving soil ecosystems. Future research should focus on long-term monitoring of soil communities in invaded landscapes, evaluating the effectiveness of different management techniques and developing innovative solutions to restore soil biodiversity in affected areas.

Conclusion

The studies reviewed here consistently show that invasive plants can have substantial negative impacts on soil macrofauna communities – including arthropods, gastropods and annelids – by disrupting microhabitat conditions, altering litter quality and changing soil chemistry. These disruptions often reduce soil biodiversity, impair nutrient cycling and compromise overall ecosystem resilience. However, it is noteworthy that not all effects of invasive plants on soil macrofauna are negative. Some studies have demonstrated that certain invasive plants may actually enhance the abundance and diversity of soil macrofauna in particular ecosystems. This highlights the complex and context-dependent nature of invasive plant impacts on soil communities. Overall, it is essential to recognize that invasive plants can exert both positive and negative influences on local soil macrofauna assemblages. Future research should adopt a multi-trophic and long-term perspective, with particular emphasis on understudied ecosystems and taxonomic groups, to improve our understanding and inform effective management and restoration strategies. Importantly, a deeper understanding of how native soil macrofauna assemblages respond to these changes is vital for managing these novel ecosystems and maximizing their potential biodiversity benefits.

References

Afzal, M.R., Naz, M., Ashraf, W. and Du, D. (2023) The legacy of plant invasion: Impacts on soil nitrification and management implications. *Plants (Basel, Switzerland)* 12(16), 1–14. DOI: 10.3390/plants12162980.

Ali, I.G., Sheridan, G. and French, J.R.J. (2013) Ecological benefits of termite soil interaction and microbial symbiosis in the soil ecosystem. *Journal of Earth Sciences and Geotechnical Engineering* 3(4), 63–85.

Aloo, P., Ojwang, W., Omondi, R., Njiru, J.M. and Oyugi, D. (2013) A review of the impacts of invasive aquatic weeds on the biodiversity of some tropical water bodies with special reference to Lake Victoria (Kenya). *Biodiversity Journal* 4(4), 471–482.

Banks-Leite, C., Ewers, R.M., Folkard-Tapp, H. and Fraser, A. (2020) Countering the effects of habitat loss, fragmentation, and degradation through habitat restoration. *One Earth* 3(6), 672–676. DOI: 10.1016/j.oneear.2020.11.016.

Black, H.I.J. and Okwakol, M.J.N. (1997) Agricultural intensification, soil biodiversity and agroecosystem function in the tropics: The role of termites. *Applied Soil Ecology* 6(1), 37–53. DOI: 10.1016/S0929-1393(96)00153-9.

Blakeslee, A.M.H., Canning-Clode, J., Lind, E.M. and Quilez-Badia, G. (2011) Biological invasions in the 21st century: Ecological impacts, predictions, and management across land and sea. *Environmental Research* 111(7), 891–892. DOI: 10.1016/j.envres.2011.09.005.

Blossey, B., Skinner, L.C. and Taylor, J. (2001) Impact and management of Purple Loosestrife (*Lythrum salicaria*) in North America. *Biodiversity & Conservation* 10(10), 1787–1807. DOI: 10.1023/A:1012065703604.

Bobuľská, L., Demková, L., Pinčáková, G. and Lošák, T. (2025) Alien plant invasion: Are they strictly nature's enemy and how can we use their supremacy? *Land* 14(1), 107. DOI: 10.3390/land14010107.

Bohlen, P.J., Scheu, S., Hale, C.M., McLean, M.A., Migge, S. *et al.* (2004) Non-native invasive earthworms as agents of change in northern temperate forests. *Frontiers in Ecology and the Environment* 2(8), 427–435. DOI: 10.1890/1540-9295(2004)002[0427:NIEAAO]2.0.CO;2.

Braschi, J., Hélard, O., Mazzia, C., Oger, P., Ponel, P. *et al.* (2021) Impacts of the removal of invasive *Carpobrotus* on spider assemblage dynamics. *Biodiversity and Conservation* 30(2), 497–518. DOI: 10.1007/s10531-020-02102-6.

Bugiel, L.N., Livingstone, S.W., Isaac, M.E., Fulthorpe, R.R. and Martin, A.R. (2018) Impacts of invasive plant species on soil biodiversity: A case study of dog-strangling vine (*Vincetoxicum rossicum*) in a Canadian national park. *Canadian Journal of Soil Science* 98(4), 716–723. DOI: 10.1139/cjss-2018-0093.

Casati, M., Kichey, T., Closset, D., Spicher, F. and Decocq, G. (2023) Is the invasive *Rhododendron ponticum* L. An emergent threat to mainland Atlantic forests? A population dynamics approach. *Forest Ecology and Management* 549(2), 121463. DOI: 10.1016/j.foreco.2023.121463.

Coakley, S. and Petti, C. (2021) Impacts of the invasive impatiens *Glandulifera*: Lessons learned from one of Europe's top invasive species. *Biology* 10(7), 619. DOI: 10.3390/biology10070619.

Dhiman, V., Pant, D., Sharma, D.K. and Prakash, P. (2020) A review on persisting threats to snail's diversity and its conservation approaches. *Archives of Agriculture and Environmental Science* 5(2), 205–217. DOI: 10.26832/24566632.2020.0502019.

Doherty, T.S., Glen, A.S., Nimmo, D.G., Ritchie, E.G. and Dickman, C.R. (2016) Invasive predators and global biodiversity loss. *Proceedings of the National Academy of Sciences USA* 113(40), 11261–11265. DOI: 10.1073/pnas.1602480113.

Ehrenfeld, J.G. (2010) Ecosystem consequences of biological invasions. *Annual Review of Ecology, Evolution, and Systematics* 41(1), 59–80. DOI: 10.1146/annurev-ecolsys-102209-144650.

Ehrenfeld, J.G. and Scott, N. (2001) Invasive species and the soil: Effects on organisms and ecosystem processes. *Ecological Applications* 11(5), 1259–1260. DOI: 10.1890/1051-0761(2001)011[1259:ISATSE]2.0.CO;2.

Flaherty, L., Hills, M., Giacobbo, V., Kuczmarski, P., Momborquette, M. *et al.* (2024) Impacts of garlic mustard (*Alliaria petiolata*, Brassicaceae) invasion on oribatid mites in urban forest soils vary with the size of the invaded patch. *Pedobiologia* 103, 150933. DOI: 10.1016/j.pedobi.2024.150933.

Gallé, R., Erdélyi, N., Szpisjak, N., Tölgyesi, C. and Maák, I. (2015) The effect of the invasive *Asclepias syriaca* on the ground-dwelling arthropod fauna. *Biologia* 70(1), 104–111. DOI: 10.1515/biolog-2015-0011.

Gaskin, J.F., Espeland, E., Johnson, C.D., Larson, D.L., Mangold, J.M. *et al.* (2021) Managing invasive plants on Great Plains grasslands: A discussion of current challenges. *Rangeland Ecology & Management* 78, 235–249. DOI: 10.1016/j.rama.2020.04.003.

Gerber, E., Krebs, C., Murrell, C., Moretti, M., Rocklin, R. *et al.* (2008) Exotic invasive knotweeds (*Fallopia* spp.) negatively affect native plant and invertebrate assemblages in European riparian habitats. *Biological Conservation* 141(3), 646–654. DOI: 10.1016/j.biocon.2007.12.009.

Getahun, S. and Kefale, H. (2023) Problem of water hyacinth (*Eichhornia crassipes* (Mart.)) in Lake Tana (Ethiopia): Ecological, economic, and social implications and management options. *International Journal of Ecology* 2023, 1–9. DOI: 10.1155/2023/4618069.

Greenleaf, J., Holásková, I., Rowen, E., Gutensohn, M., Turcotte, R. *et al.* (2023) Arthropods associated with invasive *Frangula alnus* (Rosales: Rhamnaceae): Implications for invasive plant and insect management. *Insects* 14(12), 1–18. DOI: 10.3390/insects14120913.

Gwate, O., Mantel, S.K., Gibson, L.A., Munch, Z., Gusha, B. *et al.* (2021) The effects of *Acacia mearnsii* (black wattle) on soil chemistry and grass biomass production in a South African semi-arid rangeland: Implications for rangeland rehabilitation. *African Journal of Range & Forage Science* 38(4), 270–280. DOI: 10.2989/10220119.2021.1875048.

Hale, C.M., Frelich, L.E., Reich, P.B. and Pastor, J. (2005) Effects of European earthworm invasion on soil characteristics in northern hardwood forests of Minnesota, USA. *Ecosystems* 8(8), 911–927. DOI: 10.1007/s10021-005-0066-x.

Hendrix, P.F., Callaham, M.A., Jr., Drake, J.M., Huang, C.-Y., James, S.W. *et al.* (2008) Pandora's box contained bait: The global problem of introduced earthworms. *Annual Review of Ecology, Evolution, and Systematics* 39(1), 593–613. DOI: 10.1146/annurev.ecolsys.39.110707.173426.

Herrera, A.M. and Dudley, T.L. (2003) Reduction of riparian arthropod abundance and diversity as a consequence of giant reed (arundo donax) invasion. *Biological Invasions* 5(3), 167–177. DOI: 10.1023/A:1026190115521.

Huang, X., Duan, Y., Tao, Y., Wang, X., Long, H. *et al.* (2022) Effects of *Spartina alterniflora* invasion on soil organic carbon storage in the Beihai coastal wetlands of China. *Frontiers in Marine Science* 9, 1–10. DOI: 10.3389/fmars.2022.890811.

Huerta, E. and van der Wal, H. (2012) Soil macroinvertebrates' abundance and diversity in home gardens in Tabasco, Mexico, vary with soil texture, organic matter and vegetation cover. *European Journal of Soil Biology* 50, 68–75. DOI: 10.1016/j.ejsobi.2011.12.007.

Ibrahima, A., Magouo, D.A., Ibrahim, O. and Hassana, B. (2017) Synergistic effects of earthworms and soil microorganisms on litter decomposition in Sudano-Guinea Savannah zone of Ngaoundere, Cameroon. *International Journal of Agricultural Research, Sustainability, and Food Sufficiency* 4(1), 133–150.

Jesse, W.A.M., Molleman, J., Franken, O., Lammers, M., Berg, M.P. *et al.* (2020) Disentangling the effects of plant species invasion and urban development on arthropod community composition. *Global Change Biology* 26(6), 3294–3306. DOI: 10.1111/gcb.15091.

Jordan, N.R., Larson, D.L. and Huerd, S.C. (2008) Soil modification by invasive plants: Effects on native and invasive species of mixed-grass prairies. *Biological Invasions* 10(2), 177–190. DOI: 10.1007/s10530-007-9121-1.

Jouquet, P., Chaudhary, E. and Kumar, A.R.V. (2018) Sustainable use of termite activity in agro-ecosystems with reference to earthworms. A review. *Agronomy for Sustainable Development* 38(1), 1–11. DOI: 10.1007/s13593-017-0483-1.

Jouquet, P., Traoré, S., Choosai, C., Hartmann, C. and Bignell, D. (2011) Influence of termites on ecosystem functioning. Ecosystem services provided by termites. *European Journal of Soil Biology* 47(4), 215–222. DOI: 10.1016/j.ejsobi.2011.05.005.

Keller, R.P., Geist, J., Jeschke, J.M. and Kühn, I. (2011) Invasive species in Europe: Ecology, status, and policy. *Environmental Sciences Europe* 23(1), 1–17. DOI: 10.1186/2190-4715-23-23.

Kiełtyk, P. and Delimat, A. (2019) Impact of the alien plant impatiens *Glandulifera* on species diversity of invaded vegetation in the northern foothills of the Tatra mountains, Central Europe. *Plant Ecology* 220(1), 1–12. DOI: 10.1007/s11258-018-0898-z.

Lazzaro, L., Giuliani, C., Fabiani, A., Agnelli, A.E., Pastorelli, R. *et al.* (2014) Soil and plant changes after invasion: The case of *Acacia dealbata* in a Mediterranean ecosystem. *Science of The Total Environment* 497–498, 491–498. DOI: 10.1016/j.scitotenv.2014.08.014.

Liu, J., Xu, G., Yin, L., Xu, X., Armitage, D.W. *et al.* (2020) Invasive plants exert disproportionately negative allelopathic effects on the growth and physiology of the earthworm *Eisenia fetida*. *Science of The Total Environment* 747, 141534. DOI: 10.1016/j.scitotenv.2020.141534.

Lusizi, Z., Motsi, H., Nyambo, P. and Elephant, D.E. (2024) Black (*Acacia mearnsii*) and silver wattle (*Acacia dealbata*) invasive tree species impact on soil physicochemical properties in South Africa: A systematic literature review. *Heliyon* 10(2), 1–9. DOI: 10.1016/j.heliyon.2024.e24102.

Maceda-Veiga, A., Basas, H., Lanzaco, G., Sala, M., de Sostoa, A. *et al.* (2016) Impacts of the invader giant reed (*Arundo donax*) on riparian habitats and ground arthropod communities. *Biological Invasions* 18(3), 731–749. DOI: 10.1007/s10530-015-1044-7.

Mamabolo, E., Pryke, J.S. and Gaigher, R. (2024) Soil macrofauna are important bioindicators of soil quality in agroecosystems under different management. *Ecological Indicators* 167, 112723. DOI: 10.1016/j.ecolind.2024.112723.

Mantoani, M.C., Alhakami, F.T., Fearon, H., Gioria, M., Schmidt, O. *et al.* (2022) *Gunnera tinctoria* invasions increase, not decrease, earthworm abundance and diversity. *Biological Invasions* 24(12), 3721–3734. DOI: 10.1007/s10530-022-02873-9.

Martin, M.R., Tipping, P.W., Reddy, K.R., Madeira, P.T. and Fitzgerald, D. (2011) An evaluation of the impact of *Melaleuca quinquenervia* invasion and management on plant community structure after fire. *Aquatic Botany* 95(4), 287–291. DOI: 10.1016/j.aquabot.2011.08.004.

Mavimbela, L.Z., Sieben, E.J.J. and Procheș, Ș. (2018) Invasive alien plant species, fragmentation and scale effects on urban forest community composition in Durban, South Africa. *New Zealand Journal of Forestry Science* 48(1), 1–19. DOI: 10.1186/s40490-018-0124-8.

Mgobozi, M.P., Somers, M.J. and Dippenaar-Schoeman, A.S. (2008) Spider responses to alien plant invasion: The effect of short- and long-term *Chromolaena odorata* invasion and management. *Journal of Applied Ecology* 45(4), 1189–1197. DOI: 10.1111/j.1365-2664.2008.01486.x.

Mohammadi, S. and Ahmadzadeh, F. (2024) Revealing hidden diversity and community dynamics of land snails through DNA barcoding: Implications for conservation and ecological studies. *Frontiers in Ecology and Evolution* 12, 1–14. DOI: 10.3389/fevo.2024.1329581.

Moran, V.C., Hoffmann, J.H. and Zimmermann, H.G. (2005) Biological control of invasive alien plants in South Africa: Necessity, circumspection, and success. *Frontiers in Ecology and the Environment* 3(2), 77. DOI: 10.2307/3868513.

Müller-Schärer, H., Schaffner, U. and Steinger, T. (2004) Evolution in invasive plants: Implications for biological control. *Trends in Ecology & Evolution* 19(8), 417–422. DOI: 10.1016/j.tree.2004.05.010.

Nunez-Mir, G.C. and McCary, M.A. (2024) Invasive plants and their root traits are linked to the homogenization of soil microbial communities across the United States. *Proceedings of the National Academy of Sciences USA* 121(44), 1–9. DOI: 10.1073/pnas.2418632121.

Ogura-Yamada, C.S. and Krushelnycky, P.D. (2020) The effects of the invasive thief ant, *Solenopsis papuana*, on ground-dwelling invertebrates in Mesic forests of Hawaiʻi. *Journal of Insect Conservation* 24(1), 151–162. DOI: 10.1007/s10841-019-00185-3.

Ojija, F. (2024) Perennial grasses: Natural allies for soil health and biodiversity, climate change mitigation, and invasive plant management. *Grass Research* 4(1), 1–6. DOI: 10.48130/grares-0024-0019.

Ojija, F. and Manyanza, N.M. (2021) Distribution and impact of invasive *Parthenium hysterophorus* on soil around Arusha national park, Tanzania. *Ecology and Evolutionary Biology* 6(1), 8. DOI: 10.11648/j.eeb.20210601.13.

Ojija, F. and Ngimba, C. (2021) Suppressive abilities of legume fodder plants against the invasive weed *Parthenium hysterophorus* (Asteraceae). *Environmental and Sustainability Indicators* 10, 100111. DOI: 10.1016/j.indic.2021.100111.

Ojija, F., Arnold, S.E.J. and Treydte, A.C. (2019) Impacts of alien invasive *Parthenium hysterophorus* on flower visitation by insects to co-flowering plants. *Arthropod-Plant Interactions* 13(5), 719–734. DOI: 10.1007/s11829-019-09701-3.

Ojija, F., Petruzzellis, F. and Bacaro, G. (2024) Review of invasive plant functional traits and management using remote sensing in sub-Saharan Africa. *International Journal of Plant Biology* 15(2), 358–374. DOI: 10.3390/ijpb15020029.

O'Reilly-Nugent, A., Palit, R., Lopez-Aldana, A., Medina-Romero, M., Wandrag, E. *et al.* (2016) Landscape effects on the spread of invasive species. *Current Landscape Ecology Reports* 1(3), 107–114. DOI: 10.1007/s40823-016-0012-y.

O'Sullivan, M., Dorken, M.E. and Freeland, J.R. (2019) Garlic mustard (*Alliaria petiolata*) is associated with an overall reduction in plant diversity, but is more likely to co-exist with native than alien species. *Plant Ecology & Diversity* 12(5), 427–439. DOI: 10.1080/17550874.2019.1626510.

Ovando, X.M.C., Miranda, M.J., Loyola, R. and Cuezzo, M.G. (2019) Identifying priority areas for invertebrate conservation using land snails as models. *Journal for Nature Conservation* 50, 1–10. DOI: 10.1016/j.jnc.2019.04.004.

Pearson, B.M., Minor, M.A., Robertson, A.W. and Clavijo McCormick, A.L. (2024) Plant invasion down under: Exploring the below-ground impact of invasive plant species on soil properties and invertebrate communities in the central plateau of New Zealand. *Biological Invasions* 26(12), 4215–4228. DOI: 10.1007/s10530-024-03441-z.

Pritekel, C., Whittemore-Olson, A., Snow, N. and Moore, J.C. (2006) Impacts from invasive plant species and their control on the plant community and belowground ecosystem at rocky mountain national park, USA. *Applied Soil Ecology* 32, 132–141. DOI: 10.1016/j.apsoil.2005.01.010.

Pyšek, P., Hulme, P.E., Simberloff, D., Bacher, S., Blackburn, T.M. *et al.* (2020) Scientists' warning on invasive alien species. *Biological Reviews* 95(6), 1511–1534. DOI: 10.1111/brv.12627.

Rodgers, V.L., Wolfe, B.E., Werden, L.K. and Finzi, A.C. (2008) The invasive species *Alliaria petiolata* (garlic mustard) increases soil nutrient availability in northern hardwood-conifer forests. *Oecologia* 157(3), 459–471. DOI: 10.1007/s00442-008-1089-8.

Rousseau, L., Fonte, S.J., Téllez, O., van der Hoek, R. and Lavelle, P. (2013) Soil macrofauna as indicators of soil quality and land use impacts in smallholder agroecosystems of Western Nicaragua. *Ecological Indicators* 27, 71–82. DOI: 10.1016/j.ecolind.2012.11.020.

Ruckli, R., Rusterholz, H.-P. and Baur, B. (2013) Invasion of impatiens *Glandulifera* affects terrestrial gastropods by altering microclimate. *Acta Oecologica* 47, 16–23. DOI: 10.1016/j.actao.2012.10.011.

Schirmel, J. (2020) Differential effects of American pokeweed (*Phytolacca americana*) invasion on ground-dwelling forest arthropods in southwest Germany. *Biological Invasions* 22(4), 1289–1298. DOI: 10.1007/s10530-019-02180-w.

Schweizer, M., Triebskorn, R. and Köhler, H.R. (2019) Snails in the sun: Strategies of terrestrial gastropods to cope with hot and dry conditions. *Ecology and Evolution* 9(22), 12940–12960. DOI: 10.1002/ece3.5607.

Seeney, A., Eastwood, S., Pattison, Z., Willby, N.J. and Bull, C.D. (2019) All change at the water's edge: Invasion by non-native riparian plants negatively impacts terrestrial invertebrates. *Biological Invasions* 21(6), 1933–1946. DOI: 10.1007/s10530-019-01947-5.

Sharma, G.P., Raghubanshi, A.S. and Singh, J.S. (2005) Lantana invasion: An overview. *Weed Biology and Management* 5(4), 157–165. DOI: 10.1111/j.1445-6664.2005.00178.x.

Shuvar, I., Korpita, H., Shuvar, A., Shuvar, B. and Kropyvnytskyi, R. (2021) Invasive plant species and the consequences of its prevalence in biodiversity. *BIO Web of Conferences* 31, 00024. DOI: 10.1051/bioconf/20213100024.

Sileshi, G.W., Arshad, M.A., Konaté, S. and Nkunika, P.O.Y. (2010) Termite-induced heterogeneity in African savanna vegetation: Mechanisms and patterns. *Journal of Vegetation Science* 21(5), 923–937. DOI: 10.1111/j.1654-1103.2010.01197.x.

Simberloff, D., Martin, J.-L., Genovesi, P., Maris, V., Wardle, D.A. *et al.* (2013) Impacts of biological invasions: What's what and the way forward. *Trends in Ecology & Evolution* 28(1), 58–66. DOI: 10.1016/j.tree.2012.07.013.

Singh, J., Schädler, M., Demetrio, W., Brown, G.G. and Eisenhauer, N. (2019) Climate change effects on earthworms – a review. *Soil Organisms* 91(3), 114–138. DOI: 10.25674/SO91ISS3PP114.

Sofo, A., Mininni, A.N. and Ricciuti, P. (2020) Soil macrofauna: A key factor for increasing soil fertility and promoting sustainable soil use in fruit orchard agrosystems. *Agronomy* 10(4), 1–20. DOI: 10.3390/agronomy10040456.

Srivastava, J., Kalra, S.J.S. and Naraian, R. (2014) Environmental perspectives of *Phragmites australis* (Cav.) Trin. Ex. Steudel. *Applied Water Science* 4(3), 193–202. DOI: 10.1007/s13201-013-0142-x.

Stiers, I., Crohain, N., Josens, G. and Triest, L. (2011) Impact of three aquatic invasive species on native plants and macroinvertebrates in temperate ponds. *Biological Invasions* 13(12), 2715–2726. DOI: 10.1007/s10530-011-9942-9.

Stinson, K.A., Campbell, S.A., Powell, J.R., Wolfe, B.E., Callaway, R.M. *et al.* (2006) Invasive plant suppresses the growth of native tree seedlings by disrupting belowground mutualisms. *PLoS Biology* 4(5), e140. DOI: 10.1371/journal.pbio.0040140.

Strickland, M.S., DeVore, J.L., Maerz, J.C. and Bradford, M.A. (2011) Loss of faster-cycling soil carbon pools following grass invasion across multiple forest sites. *Soil Biology and Biochemistry* 43(2), 452–454. DOI: 10.1016/j.soilbio.2010.10.006.

Strong, E.E., Gargominy, O., Ponder, W.F. and Bouchet, P. (2008) Global diversity of gastropods (Gastropoda; Mollusca) in freshwater. *Hydrobiologia* 595(1), 149–166. DOI: 10.1007/s10750-007-9012-6.

Tlou, S. and Duffy, K. (2022) The impact of *Lantana camara* on invertebrates and plant species of the Groenkloof nature reserve, South Africa. *Zoological Studies* 61(33), 1–10. DOI: 10.6620/ZS.2022.61-33.

Torres, N., Herrera, I., Fajardo, L. and Bustamante, R.O. (2021) Meta-analysis of the impact of plant invasions on soil microbial communities. *BMC Ecology and Evolution* 21(1), 1–6. DOI: 10.1186/s12862-021-01899-2.

Vacek, Z., Vacek, S., Eşen, D., Yildiz, O., Král, J. *et al.* (2020) Effect of invasive *Rhododendron ponticum* L. on natural regeneration and structure of *Fagus orientalis* Lipsky forests in the black sea region. *Forests* 11(5), 603. DOI: 10.3390/f11050603.

van Hengstum, T., Hooftman, D.A.P., Oostermeijer, J.G.B. and van Tienderen, P.H. (2014) Impact of plant invasions on local arthropod communities: A meta-analysis. *Journal of Ecology* 102(1), 4–11. DOI: 10.1111/1365-2745.12176.

Vilà, M., Espinar, J.L., Hejda, M., Hulme, P.E., Jarošík, V. *et al.* (2011) Ecological impacts of invasive alien plants: A meta-analysis of their effects on species, communities and ecosystems. *Ecology Letters* 14(7), 702–708. DOI: 10.1111/j.1461-0248.2011.01628.x.

Watz, J. and Nyqvist, D. (2022) Interspecific competition among terrestrial slugs. *Journal of Molluscan Studies* 88(2), 1–6. DOI: 10.1093/mollus/eyac007.

Wekhanya, M.N., Mbugua, P.K. and Mworia, J.K. (2020) The effect of invasive species *Lantana camara* L. on soil nutrients at Ol-Donyo Sabuk national park, Kenya. *African Journal of Pure and Applied Sciences* 1(1), 25–32. DOI: 10.33886/ajpas.v1i1.169.

Wiatrowska, B., Kurek, P., Rutkowski, T., Napierała, A., Sienkiewicz, P. *et al.* (2024) Consequences of *Spiraea tomentosa* invasion in Uropodina mite (Acari: Mesostigmata) communities in wet meadows. *Experimental & Applied Acarology* 93(3), 609–626. DOI: 10.1007/s10493-024-00951-2.

Wolfe, B.E., Rodgers, V.L., Stinson, K.A. and Pringle, A. (2008) The invasive plant *Alliaria petiolata* (garlic mustard) inhibits ectomycorrhizal fungi in its introduced range. *Journal of Ecology* 96(4), 777–783. DOI: 10.1111/j.1365-2745.2008.01389.x.

Wronski, T. and Hausdorf, B. (2010) Diversity and body-size patterns of land snails in rain forests in Uganda. *Journal of Molluscan Studies* 76(1), 87–100. DOI: 10.1093/mollus/eyp048.

Yang, J., Schrader, S. and Tebbe, C.C. (2024) Legacy effects of earthworms on soil microbial abundance, diversity, and community dynamics. *Soil Biology and Biochemistry* 190, 109294. DOI: 10.1016/j.soilbio.2023.109294.

Zhou, Y., Liu, C., Ai, N., Tuo, X., Zhang, Z. *et al.* (2022) Characteristics of soil macrofauna and its coupling relationship with environmental factors in the loess area of northern Shaanxi. *Sustainability* 14(5), 2484. DOI: 10.3390/su14052484.

Zulu, S.G., Motsa, N.M., Sithole, N.J., Magwaza, L.S. and Ncama, K. (2022) Soil macrofauna abundance and taxonomic richness under long-term no-till conservation agriculture in a semi-arid environment of South Africa. *Agronomy* 12(3), 722. DOI: 10.3390/agronomy12030722.

11 Ecological Impacts of *Solidago canadensis* Invasion on Ground-Dwelling Arthropods

Ioan Tăuşan*
Lucian Blaga University of Sibiu, Faculty of Sciences, Biology and Ecology Research Centre, Sibiu, Romania

Introduction: *Solidago canadensis*, A Background for a Great Invader

The Canadian goldenrod, *Solidago canadensis*, is a well-known invasive species in Europe and Asia (Weber, 2001; Dong *et al.*, 2006; Abhilasha *et al.*, 2008). This species is a major threat to semi-natural habitats in Central Europe and has a long history on the continent (Moroń *et al.*, 2009; Fenesi *et al.*, 2015a; Świerszcz *et al.*, 2024) (Fig. 11.1).

It was introduced to Europe in 1645 when it was first planted in England as an ornamental plant (Poljuha *et al.*, 2024). The species thrives along forest edges, roadsides and meadows (Gallardo *et al.*, 2019; Poljuha *et al.*, 2024) and is widespread (Fig. 11.1). Since 2004, it has been listed on the European and Mediterranean Plant Protection Organization (EPPO) list of invasive alien plants (EPPO, n.d.).

With late-summer blooming (typically in August), the plant produces numerous yellow flowers arranged in large paniculate inflorescences (Skrzypczak *et al.*, 1999) that attract pollinators such as syrphids, beetles, solitary bees and bumblebees (Robson *et al.*, 2017; Ustinova and Lysenkov, 2020), through their carbohydrate-rich nectar containing arabinose, galactose, glucose and arabinogalactan (Šutovská *et al.*, 2013) (Fig. 11.2).

Invasive species have unique characteristics that make them thrive in many habitats. *S. canadensis* is no exception. It may be considered a model invader due to an aggressive combination of superior reproductive strategies, potent allelopathic chemicals and the resulting ability to rapidly dominate habitats by forming dense monospecific stands (Weber, 2001; Dong *et al.*, 2006; Abhilasha *et al.*, 2008). The invasion of *S. canadensis* has significant ecological consequences, including the loss of native biodiversity and consequential alterations in plant community structure (Hejda *et al.*, 2009; Liu *et al.*, 2017). These changes cascade through trophic levels, affecting herbivorous insects (Skórka *et al.*, 2010; Fenesi *et al.*, 2015b), soil fauna such as nematodes (Li *et al.*, 2021; Zhylina *et al.*, 2025) and microbial communities essential for nutrient cycling and soil health (Zhang *et al.*, 2009). Alterations in these biotic interactions can disrupt ecosystem functions like carbon sequestration and nutrient turnover. The displacement of native species and simplification of ecosystem composition diminish ecological resilience and may trigger further invasions or

*Corresponding author: ioan.tausan@ulbsibiu.ro

DOI: 10.1079/9781836991298.0011

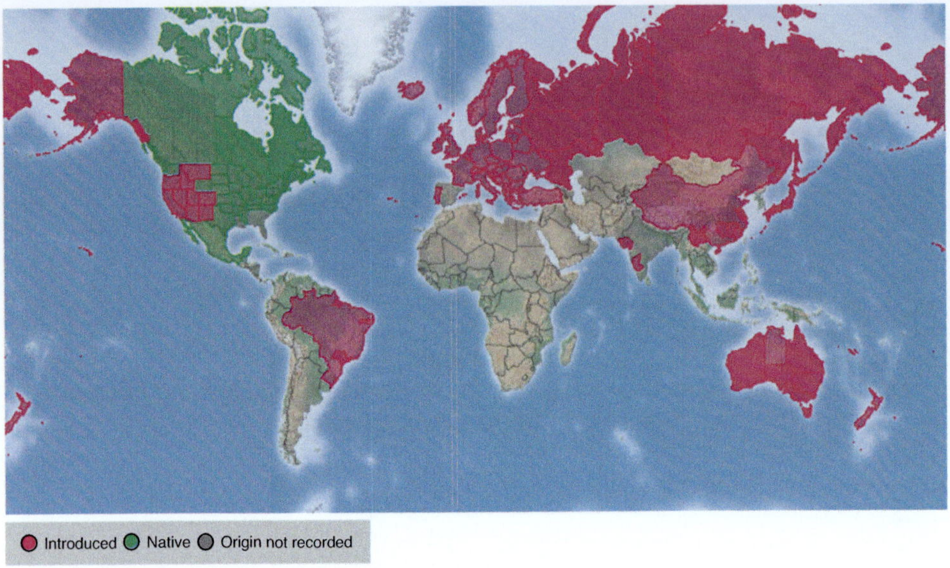

Fig. 11.1. Worldwide distribution of *Solidago canadensi*. Grey areas: origin of *Solidago* not recorded. Reproduced from Poljuha *et al*. (2024) with information from Popay and Parker (2014).

ecosystem degradation (Zhu *et al*., 2022; He *et al*., 2024; Zhylina *et al*., 2025).

These widespread effects call for urgent assessment and intervention to safeguard ecosystem services (reviewed in Table 11.1).

Soil Arthropods and Their Importance

Soil and ground-dwelling arthropods (insects, spiders, mites, centipedes, etc.) are crucial components of virtually all terrestrial ecosystems. Their activities drive processes essential for soil fertility, pH, air circulation, nutrient cycling and plant growth, effectively serving as ecosystem engineers of the soil environment (Frouz and Jilková, 2008; Culliney, 2013; Gonçalves *et al*., 2021).

In the ground-dwelling arthropods there are two groups which are an important component of terrestrial ecosystems, namely ground beetles (Coleoptera: Carabidae) and ants (Hymenoptera: Formicidae). Ground beetles are one of the most species-rich insect families globally and are ubiquitous in terrestrial habitats (Lövei and Sunderland, 1996). They are recognized for their diverse roles as highly effective generalist predators of other arthropods (including insect pests), slugs and earthworms. This predatory role provides a crucial ecosystem service of biological control in both natural and agricultural settings (Kromp, 1999; Bohan *et al*., 2000). Moreover, they are good seed dispersers, particularly in temperate regions, where they are known to consume seeds or nutrient-rich attachments to seeds (Honek *et al*., 2003; Kulkarni *et al*., 2015). Carabid beetles are reliable bioindicators which are sensitive to environmental changes, habitat disturbance and pollution, making them excellent bioindicators of ecosystem health (Rainio and Niemelä, 2003). Changes in their species' richness, abundance or community composition often reflect alterations in soil quality, ground cover or land-use intensity (Lock *et al*., 2001; Kotze *et al*., 2011; Borchard *et al*., 2014).

Ants are arguably the most successful insect group on Earth, often dominating the biomass of terrestrial invertebrate communities. Their high abundance and diverse behaviours enhance their status as keystone species in many ecosystems (Schultheiss *et al*., 2022). Ants impact a vast number of other species and ecological processes disproportionately to their biomass. They influence nutrient dynamics, prey populations and plant distribution, often being identified as keystone species due to their profound ecosystem influence (Folgarait,

Fig. 11.2. *Solidago canadensis ex situ.*

1998). They are known as soil engineers via their extensive nest-building and tunnelling activities. They turn over massive amounts of soil (bioturbation), thereby improving soil aeration, water infiltration and nutrient mixing, which benefits plant growth (Frouz and Jilková, 2008; Zelikova *et al.*, 2011). Moreover, many plant species produce seeds with elaiosomes specifically to attract ants. Ants collect these seeds, transport them back to the nest, consume the elaiosome and discard the intact seed in a nutrient-rich, protected environment, making them primary agents of seed dispersal in many forests and scrublands (Beattie and Hughes, 2002). Ants are highly effective and often cooperative predators, preying on a wide range of invertebrates, including the eggs and larvae of pest species. This continuous predation is vital for regulating local invertebrate populations (Davidson *et al.*, 2003; Rosumek *et al.*, 2009). Similar to ground beetles, ants are sensitive to habitat fragmentation, vegetation structure and microclimate, making them valuable bioindicators for monitoring biodiversity and ecological restoration success (Andersen and Majer, 2004; Lawes *et al.*, 2017).

Solidago canadensis Invasion and Ground-Dwelling Invertebrates

As already discussed, *S. canadensis* has a plethora of characteristics that make it an aggressive invasive plant species. Yet, ground beetles do not have the same responses in terms of community features. Results from studies are inconsistent in terms of the effects of invasive species on ground-dwelling invertebrates. In general, there is the assumption that an invasive species has negative effects on community richness and diversity (Hejda *et al.*, 2009). Moreover, the invasion of grasslands by *S. canadensis* often leads to the formation of large, dense stands that alter the habitat (Szymura *et al.*, 2016), which in turn may impact ground-dwelling invertebrates.

Table 11.1. *Solidago canadensis*: key features of a successful invasive species.

Feature	Description	References
Rapid clonal growth	*S. canadensis* possesses an aggressive, robust system of creeping rhizomes (underground stems). This allows it to reproduce asexually and spread laterally, forming dense, expansive patches (clones colonies) quickly. This strategy enables rapid establishment and persistence, especially in disturbed or abandoned areas.	Baranová *et al*. (2022), USDA Forest Service (n.d.)
High fecundity	The plant produces a large number of tiny seeds (achenes) per plant, leading to high propagule pressure. The seeds are equipped with a pappus, making them highly wind dispersed. This ensures wide dispersal and the colonization of new, distant sites, compounding its invasive potential.	Popay and Parker (2014), USDA Forest Service (n.d.)
Fast growth and high biomass	It exhibits a significantly higher relative growth rate and achieves greater aboveground biomass than many native species in invaded regions. The plant can grow tall (up to 2 m), effectively shading and suppressing shorter native vegetation, thereby reducing its photosynthetic capacity.	Baranová *et al*. (2022), Judžentienė *et al*. (2023)
Phenological advantage	In many invaded regions, *S. canadensis* has an earlier and longer growing season compared to native plants. This gives it a start in acquiring resources (light, water, nutrients) and allows it to dominate the growing season, further marginalizing native competitors.	Fenesi *et al*. (2015a)
Strong allelopathy	A critical weapon is the release of allelopathic chemicals (e.g. polyphenolic acids, flavonoids and essential oils) from its roots and leaves. These compounds actively inhibit the germination, root development and growth of neighbouring native plants, establishing a 'chemical exclusion zone'.	Judžentienė *et al*. (2023), Poljuha *et al*. (2024)
Formation of monospecific stands	The combined effect of clonal expansion, high biomass, shading and allelopathy results in the rapid formation of large, monodominant stands or 'thickets'. This reduces the structural and compositional diversity of the habitat, leading to the loss of native plant and animal species richness (i.e. ecological homogenization).	Kajzer-Bonk *et al*. (2016), Szitár *et al*. (2024)
Positive plant-soil feedback	Its dense stands alter the soil environment, often by creating a unique litter composition and affecting microbial communities. This creates soil conditions more favourable to itself (e.g. altered nutrient pools, lower soil moisture) than to native species, effectively promoting and reinforcing its own invasion a process known as positive plant–soil feedback.	Baranová *et al*. (2022); Zubek *et al*. (2020)
Generalist ecology	*S. canadensis* is highly adaptable and can tolerate a wide range of soil and moisture conditions (e.g. from dry open fields to damp meadows), although it prefers moist, nutrient-rich soils.	Fenesi *et al*. (2015a)

Canadian goldenrod versus ground-dwelling beetles

Baranová *et al.* (2014) showed that in eastern Slovakia invasion significantly alters the ground-dwelling beetle community, leading to a complex restructuring of species distribution and abundance. While the invaded areas exhibited greater overall diversity and evenness of ground-dwelling beetles, along with a higher abundance of rove (Staphylinidae) and carrion beetles (Silphidae), they simultaneously show a clear decline in the taxonomic richness and abundance of ground beetles. This shift is accompanied by the observation of less-common Carabidae species and a significantly higher representation of stenotopic brachypterous habitat specialists, within goldenrod stands, suggesting that the altered environment, while generally detrimental to most ground beetles, provides specialized niches for certain rare or habitat-restricted species.

In Slovenia, de Groot *et al.* (2007) showed that insect species groups most closely related to plant species composition are most sensitive to changes in the environment caused by the invasion of *Solidago*. The richness of the species group that relies only on vegetation to provide cover and prey, the ground beetles, was not affected, although their abundance decreased. In Romania, field work was carried out during August 2018 on the Târnavelor plateau (Transylvania, Romania) in 12 distinct meadows (invaded and non-invaded areas). The experimental design included two types of exposure: northern and southern. Overall, 12 sampling sites were investigated with six of them being invaded with *S. canadensis* and six non-invaded. In each invaded and non-invaded area, three sites had northern exposures and three had southern exposures. We applied the most common technique for collecting ground-dwelling invertebrates: the use of Barber pitfall traps. Each sampling site consisted of 15 pitfall traps in a 3 × 5 network with 5m between the traps. The experimental design included two types of habitats: invaded and non-invaded, and two types of exposure: northern and southern. Principal component analysis (PCA) showed that the first two canonical axes (habitat type and exposure) accounted for 58.6% of the variability in Carabidae species composition (Fig. 11.3). However, there did not appear to be an evident correlation between Carabidae species preferences and the presence or absence of *S. canadensis* in the habitats. Thus, this indicates that the Carabidae species' habitat preferences are independent of the presence of *S. canadensis*. The observations indicated by the PCA show that the Carabidae species exhibit a tendency to be found in certain habitats, especially in northern-exposed sites, where *Amara* sp., *Anchamenus dorsalis*, *Harpalus* sp. and *Pterostichus melas* are found, and in some southern-exposed sites, which hosts *Carabus violaceus*, *Pterostichus melanarius*, *Carabus cancellatus* and *Carabus arvensis*.

The results from ground-dwelling beetles show a complex and inconsistent picture, with effects being highly taxon-dependent and varying with habitat type and degree of invasion. Invasion has been shown to increase the abundance and diversity of some other beetle families (e.g. rove and carrion beetles) in invaded grasslands, suggesting a shift in the overall beetle community structure. Tóth *et al.* (2025) showed that impacts are taxon-dependent, where certain myriapod taxa, such as the centipedes, appeared to benefit, likely due to a positive plant–soil feedback mechanism.

Ant assemblages' response to *Solidago canadensis*

If other invertebrate groups have contrasting responses to the invasion of the Canadian goldenrod, ant communities generally exhibit a negative response. In Europe, there are several case studies where *S. canadensis* imposed a dramatic effect on ant assemblages. In Poland, the invasion of *Solidago* plants led to a significant reduction in the abundance and species richness of ant communities in invaded meadows compared to non-invaded ones. The most important environmental factor influencing the ant community structure was soil moisture. Lower moisture in the invaded plots was strongly and negatively correlated with the abundance of ants, which are critical host species for endangered butterflies. *Myrmica* species were abundant in uninvaded meadows; however, *Lasius niger* became the dominant species in the invaded plots, suggesting that this generalist species is more tolerant to the drier, altered conditions. The overall dry mass of epigean invertebrates (a measure of protein food availability for ants) was significantly lower in invaded meadows compared to uninvaded ones. This loss of prey is a major consequence of the invasion that contributes to the decline in ant diversity (Trigos-Peral *et al.*, 2018). These results are similar to those of Kajzer-Bonk *et al.* (2016), who provided evidence of the negative effects of invasion. This study was designed to assess the direct impact of invasive goldenrods (*S. canadensis* and *Solidago gigantea*) on ant communities in semi-natural wet grasslands. Regarding abundance and nest densities there was a strong and negative dependence between the presence

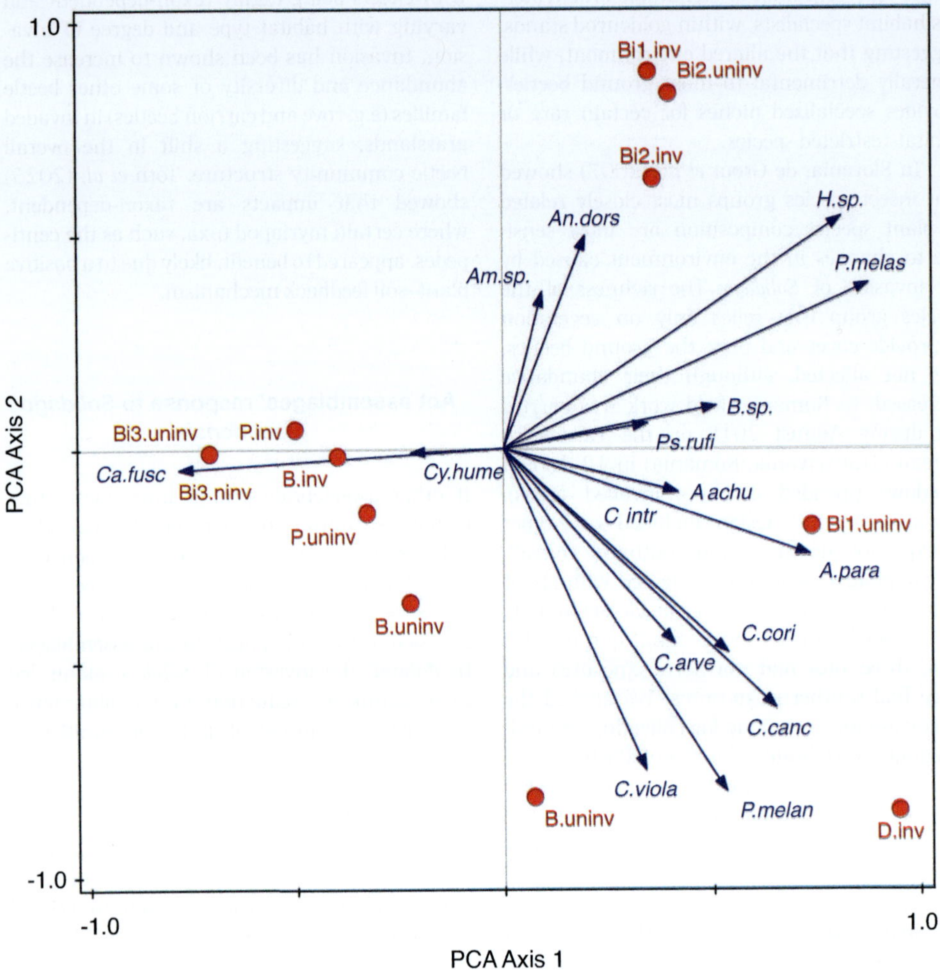

Fig. 11.3. PCA analysis of ground beetle communities from Romania. Ca.fusc, *Calathus fuscipes*; Am.sp., *Amara* sp.; An.dors, *Anchamenus dorsalis*; H.sp., *Harpalus* sp.; P.melas, *Pterostichus melas*; B.sp., *Bembidion* sp.; Ps.rufi, *Pseudophonus rufipes*; A.schu, *Abax schueppeli*; C.intr, *Carabus intricatus*; Cy.hume, *Cymindis humeralis*; C.cori, *Carabus coriaceus*; C.canc, *Carabus cancellatus*; C.viola, *Carabus violoaceus*; C.arve, *Carabus arvensis*; P.melan, *Pterostichus melanarius*; Bi1.inv, Bi2.inv, Bi3.inv, northern-exposed sites with *S. canadensis*; Bi1.uninv, Bi2.univ, Bi3.uninv, northern-exposed sites without *S. canadensis*; B.uninv, D.uninv, P.uninv, southern-exposed sites without *S. canadensis*; B.inv, D.inv, P.inv, southern-exposed sites with *S. canadensis*.

of goldenrod cover and the above parameters. Moreover, the number of ant nests was reduced by more than 50% in the plots invaded by goldenrods compared to control plots. The invasion caused a significant reduction in ant species richness and changes in species composition. *Myrmica* ants, which are considered keystone species and essential hosts for endangered *Phengaris* (*Maculinea*) butterflies, were among the most affected species, showing significantly lower nest numbers in invaded areas.

Further research indicates that ant colonies in goldenrod-invaded grasslands show smaller colony sizes and differences in brood investment compared to those in uninvaded areas. For example, *Myrmica rubra* colonies in invaded meadows are smaller but allocate reproductive output differently, possibly as a colony-level adaptive response to habitat stress around high goldenrod cover (Grześ *et al.*, 2018). The presence of *S. canadensis* seems to induce these changes through environmental stressors such as altered competition, reduced habitat heterogeneity and vegetation structural shifts. In the study of Lenda *et al.* (2013), the spread of Canadian goldenrod was associated with lower specific diversity of ants, and even with a reduced number of anthills, which were also smaller than normal, including species of the genus *Myrmica*, *Lasius* and *Formica* (Lenda *et al.*, 2013).

In more urbanized habitats, the results are even more pronounced. In Romania, the ant population is sparse and consists primarily of resilient, tolerant species, except for *Formica clara*. While invaded and non-invaded areas still show significant differences in species richness, the overall picture is bleak: an invasion's effects are so severe that it impoverishes even adjacent, non-invaded sites (Mirică and Tăuşan, 2019).

Trophic Cascade Effects

S. canadensis is a dominant and often invasive plant species with pronounced effects on ecosystem structure and function across multiple trophic levels, producing both direct and indirect (trophic cascade) effects in invaded habitats.

Within terrestrial ecosystems, species groups occupying different trophic levels responded differently to invasion, with the severity of the negative impact correlating with the group's dependence on the native plant species composition. The most affected group is, as expected, butterfly communities, for which species richness, abundance and diversity are shown to be consistently low in response to invasion. Yet, for other groups such as the pollinators, although the initial effect is negative, the impact can shift to positive during the peak flowering season of *S. canadensis*, typically in August (de Groot *et al.*, 2007).

S. canadensis often outcompetes native vegetation, primarily by altering light availability and nutrient cycling. In old-field ecosystems, a high abundance of *S. canadensis* typically correlates with reduced richness and biomass of other plant species, thereby restructuring the base of the ecosystem (Eckberg *et al.*, 2023). Its intensive litter production and rapid decomposition introduce allelopathic compounds and altered nutrient flows, which further shape soil chemical properties and plant community dynamics (Sun *et al.*, 2024). These early allelopathic effects, followed by nutrient enrichment, can facilitate further goldenrod invasion at the expense of native plants.

Arthropod communities show strong responses to *S. canadensis* dominance. Invaded patches usually support a lower diversity and total biomass of native arthropod communities, with detritivores and predators particularly suppressed due to altered plant-derived resources and habitat changes (Eckberg *et al.*, 2023). Herbivores and omnivores may be less affected directly but shifts in plant community composition limit available resources for specialist feeders. In addition, the altered chemical and structural complexity of habitats resulting from invasion can lead to changes in insect behaviour, abundance and sometimes host shifts in herbivores and their natural enemies, further changing the trophic web (Bezemer *et al.*, 2014).

Mechanism of *S. canadensis* Invasion

There are certain mechanisms that enhance *S. canadensis* to negatively affect ants and ground beetles by reducing plant and habitat diversity, leading to landscape homogenization (Zhang *et al.*, 2025) and by changing microclimatic and soil conditions, such as reducing soil fertility, altering moisture and pH, and affecting detrital food webs, which can cascade through soil and surface-active arthropod communities (Li *et al.*, 2021). Moreover, the mechanisms include the displacing of keystone species, whose ecological roles are not fully replaced, resulting in decreased functional richness of arthropods in invaded areas (Eckberg *et al.*, 2023).

Soil microbial and mycorrhizal changes initiated by *S. canadensis* may also further disadvantage ant and ground beetle populations by altering the foundational trophic structures of the ecosystem (Zhang *et al.*, 2010).

Table 11.2. The mechanism of *Solidago canadensis* affecting ant and ground beetle communities.

Mechanism	Description	Relevance to Carabidae and Formicidae	References
Habitat structure change	Formation of dense, tall and extensive clonal monocultures (monospecific stands) that shade out native low-growing plants.	Alters the microclimate (light, temperature, humidity) and physical structure of the ground layer, which is crucial for the foraging and nesting of both groups.	Baranová *et al.* (2014)
Soil alterations	Changes in soil physicochemical properties (e.g. reduced soil moisture, altered organic carbon and nitrogen levels, shifts in enzyme activity).	Influences the availability of soil-dwelling prey, nesting conditions for ants, and the overall soil food web structure. Carabid beetles and centipede species composition, for instance, are strongly influenced by soil moisture.	Li *et al.* (2021), Tóth *et al.* (2025)
Allelopathy	Release of bioactive compounds (allelochemicals such as polyphenolic acids, flavonoids, and essential oils) from *S. canadensis* roots and leaves.	These chemicals can directly inhibit other plants, reducing the diversity of native flora, which in turn reduces the prey and habitat resources available to ground-dwelling predators and omnivores.	Abhilasha *et al.* (2008), Kato-Noguchi and Kato (2022), Yuan *et al.* (2013)
Food web disruption	Reduction in the diversity of native plants leads to a decline in herbivorous insect populations (key food source for predators like carabids and some ants) and alters the quality and quantity of detritus entering the soil food web.	Primary consumers (e.g. millipedes, symphylans) are often more affected than predators (centipedes), suggesting a bottom-up trophic effect on invertebrate communities.	Eckberg *et al.* (2023), Korell *et al.* (2019), Zhylina *et al.* (2025)

The negative and shifting effects on ground-dwelling invertebrates are likely driven by a combination of factors related to the changes that *S. canadensis* induces in invaded habitats. The mechanism underlying the impacts of *S. canadensis* invasion on ground beetles and ants are summarized in the Table 11.2.

Research Gaps and Future Directions

Despite extensive research on *S. canadensis*, important gaps remain in understanding the complex biotic and abiotic interactions that govern its invasion success. Experimental studies often focus on specific stages of invasion or isolated environmental factors; however, long-term, multi-scale and integrative studies are needed to capture the dynamic processes interacting, particularly under global change scenarios that include climate change, altered precipitation patterns and nitrogen deposition (Ren *et al.*, 2022). Additionally, the combined impacts of invasive species and emerging pollutants such as microplastics on agroecosystem health require further testing and elucidation to inform management (Li *et al.*, 2023).

Further advances in environmentally friendly and sustainable control methods represent an important frontier. The allelopathic compounds produced by *S. canadensis* and their microbial interactions present avenues for developing biocontrol agents or natural product-based management tools (Elshafie *et al.*, 2019). Research into the roles of rhizosphere bacteria and microbial communities in modulating invasion dynamics suggests novel microbiome-based interventions that could support invasive weed suppression and ecosystem restoration (Huang *et al.*, 2025).

Technological progress in monitoring, modelling and predictive mapping of invasive species distributions holds significant potential for improving management strategies. The application of species distribution models using the propagule pressure–abiotic–biotic (PAB) framework has demonstrated effectiveness in understanding invasion patterns (Perera *et al.*, 2021). Remote sensing technologies combined with machine learning algorithms, such as Support Vector Machines and Random Forests applied to hyperspectral or multispectral data, offer powerful tools for detecting, mapping and quantifying *S. canadensis* populations at landscape scales (Sabat-Tomala *et al.*, 2020). Moreover, the role of protected areas in mitigating invasion risk demands further assessment, particularly by integrating climate and land-use change projections with invasion biology and ecosystem resilience frameworks (Liu *et al.*, 2024).

Conclusion

The invasion of *S. canadensis* represents a significant ecological threat to the biodiversity of ground-dwelling invertebrate communities, particularly in European grasslands and meadows.

S. canadensis consistently alters habitat structure, soil conditions and plant community composition, typically leading to a reduction in plant species richness. These habitat changes frequently, but not universally, lead to declines in many ground-specialist taxa, namely some ground beetles (Carabidae) and ants (Formicidae). Simultaneously, some detritivore and generalist taxa (e.g. several Staphylinidae and Silphidae, and sometimes spiders or centipedes) may increase in abundance resulting in taxon-specific and context-dependent responses. The overall outcome depends strongly on invasion density, habitat type, seasonality and management history, emphasizing the need for multi-taxon, multi-season monitoring combined with functional and structural trait data.

The effect on ground beetles is complex, leading to shifts in assemblage composition, but not to a consistent loss of overall diversity, as some species may benefit from the altered habitat structure. For ground beetles and similar predatory arthropods, the invasion leads to habitat simplification and trophic disruption with likely negative consequences for their populations (Eckberg *et al.*, 2023; Tóth *et al.*, 2025).

S. canadensis exerts strong negative effects on ant communities, reducing abundance and diversity and altering life-history strategies in invaded habitats (Kajzer-Bonk *et al.*, 2016; Grześ *et al.*, 2018). This results in a substantial reduction in nest abundance and species richness, with keystone genera like *Myrmica* being particularly vulnerable. These effects are complex, context-dependent, and can cascade across ecosystem functions.

The mechanisms driving these changes are multifaceted, involving alterations in physical habitat structure, soil chemistry and allelopathic effects, and resulting in disruptions to the lower trophic levels of the food web. Immediate management efforts are strongly recommended for biodiversity hotspots affected by goldenrod invasion to protect vulnerable invertebrate taxa, especially ant species that are essential hosts for myrmecophilous insects (Kajzer-Bonk *et al.*, 2016; Eckberg *et al.*, 2023).

In summary, *S. canadensis* exemplifies a highly successful invasive plant species whose global spread has far-reaching ecological, economic and social implications. Its complex reproductive strategies, allelopathic interactions and dynamic responses to environmental factors enable it to dominate diverse habitats and influence ecosystem-level processes. Despite advances in understanding its biology and impacts, integrated and innovative approaches are necessary to manage its invasion and mitigate associated risks.

References

Abhilasha, D., Quintana, N., Vivanco, J.M. and Joshi, J. (2008) Do allelopathic compounds in invasive *Solidago canadensis* s.l. restrain the native European flora? *Journal of Ecology* 96(5), 993–1001. DOI: 10.1111/j.1365-2745.2008.01413.x.

Andersen, A.N. and Majer, J.D. (2004) Ants show the way down under: The use of functional groups in ant ecology and biodiversity studies in Australia. *Biodiversity and Conservation* 13(13), 2523–2534.

Baranová, B., Manko, P. and Jászay, T. (2014) Differences in surface-dwelling beetles of grasslands invaded and non-invaded by goldenrods (*Solidago canadensis*, *S. gigantea*) with special reference to carabidae. *Journal of Insect Conservation* 18(4), 623–635. DOI: 10.1007/s10841-014-9666-0.

Baranová, B., Troščáková-Kerpčárová, E. and Gruľová, D. (2022) Survey of the *Solidago canadensis* 1. morphological traits and essential oil production: Aboveground biomass growth and abundance of the invasive goldenrod appears to be reciprocally enhanced within the invaded stands. *Plants (Basel, Switzerland)* 11(4), 535. DOI: 10.3390/plants11040535.

Beattie, A.J. and Hughes, L. (2002) Ant–plant interactions. In: Herrera, C. and Pellmyr, O. (eds) *Plant–Animal Interactions: An Evolutionary Approach*. Blackwell Science, Oxford, pp. 211–235.

Bezemer, T.M., Harvey, J.A. and Cronin, J.T. (2014) Response of native insect communities to invasive plants. *Annual Review of Entomology* 59(1), 119–141. DOI: 10.1146/annurev-ento-011613-162104.

Bohan, D.A., Bohan, A.C., Glen, D.M., Symondson, W.O.C., Wiltshire, C.W. *et al.* (2000) Spatial dynamics of predation by carabid beetles on slugs. *Journal of Animal Ecology* 69(3), 367–379. DOI: 10.1046/j.1365-2656.2000.00399.x.

Borchard, F., Buchholz, S., Helbing, F. and Fartmann, T. (2014) Carabid beetles and spiders as bioindicators for the evaluation of montane heathland restoration on former spruce forests. *Biological Conservation* 178, 185–192. DOI: 10.1016/j.biocon.2014.08.006.

Culliney, T.W. (2013) Role of arthropods in maintaining soil fertility. *Agriculture* 3(4), 629–659. DOI: 10.3390/agriculture3040629.

Davidson, D.W., Cook, S.C., Snelling, R.R. and Chua, T.H. (2003) Explaining the abundance of ants in lowland tropical rainforest canopies. *Science* 300(5621), 969–972. DOI: 10.1126/science.1082074.

de Groot, M., Kleijn, D. and Jogan, N. (2007) Species groups occupying different trophic levels respond differently to the invasion of semi-natural vegetation by *Solidago canadensis*. *Biological Conservation* 136(4), 612–617. DOI: 10.1016/j.biocon.2007.01.005.

Dong, M., Lu, J.-Z., Zhang, W.-J., Chen, J.-K. and Li, B. (2006) Canada goldenrod (*Solidago canadensis*): An invasive alien weed rapidly spreading in China. *Acta Phytotaxonomica Sinica* 44(1), 72. DOI: 10.1360/aps050068.

Eckberg, J.N., Hubbard, A., Schwarz, E.T., Smith, E.T. and Sanders, N.J. (2023) The dominant plant species *Solidago canadensis* structures multiple trophic levels in an old-field ecosystem. *Ecosphere* 14(1), e4393. DOI: 10.1002/ecs2.4393.

Elshafie, H.S., Gruľová, D., Baranová, B., Caputo, L., De Martino, L. *et al.* (2019) Antimicrobial activity and chemical composition of essential oil extracted from *Solidago canadensis* L. *Molecules (Basel, Switzerland)* 24(7), 1206. DOI: 10.3390/molecules24071206.

EPPO (n.d.) Lists of invasive alien plants. Available at: https://www.eppo.int/ACTIVITIES/invasive_alien_plants/iap_lists#iap (accessed 2 August 2025).

Fenesi, A., Geréd, J., Meiners, S.J., Tóthmérész, B., Török, P. *et al.* (2015a) Does disturbance enhance the competitive effect of the invasive *Solidago canadensis* on the performance of two native grasses? *Biological Invasions* 17(11), 3303–3315. DOI: 10.1007/s10530-015-0954-8.

Fenesi, A., Vágási, C.I., Beldean, M., Földesi, R., Kolcsár, L.-P. *et al.* (2015b) *Solidago canadensis* impacts on native plant and pollinator communities in different-aged old fields. *Basic and Applied Ecology* 16(4), 335–346. DOI: 10.1016/j.baae.2015.03.003.

Folgarait, P.J. (1998) Ant biodiversity and its relationship to ecosystem functioning: A review. *Biodiversity & Conservation* 7(9), 1221–1244. DOI: 10.1023/A:1008891901953.

Frouz, J. and Jilková, V. (2008) The effect of ants on soil properties and processes (Hymenoptera: Formicidae). *Myrmecological News* 11(11), 191–199.

Gallardo, B., Bacher, S., Bradley, B., Comín, F.A., Gallien, L. *et al.* (2019) InvasiBES: Understanding and managing the impacts of invasive alien species on biodiversity and ecosystem services. *NeoBiota* 50, 109–122. DOI: 10.3897/neobiota.50.35466.

Gonçalves, F., Carlos, C., Crespo, L., Zina, V., Oliveira, A. *et al.* (2021) Soil arthropods in the douro demarcated region vineyards: General characteristics and ecosystem services provided. *Sustainability* 13(14), 7837. DOI: 10.3390/su13147837.

Grześ, I.M., Ślipiński, P., Babik, H., Moroń, D., Walter, B. *et al.* (2018) Colony size and brood investment of *Myrmica rubra* ant colonies in habitats invaded by goldenrods. *Insectes Sociaux* 65(2), 275–280. DOI: 10.1007/s00040-018-0612-0.

He, M., Liu, T., Liu, Y., Long, Y., Li, Y. *et al.* (2024) Impacts of *Solidago canadensis* and *Erigeron canadensis* co-invasion on soil microbial communities. *Conservation Science and Practice* 6(9), e13202. DOI: 10.1111/csp2.13202.

Hejda, M., Pyšek, P. and Jarošík, V. (2009) Impact of invasive plants on the species richness, diversity and composition of invaded communities. *Journal of Ecology* 97(3), 393–403. DOI: 10.1111/j.1365-2745.2009.01480.x.

Honek, A., Martinkova, Z. and Jarosik, V. (2003) Ground beetles (Carabidae) as seed predators. *European Journal of Entomology* 100(4), 531–544. DOI: 10.14411/eje.2003.081.

Huang, Z.-Y., Li, Y., Xiong, H.-A., Naz, M., Yan, M.-T. *et al.* (2025) Rhizobacteria's effects on the growth and competitiveness of *Solidago canadensis* under nutrient limitation. *Agriculture* 15(15), 1646. DOI: 10.3390/agriculture15151646.

Judžentienė, A., Būdienė, J., Labanauskas, L., Stancelytė, D. and Nedveckytė, I. (2023) Allelopathic activity of Canadian goldenrod (*Solidago canadensis* L.) extracts on seed germination and growth of lettuce (*Lactuca sativa* L.). *Plants (Basel, Switzerland)* 12(7), 1421. DOI: 10.3390/plants12071421.

Kajzer-Bonk, J., Szpiłyk, D. and Woyciechowski, M. (2016) Invasive goldenrods affect abundance and diversity of grassland ant communities (Hymenoptera: Formicidae). *Journal of Insect Conservation* 20(1), 99–105. DOI: 10.1007/s10841-016-9843-4.

Kato-Noguchi, H. and Kato, M. (2022) Allelopathy and allelochemicals of *Solidago canadensis* L. and *S. altissima* L. for their naturalization. *Plants (Basel, Switzerland)* 11(23), 3235. DOI: 10.3390/plants12071421.

Korell, L., Schädler, M., Brandl, R., Schreiter, S. and Auge, H. (2019) Release from above- and belowground insect herbivory mediates invasion dynamics and impact of an exotic plant. *Plants (Basel, Switzerland)* 8(12), 544. DOI: 10.3390/plants8120544.

Kotze, D.J., Brandmayr, P., Casale, A., Dauffy-Richard, E., Dekoninck, W. *et al.* (2011) Forty years of carabid beetle research in Europe: From taxonomy, biology, ecology and population studies to bioindication, habitat assessment and conservation. *ZooKeys* 100(100), 55–148. DOI: 10.3897/zookeys.100.1523.

Kromp, B. (1999) Carabid beetles in sustainable agriculture: A review on pest control efficacy, cultivation impacts and enhancement. *Agriculture, Ecosystems & Environment* 74(1–3), 187–228. DOI: 10.1016/S0167-8809(99)00037-7.

Kulkarni, S.S., Dosdall, L.M. and Willenborg, C.J. (2015) The role of ground beetles (Coleoptera: Carabidae) in weed seed consumption: A review. *Weed Science* 63(2), 355–376. DOI: 10.1614/WS-D-14-00067.1.

Lawes, M.J., Moore, A.M., Andersen, A.N., Preece, N.D. and Franklin, D.C. (2017) Ants as ecological indicators of rainforest restoration: Community convergence and the development of an Ant Forest Indicator Index in the Australian wet tropics. *Ecological Indicators* 83, 44–55. DOI: 10.1002/ece3.2992.

Lenda, M., Witek, M., Skórka, P., Moroń, D. and Woyciechowski, M. (2013) Invasive alien plants affect grassland ant communities, colony size and foraging behaviour. *Biological Invasions* 15(11), 2403–2414. DOI: 10.1007/s10530-013-0461-8.

Li, G., Wang, J., Zhang, J., Li, Y., Liu, E. *et al.* (2021) Effects of experimental warming and Canada goldenrod invasion on the diversity and function of the soil nematode community. *Sustainability* 13(23), 13145. DOI: 10.3390/su132313145.

Li, G., Zhao, X., Iqbal, B., Zhao, X., Liu, J. *et al.* (2023) The effect of soil microplastics on *Oryza sativa* L. root growth traits under alien plant invasion. *Frontiers in Ecology and Evolution* 11, 1172093. DOI: 10.3389/fevo.2023.1172093.

Liu, X., Huang, W., Liu, Y. and Zhan, A. (2024) Perspectives of invasive alien species management in China. *Ecological Applications* 34(1), e2926. DOI: 10.1002/eap.2926.

Liu, Y., Oduor, A.M.O., Zhang, Z., Manea, A., Tooth, I.M. *et al.* (2017) Do invasive alien plants benefit more from global environmental change than native plants? *Global Change Biology* 23(8), 3363–3370. DOI: 10.1111/gcb.13579.

Lock, K., Desender, K. and Janssen, C.R. (2001) Effects of metal contamination on the activity and diversity of carabid beetles in an ancient Pb-Zn mining area at Plombières (Belgium). *Entomologia Experimentalis et Applicata* 99(3), 355–360. DOI: 10.1046/j.1570-7458.2001.00835.x.

Lövei, G.L. and Sunderland, K.D. (1996) Ecology and behavior of ground beetles (Coleoptera: Carabidae). *Annual Review of Entomology* 41(1), 231–256. DOI: 10.1146/annurev.en.41.010196.001311.

Mirică, R. and Tăuşan, I. (2019) Impact of goldenrod invasion (*Solidago canadensis*) on ant assemblages (Hymenoptera: Formicidae) from urban habitats in southern Transylvania (Romania). *Acta Musei Brukenthal* 14(3), 621–628.

Moroń, D., Lenda, M., Skórka, P., Szentgyörgyi, H., Settele, J. *et al.* (2009) Wild pollinator communities are negatively affected by invasion of alien goldenrods in grassland landscapes. *Biological Conservation* 142(7), 1322–1332. DOI: 10.1016/j.biocon.2008.12.036.

Perera, P.C.D., Szymura, T.H., Zając, A., Chmolowska, D. and Szymura, M. (2021) Drivers of *Solidago* species invasion in Central Europe - Case study in the landscape of the Carpathian mountains and their foreground. *Ecology and Evolution* 11(18), 12429–12444. DOI: 10.1002/ece3.7989.

Poljuha, D., Sladonja, B., Uzelac Božac, M., Šola, I., Damijanić, D. *et al.* (2024) The invasive alien plant *Solidago canadensis*: Phytochemical composition, ecosystem service potential, and application in bioeconomy. *Plants (Basel, Switzerland)* 13(13), 1745. DOI: 10.3390/plants13131745.

Popay, I. and Parker, C. (2014) *Solidago canadensis* (Canadian goldenrod). In: *CABI Compendium*. Available at: https://www.cabidigitallibrary.org/doi/10.1079/cabicompendium.50599 (accessed 22 January 2026).

Rainio, J. and Niemelä, J. (2003) Ground beetles (Coleoptera: Carabidae) as bioindicators. *Biodiversity & Conservation* 12(3), 487–506. DOI: 10.1023/A:1022412617568.

Ren, G., Yang, B., Cui, M., Dai, Z., Xiang, Y. *et al.* (2022) Warming and elevated nitrogen deposition accelerate the invasion process of *Solidago canadensis* L. *Ecological Processes* 11(1), 62. DOI: 10.1186/s13717-022-00407-8.

Robson, D.B., Hamel, C. and Neufeld, R. (2017) Identification of plant species for pollinator restoration in the northern prairies. *Journal of Pollination Ecology* 21(6), 98–108. DOI: 10.26786/1920-7603(2017)429.

Rosumek, F.B., Silveira, F.A.O., de S Neves, F., de U Barbosa, N.P., Diniz, L. *et al.* (2009) Ants on plants: A meta-analysis of the role of ants as plant biotic defenses. *Oecologia* 160(3), 537–549. DOI: 10.1007/s00442-009-1309-x.

Sabat-Tomala, A., Raczko, E. and Zagajewski, B. (2020) Comparison of support vector machine and random forest algorithms for invasive and expansive species classification using airborne hyper-spectral data. *Remote Sensing* 12(3), 516. DOI: 10.3390/rs12030516.

Schultheiss, P., Nooten, S.S., Wang, R., Wong, M.K.L., Brassard, F. *et al.* (2022) The abundance, biomass, and distribution of ants on earth. *Proceedings of the National Academy of Sciences USA* 119(40), e2201550119. DOI: 10.1073/pnas.2201550119.

Skórka, P., Lenda, M. and Tryjanowski, P. (2010) Invasive alien goldenrods negatively affect grassland bird communities in Eastern Europe. *Biological Conservation* 143(4), 856–861. DOI: 10.1016/j.biocon.2009.12.030.

Skrzypczak, L., Wesołowska, M., Thiem, B. and Budzianowski, J. (1999) *Solidago* L. species (Goldenrod): *In vitro* regeneration and biologically active secondary metabolites. *Biotechnology in Agriculture and Forestry* 43, 384–403. DOI: 10.1007/978-3-662-08614-8_23.

Sun, J., Fu, Y., Hu, W., Bo, Y., Nawaz, M. *et al.* (2024) Early allelopathic input and later nutrient addition mediated by litter decomposition of invasive *Solidago canadensis* affect native plant and facilitate its invasion. *Frontiers in Plant Science* 15, 1503203. DOI: 10.3389/fpls.2024.1503203.

Šutovská, M., Capek, P., Kocmálová, M., Fraňová, S., Pawlaczyk, I. *et al.* (2013) Characterization and biological activity of *Solidago canadensis* complex. *International Journal of Biological Macromolecules* 52, 192–197. DOI: 10.1016/j.ijbiomac.2012.09.021.

Świerszcz, S., Czarniecka-Wiera, M., Szymura, T.H. and Szymura, M. (2024) From invasive species stand to species-rich grassland: Long-term changes in plant species composition during *Solidago* invaded site restoration. *Journal of Environmental Management* 353, 120216. DOI: 10.1016/j.jenvman.2024.120216.

Szitár, K., Kabai, M., Zimmermann, Z., Szabó, G., Reis, B.P. *et al.* (2024) Effects of goldenrod (*Solidago gigantea* Aiton and *S. canadensis* L.) invasion on the wet meadow vegetation composition of Somogyfajsz wood pasture. *Ecology*. DOI: 10.1101/2024.05.10.593301.

Szymura, M., Szymura, T.H. and Wolski, K. (2016) Invasive *Solidago* species: How large area do they occupy and what would be the cost of their removal? *Polish Journal of Ecology* 64(1), 25–34. DOI: 10.3161/15052249PJE2016.64.1.003.

Tóth, Z., Novák, J., Sipőcz, L. and Dombos, M. (2025) Canadian goldenrod (*Solidago canadensis*) invasion affects millipedes and symphylans more than centipedes in an urban meadow. *Insect Conservation and Diversity* 18(2), 260–274. DOI: 10.1111/icad.12802.

Trigos-Peral, G., Casacci, L.P., Ślipiński, P., Grześ, I.M., Moroń, D. *et al.* (2018) Ant communities and *Solidago* plant invasion: Environmental properties and food sources. *Entomological Science* 21(3), 270–278. DOI: 10.1111/ens.12304.

USDA Forest Service (n.d.) *Solidago canadensis*, Canada goldenrod. Available at: https://research.fs.usda.gov/feis/species-reviews/solcan (accessed 28 January 2026).

Ustinova, E.N. and Lysenkov, S.N. (2020) Comparative study of the insect community visiting flowers of invasive goldenrods (*Solidago canadensis* and *S. gigantea*). *Arthropod-Plant Interactions* 14(6), 825–837. DOI: 10.1007/s11829-020-09780-7.

Weber, E. (2001) Current and potential ranges of three exotic goldenrods (*Solidago*) in Europe. *Conservation Biology* 15(1), 122–128. DOI: 10.1111/j.1523-1739.2001.99424.x.

Yuan, Y., Wang, B., Zhang, S., Tang, J., Tu, C. *et al.* (2013) Enhanced allelopathy and competitive ability of invasive plant *Solidago canadensis* in its introduced range. *Journal of Plant Ecology* 6(3), 253–263. DOI: 10.1093/jpe/rts033.

Zelikova, T.J., Sanders, N.J. and Dunn, R.R. (2011) The mixed effects of experimental ant removal on seedling distribution, belowground invertebrates, and soil nutrients. *Ecosphere* 2(5), art63. DOI: 10.1890/ES11-00073.1.

Zhang, C.B., Wang, J., Qian, B.Y. and Li, W.H. (2009) Effects of the invader *Solidago canadensis* on soil properties. *Applied Soil Ecology* 43(2–3), 163–169. DOI: 10.1016/j.apsoil.2009.07.001.

Zhang, Q., Yang, R., Tang, J., Yang, H., Hu, S. *et al.* (2010) Positive feedback between mycorrhizal fungi and plants influences plant invasion success and resistance to invasion. *PLoS ONE* 5(8), e12380. DOI: 10.1371/journal.pone.0012380.

Zhang, R.-F., Zhang, X., Dang, L., Guo, Y.-J., Xu, Y.-H. *et al.* (2025) Species diversity mediates negative impacts of plant invasion on native-community flowering synchrony. *The New Phytologist* 248(1), 389–400. DOI: 10.1111/nph.70401.

Zhu, X., Li, W., Shao, H. and Tang, S. (2022) Selected aspects of invasive *Solidago canadensis* with an emphasis on its allelopathic abilities: A review. *Chemistry & Biodiversity* 19(10), e202200728. DOI: 10.1002/cbdv.202200728.

Zhylina, T.M., Lukash, O.V., Yakovenko, O.I. and Shevchenko, V.L. (2025) The invasive effect of *Solidago canadensis* L. on the structural characteristics of soil nematode communities in the ecosystems of the eu-mesophytic meadows. *Agricultural Science and Practice* 12(1), 46–62. DOI: 10.15407/agrisp12.01.046.

Zubek, S., Majewska, M.L., Kapusta, P., Stefanowicz, A.M., Błaszkowski, J. *et al.* (2020) *Solidago canadensis* invasion in abandoned arable fields induces minor changes in soil properties and does not affect the performance of subsequent crops. *Land Degradation & Development* 31(3), 334–345. DOI: 10.1002/ldr.3452.

12 Direct and Indirect Effects of Invasive Plants on Soil Invertebrate Communities

Viktoriia O. Komlyk and Viktor V. Brygadyrenko*
Department of Biodiversity and Ecology, Faculty of Biology and Ecology, Oles Honchar Dnipro National University, Dnipro, Ukraine

Introduction

Invasive plant species are defined as organisms that are introduced into new regions, where they successfully reproduce and occupy territories. Most invasive species cause significant damage to ecosystems, human health and the economic well-being of societies (Kumar and Singh, 2020). In recent decades, due to the growth of global trade and travel, alien plant species have become a serious threat to local ecosystems, causing irreversible biological processes such as displacement of native flora and fauna, biodiversity loss, changes in community structure across several trophic levels, disruption of food chains and deterioration of environmental quality (Murrell *et al.*, 2010; Turbelin *et al.*, 2016; Roberge *et al.*, 2024). The majority of all literature (about two-thirds) on biological invasions is devoted to the introduction and establishment of an alien species into an ecosystem. Only about one-third focuses on the deeper ecological consequences of invasions (Lowry *et al.*, 2013).

Despite the increasing frequency of invasions of alien plant species and their significant ecological and economic consequences, the impact of these processes on the animal communities of the soil and litter remains unclear. Food webs involving soil organisms are among the most diverse in terrestrial ecosystems. Microorganisms, which act as primary consumers (Coleman and Hendrix, 2000), occupy a central place in the functioning of these webs. This group includes archaea, bacteria, fungi and arthropods belonging to the microfauna group (Acari, Collembola). Soil microorganisms regulate nutrient cycling and soil formation processes. Arthropods of microfauna are food sources for primary and secondary zoophages (Coleman and Hendrix, 2000) and thus determine the functioning and stability of soil invertebrate communities.

The impact of invasive plants on the abundance of soil fauna depends on soil characteristics, microclimatic conditions, plant community structure and the trophic specialization of consumers of different orders. The susceptibility of soil invertebrates depends on the ecosystem type (forest, meadow, desert, steppe, etc.) and its structure. Ecosystems with lower plant diversity and a simpler taxonomic structure are more susceptible to the impact of adventive plant species (McCary *et al.*, 2016). For example, Abgrall *et al.* (2019) found that the abundance of soil invertebrates increased only in forest communities (+18.2%) following the invasion of invasive plants. This effect was observed exclusively

*Corresponding author: brigad@ua.fm

DOI: 10.1079/9781836991298.0012

among primary consumers: the abundance of herbivores increased by 29.1% and detritophages by 66.7%. In contrast, predators and microbiophages showed no significant response to the presence of invasive species, regardless of the habitat (Abgrall *et al.*, 2019).

McCary *et al.* (2016) found a significant impact of adventive plant species on representatives of different levels of the soil food web. First-order consumers (e.g. saprotrophic bacteria, fungi and rhizophagous insects) are more sensitive to changes in plant species composition than consumers at higher levels. Detrital food webs encompass a large number of first-order consumer groups, creating a diversity of energy pathways. Therefore, detrital food webs are relatively more resilient to the invasion of adventive species (McCary *et al.*, 2016).

Morriën *et al.* (2011) studied the response of soil nematodes to the invasion of alien plant species. Nematodes were chosen as indicator species because they are represented at different trophic levels: some feed on plant roots while others feed on microorganisms or invertebrates. The results of these studies showed that invasive plants had a lower abundance of root-feeding nematodes per unit of root biomass than related native species (Morriën *et al.*, 2011). These plants demonstrated resistance, rather than tolerance, to native root-feeding nematodes. However, the composition of the nematode community (both taxonomic and functional) is generally determined by the plant species, not its geographic range. The influence of plant species decreased with increasing trophic level of the nematodes: the greatest impact was observed on herbivorous forms, while the impact on predators and mycophages was minimal.

However, the invasion of the adventive plant *Chromolaena odorata* (L.) R.M. King and H. Rob (Asterales, Asteraceae) into the savannahs of South Africa reduced plant taxonomic diversity and affected communities that serve as sensitive indicators of ecosystem change. With increasing duration of invasion, abundance, assemblage patterns, diversity and species richness of spiders significantly declined (Mgobozi *et al.*, 2008).

Thus, the reaction of soil fauna is determined by both habitat type and trophic level, with organisms at the bottom of the food chain being the most sensitive. When invasive plants take over large areas in new habitats, they have a significant impact on local plant species and the associated trophic networks. Numerous studies confirm their negative impact on the taxonomic diversity of native flora (Gerber *et al.*, 2008; Murrell *et al.*, 2010; Kato-Noguchi and Kato, 2023) and primary consumers (Hartley *et al.*, 2010; Bezemer *et al.*, 2014; McCary and Wise, 2019). Primary consumers, in turn, provide a significant portion of the energy flow through the ecosystem. Their decline in invasive habitats inevitably influences populations of predatory invertebrates, including soil invertebrates (Bassett *et al.*, 2011; Motard *et al.*, 2015; Clark and Seewagen, 2019). However, knowledge about how the invasion of adventive plants affects the structure, species richness and biomass of predators remains limited. In this regard, the purpose of this chapter is (1) to identify the features of the direct and indirect impact of invasive plant species on soil macrofauna; (2) to consider examples of the effect of various classes of secondary metabolites of adventive plant species on representatives of soil communities; (3) to highlight the impact of invasive plants on soil macrofauna in the context of climatic cycles; and (4) to demonstrate both positive and negative consequences of the impact of invasive plant species on soil macrofauna.

Direct and Indirect Impacts of Invasive Plants on Soil Macrofauna

Invasive plants have both direct and indirect impacts on invertebrates (Litt *et al.*, 2014). Direct impacts include changes in the food supply for primary consumers (Gerber *et al.*, 2008; McCary *et al.*, 2016). Many adventive plant species have specific chemical compositions in their leaves and roots, which can reduce the nutritional value and availability of food substrates. An additional factor is the release of secondary metabolites through leaves and roots, which in most cases are toxic to invertebrates (see next section).

A decrease in the abundance, diversity and biomass of phytophages and saprophages leads to a decline in the abundance of representatives of higher trophic levels (primary and secondary predators) (Lau, 2012). This is already a manifestation of the indirect impact of adventive species on representatives of soil invertebrates. Indirect impacts also include, for example, a decrease in the abundance and diversity of saprophages as a result of changes in the biomass and quality of leaf litter, as well as alterations in the chemical and microbial composition of the soil (McLeod *et al.*, 2016).

Effects of Secondary Metabolites of Invasive Plants on Soil Invertebrates

Main classes of secondary metabolites and their effects on soil invertebrates

Plants synthesize both primary and secondary metabolites throughout their life cycle. Primary metabolites are directly involved in metabolic processes such as growth, development and reproduction (Elshafie *et al.*, 2023). Secondary metabolites are also involved in basic metabolic processes, and some are synthesized only under specific stress conditions (Kumari *et al.*, 2024). Secondary metabolites help invasive plants adapt and survive in new environments (Chiocchio *et al.*, 2021). They mediate chemical plant defences against herbivores, fungi and bacteria, enhance adaptive properties and provide allelopathic effects (Zhai *et al.*, 2024).

Plant secondary metabolites are divided into three main groups: terpenoids, phenolic compounds and nitrogen-containing compounds (alkaloids; see Fig. 12.1). This includes a very large number of substances (at least 40,000 terpenoids, 12,000 alkaloids and 8000 phenylpropanoids; Bourgaud *et al.*, 2001; Elshafie *et al.*, 2023). The concentration of secondary metabolites in plant tissues depends on the plant's age. In the leaves of young plants, especially in the first year of life, the level of many secondary metabolites reaches a maximum and then decreases with age (Cipollini and Gruner, 2006).

Secondary metabolites synthesized by plants are often found in glycosylated forms, which increases their water solubility (Barto and Cipollini, 2009). Plant metabolites can be classified as non-volatile (liquid) and airborne (gaseous) organic compounds. The effects of gaseous secondary metabolites on other plants and insects are poorly studied (McCormick *et al.*, 2023).

Many invasive plants are of interest as sources of biologically active compounds with insecticidal and herbicidal properties. For example, *Nerium oleander* L. (Gentianales, Apocynaceae) is widespread and known for its toxic properties. It contains a wide range of secondary metabolites, such as alkaloids, cardenolides and terpenoids, which have been studied for their insecticidal activity (Al-Ansi *et al.*, 2024). Secondary metabolites of many plants can act as attractants or as toxic substances in invertebrates (Zeni *et al.*, 2021; Parhomenko *et al.*, 2024). At the molecular level, secondary metabolites disrupt cellular respiration, cause mitochondrial damage and trigger apoptosis, ultimately leading to insect death (Kumar *et al.*, 2024). They also affect the insect's chemoreceptors, reducing feeding behaviour. Essential oils affect the nervous system, causing neurotoxicity, paralysis and eventually death of insects (Kumar *et al.*, 2024).

Although the effects of secondary metabolites on invertebrates have been studied in individual species, a comprehensive understanding of their effects remains limited. Volatiles emitted by invasive plants affect the abundance and diversity of pollinators, phytophages and their parasitoids and predators (Brown and Mitchell, 2001; Seybold *et al.*, 2006; Beck *et al.*, 2013; Bezemer *et al.*, 2014). For example, Vasquez *et al.* (2024) found that aphid-infested plants emit volatile metabolites that attract aphid natural enemies. Non-volatile secondary metabolites accumulate in various plant organs, performing defensive and adaptive functions, and can also be released through the root system into the soil (Chepel *et al.*, 2020).

We conducted a literature review to summarize the effects of secondary metabolites of invasive plant species on invertebrates with particularly attention to soil organisms. A comprehensive literature search was carried out in the Scopus database using the keywords 'secondary metabolites', 'alkaloids', 'terpenoids' and 'phenols'. Forty-seven articles were identified, of which eight focused on terrestrial phytophages (Özenver *et al.*, 2022; Kato-Noguchi and Kato, 2023; Kumar *et al.*, 2024

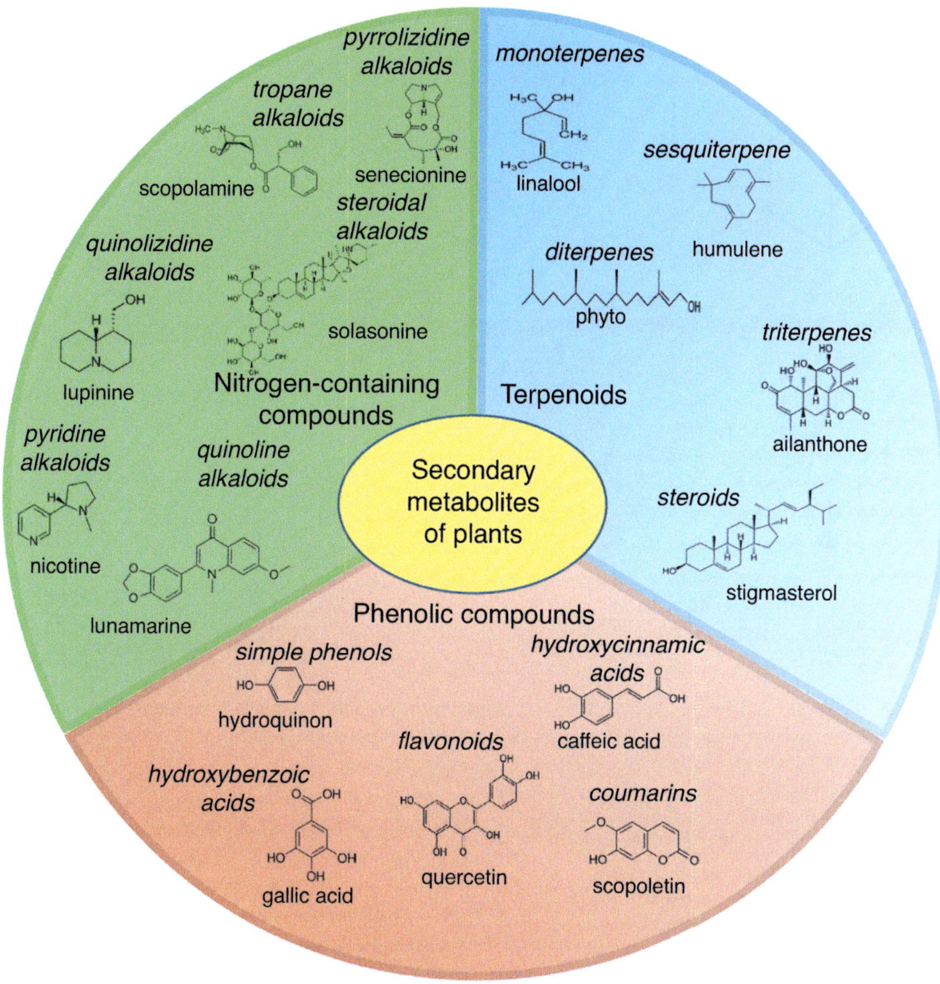

Fig. 12.1. The major types of secondary metabolite.

and others). However, the literature on soil invertebrates is significantly less represented: we found only a few sources examining the influence of invasive plants on soil organisms (Abgrall *et al.*, 2019; McCary and Wise, 2019; Al-Ansi *et al.*, 2024).

Terpenoids

Terpenoids are a large class of natural compounds formed during the biosynthesis of mevalonic acid (Fig. 12.2). They are characterized by a wide variety of chemical structures and biological functions (Chiocchio *et al.*, 2021). Their classification is based on the number of isoprene units: hemiterpenes (C5), isoprene; monoterpenes (C10), limonene (see Fig. 12.2) pinene, menthol, camphor and terpineol; sesquiterpenes (C15), farnesene, chamazulene and artemisinin; diterpenes (C20), phytol, taxol, gibberellins and labdanes; triterpenes (C30), squalene, dammarenes, lupanes, oleanes and saponins; steroids as derivatives of triterpenes, phytosterols, corticosteroids and steroid saponins (e.g. cardenolides); tetraterpenes (C40), carotenoids (β-carotene, lycopene, xanthophyll); polyterpenes (>C40), natural rubbers (e.g. from

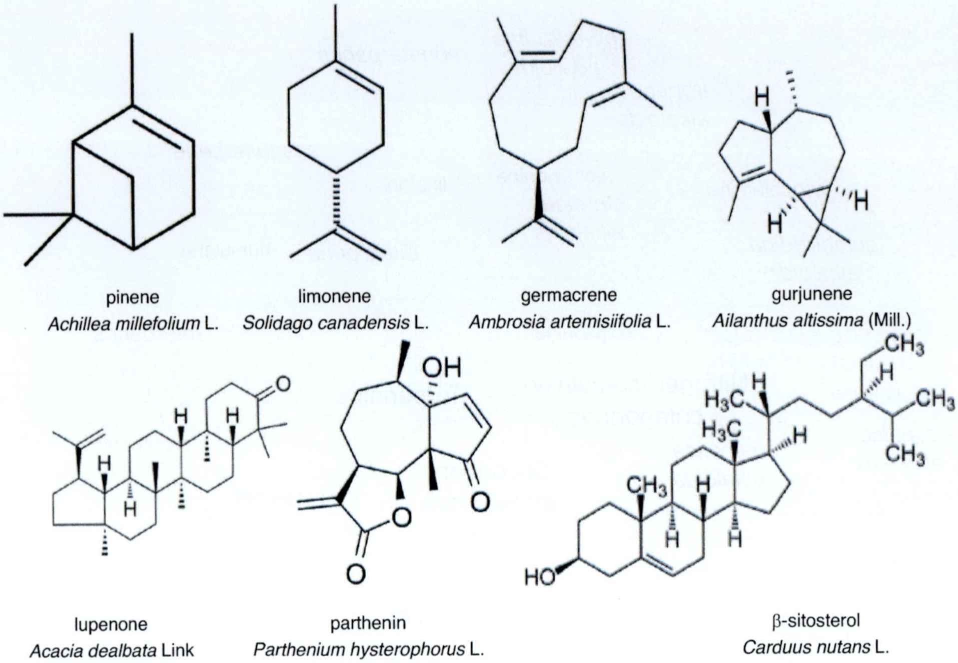

Fig. 12.2. Examples of terpenoids from invasive plants: monoterpenes (pinene, limonene), sesquiterpenes (germacrene, gurjunene, parthenin), triterpenes (lupenone) and phytosterols (β-sitosterol).

plants of the Euphorbiaceae and Sapotaceae) (Elshafie *et al.*, 2023).

In plant ecology, terpenoids perform numerous functions: they provide allelopathic properties to invasive plants, attract pollinators and protect plants from phytophages and pathogens (Zeni *et al.*, 2021). For example, the sesquiterpenes of *Ailanthus altissima* (Mill.) Swingle (Sapindales, Simaroubaceae) such as α-curcumene, α-gurjunene, γ-cadinene, α-humulene, β-caryophyllene, caryophyllene oxide and germacrene D have essential oil repellent, insecticidal and herbicidal properties (Kozuharova *et al.*, 2022) (Table 12.1). Monoterpenes have antimicrobial and insecticidal effects (Kumar *et al.*, 2024), while steroid saponins are toxic to invertebrates because they block the activity of Na^+/K^+-ATPase (Woerdenbag, 1986). Several studies have highlighted the effects of various essential oils on the physiological parameters of invertebrates (Remezok *et al.*, 2022; Kolombar *et al.*, 2025).

Many terpenoids are characteristic of invasive plant species (Table 12.1). For instance: *Ambrosia artemisiifolia* L. (Asterales, Asteraceae) contains germacrene D and β-caryophyllene (Porter and Mabry, 1969); *Bidens pilosa* L. (Asterales, Asteraceae) contains α-pinene and β-caryophyllene (Kato-Noguchi and Kurniadie, 2024); *Eupatorium cannabinum* L. (Asterales, Asteraceae) contains eupatolide (Woerdenbag, 1986); *Impatiens glandulifera* Royle (Ericales, Balsaminaceae) contains linalool (Szewczyk *et al.*, 2016); and *Lantana camara* L. (Lamiales, Verbenaceae) contains lantadene A–D, β-caryophyllene, α-humulene and linalool (Kumar *et al.*, 2024).

Phenolic compounds

Phenolic compounds are a large and diverse group of secondary plant metabolites (Fig. 12.3). They are common in nearly all higher plants, as they participate in lignin synthesis. Their classification is based on biosynthetic origin and the number of carbon atoms, aromatic rings and hydroxyl groups.

Table 12.1. Invasive plant species and their secondary metabolites.

Family	Species	Region in which the species is invasive	Example substance(s)	Source
Fabaceae	*Acacia dealbata* Link	Chile, Italy, Portugal, South Africa, Spain	Terpenoids: lupenone, stigmasterol. Phenolic compounds: methyl p-hydroxycinnamate, resorcinol	Aguilera *et al.* (2015)
Fabaceae	*Acacia saligna* (Labill.) Wendl.	Australia, Cyprus, Israel, Italy	Phenolic compounds: 2-(2-hydroxyl-3-methoxypropyl)–4,6-dibenzoylresorcinol. Nitrogen-containing compounds: 1-methylpyrazole, 1,3,5-triazine-2,4-diamine, 1-propanamine, 3-(10,11-dihydro-5H-dibenzo(a,d)cyclohepten-5-ylidene)-*N*-methyl, dimethylphosphinic azide, 2-methyleneamino-propyl	Salem *et al.* (2012)
Sapindaceae	*Acer negundo* L.	Argentina, Bosnia and Herzegovina, Brazil, Czechia, France, Germany, Greece, Italy, Latvia, Lithuania, New Zealand, Serbia, Slovakia, Slovenia, Ukraine	Phenolic acids: 4-coumaric acid, 4-hydroxybenzoic acid, caffeic acid, chlorogenic acid, ferulic acid, gallic acid, gentisic acid, protocatechuic acid, salicylic acid, sinapic acid, trans-cinnamic acid, vanillic acid. Phenolic compounds: scopoletin, vanillin. Flavonoids: catechin, epigallocatechin, epicatechin, kaempferol-3-*O*-glucoside, luteolin, luteolin-7-*O*-glucoside, naringin, quercetin-3-glucoside, quercetin-3,4′-di-*O*-glucoside, quercetin-3-D-galactoside, rutin trihydrate	Barrales-Cureño *et al.* (2020)
Asteraceae	*Achillea ageratum* L.	Europe	Terpenoids: artemisyl acetate, artemisia alcohol, santolina alcohol, stigmasterol, β-sitosterol, yomogi alcohol. Phenolic acids: caffeic acid, ferulic acid, p-coumaric acid, salicylic acid, sinapic acid, vanillic acid. Flavonoids: kaempferol, luteolin, quercetin	Grandi *et al.* (1972), El Bouzidi *et al.* (2012), Strzępek-Gomółka *et al.* (2021), Açıkgöz *et al.* (2025)

Continued

Table 12.1. Continued

Family	Species	Region in which the species is invasive	Example substance(s)	Source
Asteraceae	*Achillea millefolium* L.	Europe	Terpenoids: 1,8-cineole, β-pinene, chamazulene. Phenolic acids: caffeic acid, chlorogenic acid, dicaffeoylquinic acids. Flavonoids: apigenin, apigenin-7-*O*-glucoside, luteolin, luteolin-7-*O*-glucoside, rutin, vicenin-2	Verma *et al*. (2017), Górska-Drabik *et al*. (2025), Konarska *et al*. (2025)
Simaroubaceae	*Ailanthus altissima* (Mill.)	Europe, USA	Terpenoids: α-curcumene, α-gurjunene, α-humulene, β-caryophyllene, γ-cadinene, ailanthone, caryophyllene oxide, germacrene D. Phenolic acids: caffeic acid, chlorogenic acid, ferulic acid, gallic acid, p-coumaric acid. Flavonoids: apigenin, isoquercetin, luteolin, quercetin	Bailly (2020), Kozuharova *et al*. (2022)
Brassicaceae	*Alliaria petiolata* (Bieb.) Cavara & Grande	Canada, New Zealand, USA	Phenolic compounds: kaempferol, quercetin, rutin. Glucosinolates: glucotropaeolin, sinigrin	Vaughn and Berhow (1999), Cipollini *et al*. (2005)
Amaranthaceae	*Amaranthus retroflexus* L.	Belgium, France, USA	Terpenoids: α-tocopherol, oleanolic acid, stigmasterol. Flavonoids: kaempferol, quercetin, rutin. Alkaloids: amaranthine	Fiorito *et al*. (2017)
Asteraceae	*Ambrosia artemisiifolia* L.	Europe, North America	Terpenoids: β-caryophyllene, ambrosin, artemisiifolin, germacrene D. Flavonoids: kaempferol, quercetin	Porter and Mabry (1969), Parkhomenko *et al*. (2006), Šćepanović *et al*. (2022)
Apocynaceae	*Asclepias syriaca* L.	Bosnia and Herzegovina, Croatia, Czechia, Lithuania, Romania, Serbia, Slovakia	Phenolic acids: 4-hydroxybenzoic acid, benzoic acid, caffeic acid, chlorogenic acid, ferulic acid, gallic acid, p-coumaric acid, syringic acid, vanillic acid. Flavonoids: kaempferol, quercetin, rutin	Umiljendić *et al*. (2025)

Brassicaceae	*Brassica elongata* Ehrh.	Poland, USA	Phenolic compounds: caffeic acid, chlorogenic acid, gallic acid, p-coumaric acid. Flavonoids: quercetin, rutin. Glucosinolates: glucobrassicin, glucoiberin, sinigrin	Horn and Vaughan (1983), Muthusamy and Lee (2024)
Asteraceae	*Carduus nutans* L.	Australia, New Zealand, USA	Phytosterols: β-sitosterol, β-sitosterol 3-*O*-xyloside, taraxasterol. Phenolic acids: caffeic acid, chlorogenic acid, gallic acid, p-coumaric acid. Flavonoids: apigenin, isorhamnetin, kaempferol, luteolin, quercetin. Coumarins: esculin, herniarin, scopoletin, umbeliferon	Bain and Desrochers (1988), Kozyra *et al.* (2022)
Asteraceae	*Centaurea diffusa* Lam.	Canada, Poland, USA	Terpenoids: caryophyllene oxide, cnicin. Phenolic acids: coumaric acid	Skliar *et al.* (2005), Quintana *et al.* (2009)
Asteraceae	*Centaurea solstitialis* L.	Canada, USA	Sesquiterpene lactones: 13-acetyl solstitialin A, centaurepensin, chlorojanerin	Cheng *et al.* (1992), Özçelik *et al.* (2009)
Asteraceae	*Cirsium arvense* (L.) Scop.	Australia, Canada, India, Japan, New Zealand, Northern Africa, South America, Southern Africa, UK, USA	Glycosides: cardiac glycosides, sapogenin. Flavonoids: anthocyanin, epicatechin, flavan-3-ol, kaempferol, naringenin, naringin, rutin. Polyphenols: resveratrol. Alkaloids: lunamarin, ribalinidine, sparteine	Nwauche *et al.* (2024)
Poaceae	*Cynodon dactylon* (L.) Pers.	Australia, Bermuda, Cambodia, Chile, Christmas Island, Colombia, Ecuador, El Salvador, Fiji, French Polynesia, Guam, Guatemala, Indonesia, Kiribati, Malaysia, Marshall Islands, Mexico, New Caledonia, New Zealand, Nicaragua, Niue, Palau, Peru, Philippines, Saint Helena, Singapore, Solomon Islands, Thailand, Tonga, USA, Vietnam	Terpenoids: phytol. Phenolic compounds: hydroquinone. Phenolic acids: syringic acid, vanillic acid	Savadi *et al.* (2020)

Continued

Table 12.1. Continued

Family	Species	Region in which the species is invasive	Example substance(s)	Source
Solanaceae	*Datura stramonium* L.	Anguilla, Australia, Bolivia, Bosnia and Herzegovina, Botswana, China, Croatia, Cuba, Ethiopia, Fiji, Greece, India, Israel, Italy, Japan, Kenya, Lesotho, Madagascar, Namibia, New Zealand, Papua New Guinea, Portugal, South Africa, Sweden, Tanzania, Turkey, Uganda, UK, Zambia, Zimbabwe	Tropane alkaloids: apoatropin, aposcopolamine, hyoscyamine, scopolamine, tigloidin	Soni *et al*. (2012), Cinelli and Jones (2021)
Apiaceae	*Daucus carota* L.	Canada, China, USA	Terpenoids: α-humulene, β-caryophyllene, β-myrcene, sabinene. Phenolic acids: caffeic acid, chlorogenic acid, p-hydroxybenzoic acid	Keilwagen *et al*. (2017), Ahmad *et al*. (2019)
Poaceae	*Elymus repens* (L.) Gould	Argentina, Canada, Mexico, South Africa	Phenolic acids: chlorogenic acid, ferulic acid, p-coumaric acid, p-hydroxybenzoic acid, vanillic acid. Flavonoids: baicalein, hyperoside, luteolin, quercetin, rutin, tricin	Bortolami *et al*. (2022)
Asteraceae	*Erigeron canadensis* L.	Bosnia and Herzegovina, Bulgaria, China, Czechia, Greece, Guyana, India, Italy, Japan, Korea, Portugal, Romania, Slovenia, Turkey, Ukraine	Terpenoids: α-curcumene, β-himachalene, β-santalen, cuparene, γ-cadinene. Phenolic compounds: 2-furoic acid, 4-hydroxybenzaldehyde, gallic acid, gentisic acid. Flavonoids: genistein. Nitrogen-containing compounds: piperidine	Khare (2007), Ghosh *et al*. (2025)
Asteraceae	*Eupatorium cannabinum* L.	Albania, Algeria, Armenia, Austria, Azerbaijan Bavarian alps, Belarus, Belgium, Bulgaria, China, Cyprus, Czechia, Denmark, Estonia, Finland, France, Georgia, Germany, Hungary, India, Iran, Iraq, Israel, Italy, Latvia, Lebanon, Lithuania, Moldova, Morocco, Nepal, Netherlands, New Zealand, Norway, Pakistan, Poland, Portugal, Romania, Serbia, Spain, Sweden, Switzerland, Turkey, Turkmenistan, USA	Sesquiterpene lactones: eupatolide, eupatoriopicrin, other germacranolides. Phenolic acids: caffeic acid, chlorogenic acid, dicaffeoylquinic acids. Flavonoids: astragalin, hispidulin, hyperoside, isoquercitrin, kaempferol, kaempferol-3-rhamnoside, quercetin, rutin. Pyrrolizidine alkaloids: echinatine, intermedine, lycopsamine, supinine	Woerdenbag (1986), Ribeiro-Varandas *et al*. (2014), Piwowarski *et al*. (2018)

Polygonaceae	*Fallopia sachalinensis* (F.Schmidt) Ronse Decr.	Belgium, Czechia, Denmark, Estonia, Finland, France, Germany, Greece, Ireland, Italy, Luxembourg, Netherlands, New Zealand, Norway, Poland, Spain, Sweden, Switzerland, UK, USA	Terpenoids: betulinic acid, oleanolic acid, ursolic acid. Stilbenes: *cis*-polydatin, piceatannol 3′-*O*-glucoside, resveratroloside, trans-polydatin. Flavonoids: catechin, epicatechin-3-gallate, kaempferol derivatives, luteolin derivatives, quercetin derivatives, quercitrin	Lachowicz and Oszmiański (2019), Suprun *et al*. (2024)
Fabaceae	*Gleditsia triacanthos* L.	Argentina, Australia, South Africa, USA	Terpenoids: echinocystic acid, gleditschoside A–E, triacanthoside A, B, C. Flavonoids: isovitexin, luteolin, quercetin, vitexin	Zhang *et al*. (2016)
Asteraceae	*Helianthus tuberosus* L.	Australia, Austria, Czechia, France, Germany, Hungary, Italy, Japan, Montenegro, New Zealand, Slovakia, Spain, Switzerland, Tunisia	Phenolic compounds: chlorogenic acid, ferulic acid, vanillin. Flavonoids: kaempferol, quercetin	Tapera *et al*. (2024)
Solanaceae	*Hyoscyamus niger* L.	Canada, USA	Tropane alkaloids: hyoscyamine, scopolamine. Flavonoids: kaempferol, quercetin, rutin	Sengupta *et al*. (2010), Passos and Mironidou-Tzouveleki (2016), Fatur and Kreft (2020)
Balsaminaceae	*Impatiens glandulifera* Royle	Austria, Croatia, Czechia, Finland, Germany, Hungary, Ireland, Netherlands, Norway, Serbia, Slovakia, Sweden, Switzerland	Terpenoids: linalool. Flavonoids: kaempferol, quercetin derivatives	Szewczyk *et al*. (2016), Block *et al*. (2019), Whaley *et al*. (2023)
Brassicaceae	*Isatis tinctoria* L.	USA	Phenolic compounds: 4-hydroxyphenylacetic acids, ferulic acid, salicylic acid, sinapic acid, vanillic acid. Flavonoids: isoorientin, isoscoparin, isovitexin, quercetin, rutin, stellarin-2, vicenin-2. Alkaloids: indirubin, tryptanthrin. Glucosinolates: epipogoitrin, glucobrassicin, gluconapin, progoitrin	Speranza *et al*. (2020), Caliskan *et al*. (2022)

Continued

Table 12.1. Continued

Family	Species	Region in which the species is invasive	Example substance(s)	Source
Asteraceae	*Lactuca serriola* L.	Argentina, Brazil, Canada, China, Estonia, Lithuania,	Sesquiterpene lactones: lactucarin, lactucin, deoxylactucin. Phenolic compounds: caffiec acid, chlorogenic acid. Flavonoids: quercetin, rutin	Sessa *et al*. (2000), Ilgün *et al*. (2020), Hussien and Al-Hamashi (2024)
Verbenaceae	*Lantana camara* L.	Argentina, Australia, Bahamas, Barbados, Bermuda, Burundi, Cambodia, China, Colombia, Cook Islands, Cyprus, Dominican Republic, Ecuador, Fiji, French Polynesia, Ghana, Guam, Haiti, Hong Kong, India, Indonesia, Israel, Japan, Kenya, Kiribati, Liberia, Madagascar, Malaysia, Marshall Islands, Mauritius, Mayotte, Micronesia, Nauru, New Caledonia, New Zealand, Niue, Norfolk Island, Northern Mariana Islands, Palau, Papua New Guinea, Philippines, Pitcairn, Reunion, Rwanda, Saint Helena, Samoa, Solomon Islands, South Africa, Southern Africa, Sri Lanka, Swaziland, Thailand, Tonga, Turkey, Tuvalu, Uganda, United Arab Emirates, USA, USA Minor Outlying Islands, Vanuatu, Wallis and Futuna, Zambia, Zimbabwe	Terpenoids: α-humulene, β-caryophyllene, lantadene A–D, linalool. Flavonoids: quercetin, rutin	Sagar *et al*. (2005), Kumar *et al*. (2024)
Brassicaceae	*Lepidium virginicum* L.	Australia, Cayman Islands, China, Cuba, Ecuador, French Polynesia, Guam, Japan, New Caledonia, New Zealand, Niue, Palau, Philippines, Solomon Islands, Tonga, USA, USA minor outlying islands	Phenolic acids: caffeic acid, cinnamic acid, ferulic acid, gallic acid, p-coumaric acid, vanillic acid. Flavonoids: kaempferol, naringenin, quercetin, quercetin-3-galactoside, quercetin-3-glucoside	Gallegos-Saucedo *et al*. (2024)
Fabaceae	*Lupinus polyphyllus* L.	Argentina, Czechia, Denmark, Estonia, Finland, Japan, New Zealand, Latvia, Lithuania, Serbia, Sweden	Phenolic compounds: apigenin, diosmetin, genistein, hydroxycinnamic acid, p-hydroxybenzoic acid. Quinolizidine alkaloids: angustifoline, lupanine, lupinine, sparteine	Khan *et al*. (2015), Frick *et al*. (2017)

Solanaceae	*Nicotiana glauca* Graham.	Australia, Croatia, Egypt, France, Greece, Israel, Italy, Marshall Islands, Mexico, Morocco, New Zealand, Portugal, South Africa, Spain, Ukraine, UK, USA	Pyridine alkaloids: anabasine, nicotine, nornicotine	Massadeh *et al*. (2021)
Asteraceae	*Parthenium hysterophorus* L.	Australia, Ethiopia, India, Madagascar, Mayotte, New Caledonia, Papua New Guinea, Reunion, South Africa, Swaziland, Taiwan, USA	Sesquiterpene lactones: parthenin. Phenolic acids: aerulic acid, caffeic acid, chlorogenic acid, ferulic acid, fumaric acid, neochlorogenic acid, p-anisic acid, p-coumaric acid, p-hydroxybenzoic acid, protocatechuic acid, vanillic acid	Hassan *et al*. (2018), Bashar *et al*. (2023)
Fabaceae	*Prosopis juliflora* (Sw.) DC.	Australia, Brazil, India, Israel, Kenya, Reunion	Flavonoids: apigenin, luteolin. Alkaloids: 3-oxo-juliprosine, juliprosine, juliprosopine	Nakano *et al*. (2004), Damasceno *et al*. (2016)
Fabaceae	*Pueraria lobata* (Willd.) Ohwi	Europe, Fiji, Italy, Mexico, New Zealand, USA	Flavonoids: daidzein, genistein, puerarin	Kim *et al*. (2006)
Polygonaceae	*Reynoutria japonica* Houtt.	Australia, Canada, Czechia, Denmark, France, Germany, Ireland, Italy, Latvia, New Zealand, Poland, Romania, Slovakia, UK, USA	Phenolic compounds: catechin, chlorogenic acid, emodin, epicatechin, hyperoside, quercetin, physcion, piceatannol, polydatin, procyanidin B3, resveratrol. Phenolic acids: caffeic acid, caftaric acid, ferulic acid, protocatechic acid	Stefanowicz *et al*. (2022)
Fabaceae	*Robinia pseudoacacia* L.	Canada, Cyprus, Czechia, France, Italy, Korea, Poland, Reunion, Swaziland, Ukraine, USA	Phenolic acids: caffeic acid, chlorogenic acid, coumaric acid, ferulic acid, gallic acid, protocatechuic acid. Flavonoids: acacetin glycosides, apigenin, catechin, fustin, dihydrorobinetin, kaempferol, luteolin, myricetin, quercetin, robinetin	Sanz *et al*. (2011), Uzelac *et al*. (2023)

Continued

Table 12.1. Continued

Family	Species	Region in which the species is invasive	Example substance(s)	Source
Asteraceae	*Senecio vulgaris* L.	Cananda, USA	Terpenoids: α-humulene, caryophyllene oxide, germacrene D, humulene epoxide II, limonene, para-cymene. Pyrrolizidine alkaloids: retrorsine, senecionine, seneciphylline, senecivernine, senkirkine	Flade *et al*. (2019), Hajmoradi *et al*. (2025)
Solanaceae	*Solanum elaeagnifolium* Cav.	Australia, Greece, India, Israel	Phenolic compounds: catechin, chlorogenic acid, quercetin. Alkaloids: solamargine, solasonine	Bouslamti *et al*. (2022), Kozuharova *et al*. (2024)
Asteraceae	*Solidago canadensis* L.	Denmark, Estonia, Finland, France, Germany, Romania, Sweden, Switzerland, UK	Terpenoids: α-pinene, bornyl acetate, germacrene D, limonene. Phenolic acids: caffeic acid, chlorogenic acid, ferulic acid, synaptic, vanillin acids. Flavonoids: astragalin, hyperoside, isorhamnetin, kaempferol, quercetin, rutin	Elshafie *et al*. (2019), Poljuha *et al*. (2024)

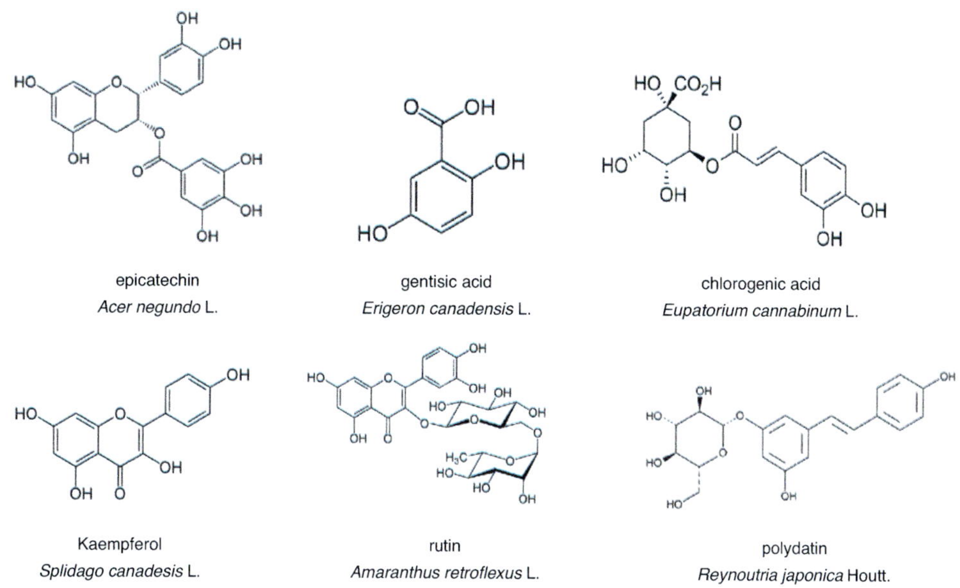

Fig. 12.3. Examples of phenolic compounds from invasive plants: flavonoids (epicatechin, kaempferol, rutin), phenolic acids (chlorogenic acid, gentisic acid) and stilbenes (polydatin).

The main subclasses include simple phenols, phenolic acids, flavonoids and tannins. Phenolic compounds are widely distributed in various plant organs but their content can vary during different stages of plant development and also depend on climatic conditions and soil composition (Liu *et al.*, 2021; Cucu *et al.*, 2022). In some plant species the greatest amounts of phenols are found in the roots during the vegetative stage, in the flowers during the flowering period and in leaves during bud formation and seed maturation (Chepel *et al.*, 2020).

Phenolic compounds perform numerous functions: flavonoids contribute to flower colour and pollinator attraction, protect plant tissues from ultraviolet radiation and act as potent antioxidants (Agati *et al.*, 2012). Certain phenolic glycosides, phenolic acids and tannins exhibit antimicrobial and insecticidal activity (Boeckler *et al.*, 2011; Zazharskyi *et al.*, 2024). Furthermore, phenolic compounds can regulate plant growth and development, influencing interactions with microorganisms and the formation of symbiotic relationships (Liu and Murray, 2016). Flavonoids play an important role in defence against phytophagous insects, regulating their growth and development, and often cause their death (Kumar *et al.*, 2024). For example, flavonoids such as quercetin and rutin affect nutrient accumulation in Lepidoptera larvae, resulting in their inability to complete the moulting process and, consequently, their death (Zhang *et al.*, 2024).

Invasive plant species contain a wide variety of phenolic compounds (Table 12.1). For example, *Ambrosia artemisiifolia* L. (Asterales, Asteraceae) contains caffeic acid and chlorogenic acid (Šćepanović *et al.*, 2022); *Carduus nutans* L. (Asterales, Asteraceae) contains gallic acid, chlorogenic acid, caffeic acid and p-coumaric acid (Lalk *et al.*, 2021; Kozyra *et al.*, 2022); *Hyoscyamus niger* L. (Solanales, Solanaceae) contains quercetin, rutin and kaempferol (Sengupta *et al.*, 2010); and *Lantana camara* L. (Lamiales, Verbenaceae) contains quercetin and rutin (Kumar *et al.*, 2024).

Alkaloids

Alkaloids are a large class of secondary plant metabolites characterized by the presence of a nitrogen atom in a heterocyclic or amine group. These compounds play an important role in

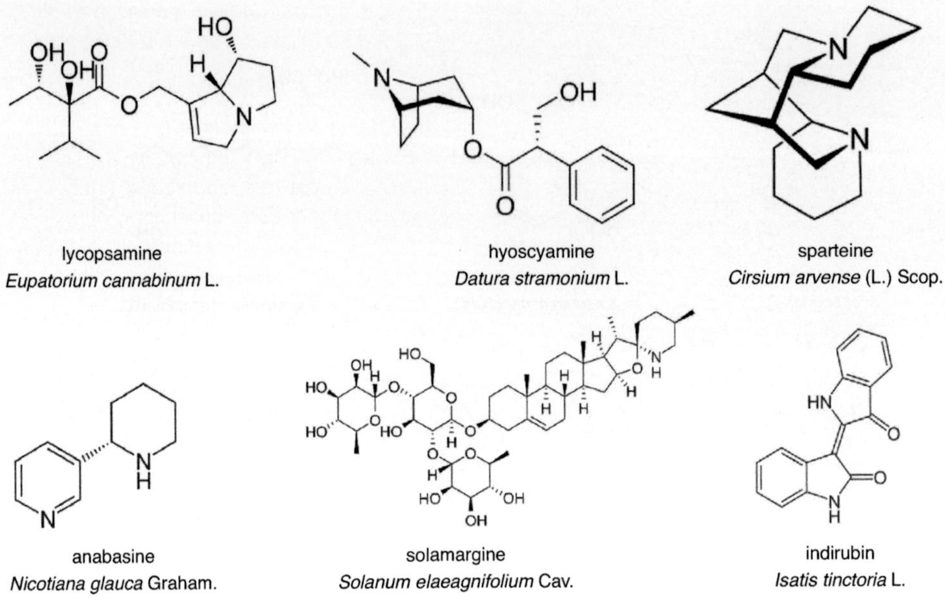

Fig. 12.4. Examples of alkaloids from invasive plants: pyrrolizidine alkaloids (lycopsamine), tropane alkaloids (hyoscyamine), quinolizidine alkaloids (sparteine), piperidine alkaloids (anabasine), steroidal glycoalkaloids (solamargine) and indole derivatives (indirubin).

plant defence against pathogens and herbivores, as well as in allelopathy and pollinator attraction (Chiocchio *et al.*, 2021) (Fig. 12.4). For example, quinolizidine alkaloids provide protection against bacteria, fungi and insects (López-Rodríguez *et al.*, 2022). Alkaloids are widespread among angiosperms and occur in nearly all plant organs: leaves, flowers, seeds, roots, stems, fruits, bark and bulbs. These substances are highly species-specific, characteristic only of certain genera and species (Bourgaud *et al.*, 2001). Their concentrations depend on the plant's developmental stage and vary significantly between species. The highest concentrations are usually found in the reproductive organs, consistent with their protective function. Many alkaloids are toxic to pathogens and non-specialized phytophages, reducing damage and increasing plant survival (Elshafie *et al.*, 2023). Some alkaloids help attract pollinators, increasing flower visitation and promoting successful reproduction. Furthermore, alkaloids can influence plant growth and development and participate in complex interactions with surrounding organisms, including soil microbiota and pollinating insects.

Adventive plant species often have increased alkaloid content, which serves as one of the key elements of their chemical defence in new environmental conditions (Fig. 12.4). For example, *Amaranthus retroflexus* L. (Caryophyllales, Amaranthaceae) contains amaranthine (Fiorito *et al.*, 2017); *Cirsium arvense* (L.) Scop. contains lunamarin, ribalinidine and sparteine (Nwauche *et al.*, 2024); *Lupinus polyphyllus* L. (Fabales, Fabaceae) contains quinolizidine alkaloids: lupanine, sparteine, lupinine and angustifoline (Frick *et al.*, 2017); and *Nicotiana glauca* Graham. (Solanales, Solanaceae) contains pyridine alkaloids: anabasine and nicotine (Massadeh *et al.*, 2021) and others (Table 12.1).

Alkaloids disrupt the functioning of the nervous system, leading to paralysis and death of insects (Kumar *et al.*, 2024). Pyrrolizidine alkaloids are highly toxic and possess strong insecticidal properties (Kato-Noguchi and Kato, 2023). Compounds such as xanthine and theobromine (methylxanthine) cause mortality in *Spodoptera frugiperda* (Smith, 1797) (Lepidoptera, Noctuidae) larvae (Kuddus *et al.*, 2023). Matrine and sophocarpine lead to the death of fire ants,

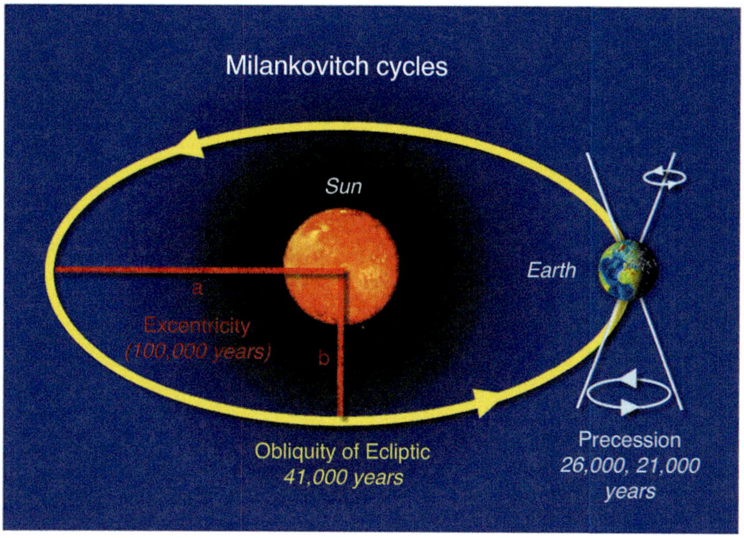

Fig. 12.5. Milankovitch cycles. Image by Grobe, H. (2007). Milankovitch cycles diagram, Wikimedia Commons, File: https://commons.wikimedia.org/wiki/File:Milankovitch-cycles_hg.png, used under CC-BY-SA-2.5 licence.

Solenopsis invicta Buren, 1972 (Hymenoptera, Formicidae) (Tian and Zhang, 2023). Thus, data on the effects of alkaloids concern mainly phytophages; information on effects at other trophic level is still lacking.

The Impact of Global Climate Change on Flora and Fauna

The history of plant dispersals is closely linked to climate fluctuations caused by Milankovitch cycles (Pälike, 2005) (Fig. 12.5). During the Pleistocene glaciations, vast areas of Europe and North America were covered by glaciers, and the flora retreated to refugia: relatively stable zones with a mild climate. With the onset of interglacial periods, plant cover was rapidly restored: species with high ecological plasticity and strong dispersal ability were the first to colonize newly available territories (Fig. 12.6). These processes can be considered historical analogues of modern biological invasions.

Glacial retreat was accompanied by primary soil succession and the formation of ecosystems. New plant species altered the chemical composition of soils, introduced organic matter and created conditions for the development of soil invertebrates. Secondary metabolites (terpenoids, phenolic compounds and alkaloids) could have both stimulating and inhibitory effects on the taxonomic structure and biomass of soil invertebrate communities. Thus, interactions between new plant communities and soil animals were observed in the paleoecological period.

Today, we are witnessing a similar scenario, albeit at a much faster pace. Climate warming is expanding the potential ranges of invasive plant species. Their ability to alter the structure of soil communities through leaf litter, root exudates and secondary metabolites affects the functioning of macrofauna.

This phenomenon is not unique in Earth's history, it has occurred at least ten times over the past million years. Currently, however, the strong effect of invasive plant species are observed not at the glacier margins, but at the transition zones between climatic regions (forests and steppes, steppes and deserts, etc.). Every 10,000 years, ranges shift by 1000–2000 km, with constant plant invasions occurring at these interfaces. Terrestrial invertebrate species, many of which have existed for hundreds of thousands

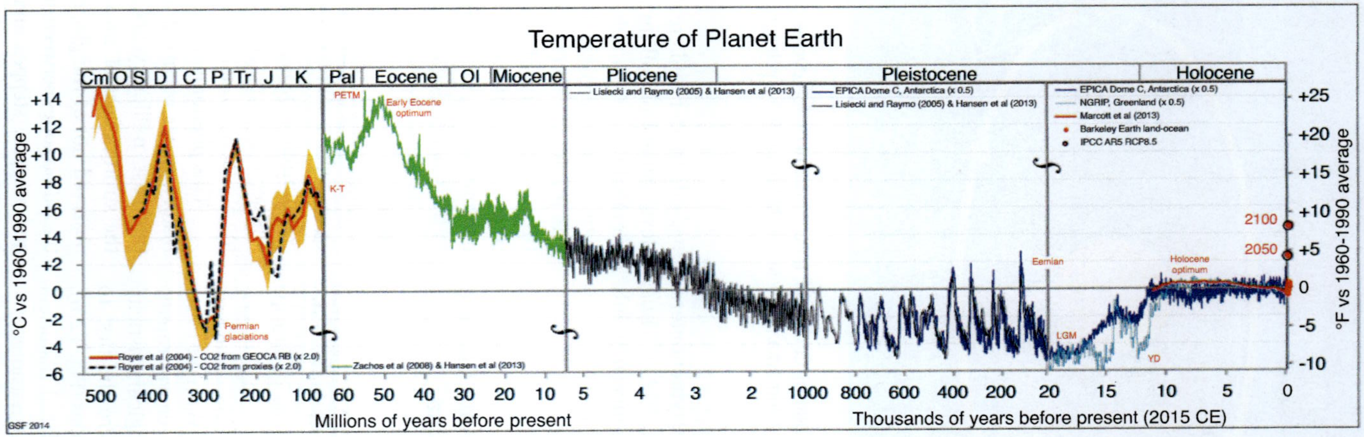

Fig. 12.6. Fluctuations in the temperature of Earth through geological time. Image by Fergus, G. (2007) All palaeotemps, Wikimedia Commons, https://commons.wikimedia.org/wiki/File:All_palaeotemps.png, used under CC-BY-3.0 licence.

of years (e.g. butterflies) to several million years (e.g. beetles), have repeatedly experienced such cycles (Schädel *et al.*, 2022; Kawahara *et al.*, 2023). Their generations have encountered shifting range boundaries, changes in community structure and fluctuations in resource availability, which has shaped their resilience, adaptive properties and distribution patterns within ecosystems.

Direction of Impact of Invasive Plants on Soil Macrofauna

The impact of invasive plant species on invertebrates is ambiguous: it can be harmful, neutral or beneficial for certain groups. The severity of the negative effect depends on the dominance of specific trophic groups in the ecosystem (Litt *et al.*, 2014).

Invasive plants influence soil invertebrates through food chains: both via living roots and through leaf litter. They can release secondary metabolites into the soil that have a toxic effect, reducing the activity and survival of certain invertebrate species (especially herbivores and saprophages) (Zeni *et al.*, 2021). Alien species also alter the quality, quantity and timing of leaf litter deposition, which affects decomposition rates, organic matter accumulation and nutrient cycling (Abgrall *et al.*, 2019). Furthermore, invasive plants may reduce the abundance of soil microorganisms and fungi by releasing secondary metabolites into the soil. A decline in microbial diversity can lead to changes in the communities of primary and secondary consumers (McCary and Wise, 2019).

Litt *et al.* (2014) reviewed existing studies on the effects of adventive plant species on arthropods and found that overall abundance and diversity generally decrease. Several studies have confirmed a decrease in predator abundance, which may be due to a decrease in primary consumer populations (Gerber *et al.*, 2008; Simao *et al.*, 2010). However, other studies show an increase in predator abundance (Bassett *et al.*, 2011; Litt *et al.*, 2014). Detritovores often show mixed responses, either increasing their abundance or remaining stable despite plant community changes (Levin *et al.*, 2006; Litt *et al.*, 2014). For example, an increase in the density of *A. altissima* (Mill.) leads to a decrease in soil microbial activity and the number of detrivores (mites, springtails), predatory beetles and the diversity of terrestrial molluscs, while simultaneously increasing the number of earthworms (Motard *et al.*, 2015).

Not all invertebrate groups respond negatively. In some cases, invasive plants provide new food resources (Rodriguez, 2006; Bassett *et al.*, 2011; Bassett, 2014; Effah *et al.*, 2020) and improve soil conditions through nutrient enrichment (Galappaththi *et al.*, 2022). Some invasions increase soil nitrate concentrations and accelerate nitrification processes, promoting the abundance of ammonia-oxidizing bacteria (Lee *et al.*, 2012; McLeod *et al.*, 2016; Tun *et al.*, 2020; Woodworth *et al.*, 2020). For example, the invasion of *C. odorata* (L.) increased soil microbial activity and available N and P thereby stimulating invertebrate populations (Yelenik *et al.*, 2004; Kato-Noguchi and Kato, 2023). Root exudates and leaf litter serve as a source of organic carbon for soil microorganisms. This increases the abundance and diversity of soil bacteria and fungi. Consequently, the abundance of invertebrates feeding on microorganisms may increase (Bais *et al.*, 2006).

Increased leaf litter during adventive plant invasion can provide additional shelter and food sources for predators (e.g. ground beetles and spiders), positively influencing their abundance and community composition (McCary *et al.*, 2016). Thickened leaf litter can modify microclimate conditions. For example, during the invasion of *Chrysanthemoides monilifera* (L.) (Asterales, Asteraceae), the environment becomes more humid and shaded, favouring millipedes, amphipods, earthworms, pseudoscorpions and isopods (Lindsay and French, 2006).

Native plant species shape the unique community structure, litter composition and food webs, which strongly influence soil invertebrate response (Belnap and Phillips, 2001). The invasion of the same plant into communities with different compositions can cause opposing outcomes: some groups may increase in one context and decline in another. These observations emphasize how the effect of invasive species depends not only on the characteristics of the invasive species itself but also on interactions with other plants, which determine the initial conditions of the ecosystem and the adaptive capacity of soil organisms (Belnap and Phillips, 2001; John *et al.*, 2006).

Woodworth *et al.* (2020) note that increases in invertebrate abundance during an invasion are often short-lived, lasting a single season until the litter decomposes. After decomposition, a sharp decline in abundance is typically observed. By contrast, in natural ecosystems, the abundance and diversity of litter invertebrates remains more stable and higher over time.

Conclusion

The impact of invasive plant species on soil invertebrates is a complex and multifaceted process. Throughout history, approximately every 10,000 years, glacial cycles have shifted plant ranges by 1000–2000 km or more. These plants penetrated ecosystems, causing changes in the structure of soil invertebrate communities; some species adapted to the changed food supply and environmental conditions while others experienced stress or were displaced.

The invasion of alien plant species is in most cases accompanied by a direct inhibitory (toxic) effect on herbivores and saprophages undergoing grazing and detrital activity. Secondary metabolites of invasive plant species often accumulate in photosynthetic organs (leaves) or fruits. These plant parts fall to the soil surface and can saturate the litter with secondary metabolites that are toxic to invertebrates. The dominant organic compounds are sterols, flavonoids, phenolic compounds, alkaloids, tannins and other classes of organic matter. The inhibitory effects on populations of primary and secondary predators are generally weaker and have remained largely unstudied.

An analysis of existing publications on the impact of invasive plant species on soil invertebrates revealed that the primary focus is on the impact on herbivores. Information on the impact on predators is extremely limited. This knowledge gap hinders the development of a comprehensive understanding of the mechanisms by which invasive plants influence the functioning of soil ecosystems and requires further targeted research.

References

Abgrall, C., Forey, E. and Chauvat, M. (2019) Soil fauna responses to invasive alien plants are determined by trophic groups and habitat structure: A global meta-analysis. *Oikos* 128(10), 1390–1401. DOI: 10.1111/oik.06493.

Açıkgöz, M.A., Ay, E.B. and Kocaman, B. (2025) Secondary metabolites and biological activities from *Achillea gypsicola* Hub-Mor. under foliar applications of chitosan and salicylic acid. *Food Science & Nutrition* 13(5), e70225. DOI: 10.1002/fsn3.70225.

Agati, G., Azzarello, E., Pollastri, S. and Tattini, M. (2012) Flavonoids as antioxidants in plants: Location and functional significance. *Plant Science* 196, 67–76. DOI: 10.1016/j.plantsci.2012.07.014.

Aguilera, N., Becerra, J., Villaseñor-Parada, C., Lorenzo, P., González, L. *et al.* (2015) Effects and identification of chemical compounds released from the invasive *Acacia dealbata* Link. *Chemistry and Ecology* 31(6), 479–493. DOI: 10.1080/02757540.2015.1050004.

Ahmad, T., Cawood, M., Iqbal, Q., Ariño, A., Batool, A. *et al.* (2019) Phytochemicals in *Daucus carota* and their health benefits - review article. *Foods (Basel, Switzerland)* 8(9), 424. DOI: 10.3390/foods8090424.

Al-Ansi, A., Dhafer, H.A., Al-Tamimi, J. and Murshed, M. (2024) Evaluation of *Nerium oleander* extracts against *Pachycondyla sennaarensis* (Hymenoptera, Formicidae) adults. *Arquivo Brasileiro de Medicina Veterinária e Zootecnia* 76(6), e13189. DOI: 10.1590/1678-4162-13189.

Bailly, C. (2020) Anticancer properties and mechanism of action of the quassinoid ailanthone. *Phytotherapy Research* 34(9), 2203–2213. DOI: 10.1002/ptr.6681.

Bain, J.F. and Desrochers, A.M. (1988) Flavonoids of *Carduus nutans* and *C. acanthoides*. *Biochemical Systematics and Ecology* 16(3), 265–268. DOI: 10.1016/0305-1978(88)90005-1.

Bais, H.P., Weir, T.L., Perry, L.G., Gilroy, S. and Vivanco, J.M. (2006) The role of root exudates in rhizosphere interactions with plants and other organisms. *Annual Review of Plant Biology* 57(1), 233–266. DOI: 10.1146/annurev.arplant.57.032905.105159.

Barrales-Cureño, H.J., Salgado-Garciglia, R., López-Valdez, L.G., Monribot-Villanueva, J.L., Guerrero-Analco, J.A. *et al.* (2020) Metabolomic data of phenolic compounds from *Acer negundo* extracts. *Data in Brief* 30, 105569. DOI: 10.1016/j.dib.2020.105569.

Barto, K.E. and Cipollini, D. (2009) Half-lives and field soil concentrations of *Alliaria petiolata* secondary metabolites. *Chemosphere* 76(1), 71–75. DOI: 10.1016/j.chemosphere.2009.02.020.

Bashar, H.M.K., Juraimi, A.S., Ahmad-Hamdani, M.S., Uddin, M.K., Asib, N. *et al.* (2023) Evaluation of allelopathic effects of *Parthenium hysterophorus* L. methanolic extracts on some selected plants and weeds. *PLoS ONE* 18(1), e0280159. DOI: 10.1371/journal.pone.0280159.

Bassett, I.E. (2014) Impacts on invertebrate fungivores: A predictable consequence of ground-cover weed invasion. *Biodiversity and Conservation* 23(4), 791–810. DOI: 10.1007/s10531-014-0634-5.

Bassett, I., Paynter, Q. and Beggs, J.R. (2011) nvasive *Alternanthera philoxeroides* (alligator weed) associated with increased fungivore dominance in Coleoptera on decomposing leaf litter. *Biological Invasions* 13(6), 1377–1385. DOI: 10.1007/s10530-010-9896-3.

Beck, J.J., Smith, L. and Baig, N. (2013) An overview of plant volatile metabolomics, sample treatment and reporting considerations with emphasis on mechanical damage and biological control of weeds. *Phytochemical Analysis* 25(4), 331–341. DOI: 10.1002/pca.2486.

Belnap, J. and Phillips, S.L. (2001) Soil biota in an ungrazed grassland: Response to annual grass (*Bromus tectorum*) invasion. *Ecological Applications* 11(5), 1261–1275. DOI: 10.1890/1051-0761(2001)011[1261:SBIAUG]2.0.CO;2.

Bezemer, T.M., Harvey, J.A. and Cronin, J.T. (2014) Response of native insect communities to invasive plants. *Annual Review of Entomology* 59(1), 119–141. DOI: 10.1146/annurev-ento-011613-162104.

Block, A.K., Yakubova, E. and Widhalm, J.R. (2019) Specialized naphthoquinones present in impatiens glandulifera nectaries inhibit the growth of fungal nectar microbes. *Plant Direct* 3(5), e00132. DOI: 10.1002/pld3.132.

Boeckler, G.A., Gershenzon, J. and Unsicker, S.B. (2011) Phenolic glycosides of the salicaceae and their role as anti-herbivore defenses. *Phytochemistry* 72(13), 1497–1509. DOI: 10.1016/j.phytochem.2011.01.038.

Bortolami, M., Matteo, P., Rocco, D., Feroci, M. and Petrucci, R. (2022) Metabolic profile of *Agropyron repens* (L.) P. Beauv. rhizome herbal tea by HPLC-PDA-ESI-MS/MS analysis. *Molecules (Basel, Switzerland)* 27(15), 4962. DOI: 10.3390/molecules27154962.

Bourgaud, F., Gravot, A., Milesi, S. and Gontier, E. (2001) Production of plant secondary metabolites: A historical perspective. *Plant Science* 161(5), 839–851. DOI: 10.1016/S0168-9452(01)00490-3.

Bouslamti, M., Metouekel, A., Chelouati, T., El Moussaoui, A., Barnossi, A.E. *et al.* (2022) *Solanum elaeagnifolium* var. obtusifolium (dunal) dunal: Antioxidant, antibacterial, and antifungal activities of polyphenol-rich extracts chemically characterized by use of *in vitro* and *in silico* approaches. *Molecules (Basel, Switzerland)* 27(24), 8688. DOI: 10.3390/molecules27248688.

Brown, B.J. and Mitchell, R.J. (2001) Competition for pollination: Effects of pollen of an invasive plant on seed set of a native congener. *Oecologia* 129(1), 43–49. DOI: 10.1007/s004420100700.

Caliskan, H., Argon, M. and Sabudak, T. (2022) The genus *Isatis* L.: A review on its flavonoid and phenolic compound profile. *Records of Agricultural and Food Chemistry* 2(2), 75–83. DOI: 10.25135/rfac.13.2209.2576.

Cheng, C.H., Costall, B., Hamburger, M., Hostettmann, K., Naylor, R.J. *et al.* (1992) Toxic effects of solstitialin a 13-acetate and cynaropicrin from *Centaurea solstitialis* L. (asteraceae) in cell cultures of foetal rat brain. *Neuropharmacology* 31(3), 271–277. DOI: 10.1016/0028-3908(92)90177-q.

Chepel, V., Lisun, V. and Skrypnik, L. (2020) Changes in the content of some groups of phenolic compounds and biological activity of extracts of various parts of heather (*Calluna vulgaris* (l.) hull) at different growth stages. *Plants (Basel, Switzerland)* (8), 926. DOI: 10.3390/plants9080926.

Chiocchio, I., Mandrone, M., Tomasi, P., Marincich, L. and Poli, F. (2021) Plant secondary metabolites: An opportunity for circular economy. *Molecules (Basel, Switzerland)* 26(2), 495. DOI: 10.3390/molecules26020495.

Cinelli, M.A. and Jones, A.D. (2021) Alkaloids of the genus *Datura*: Review of a rich resource for natural product discovery. *Molecules (Basel, Switzerland)* 26(9), 2629. DOI: 10.3390/molecules26092629.

Cipollini, D. and Gruner, B. (2006) Cyanide in the chemical arsenal of garlic mustard, *Alliaria petiolata*. *Journal of Chemical Ecology* 33(1), 85–94. DOI: 10.1007/s10886-006-9205-x.

Cipollini, D., Mbagwu, J., Barto, K., Hillstrom, C. and Enright, S. (2005) Expression of constitutive and inducible chemical defenses in native and invasive populations of *Alliaria petiolata*. *Journal of Chemical Ecology* 31(6), 1255–1267. DOI: 10.1007/s10886-005-5284-3.

Clark, R.E. and Seewagen, C.L. (2019) Invasive Japanese barberry, *Berberis thunbergii* (Ranunculales: Berberidaceae) is associated with simplified branch-dwelling and leaf-litter arthropod communities in a New York forest. *Environmental Entomology* 48(5), 1071–1078. DOI: 10.1093/ee/nvz095.

Coleman, D.C. and Hendrix, P.F. (eds) (2000) *Invertebrates as Webmasters in Ecosystems*. CABI International, Wallingford. DOI: 10.1079/9780851993942.0000.

Cucu, A.-A., Baci, G.-M., Cucu, A.-B., Dezsi, Ş., Lujerdean, C. *et al.* (2022) *Calluna vulgaris* as a valuable source of bioactive compounds: Exploring its phytochemical profile, biological activities and apitherapeutic potential. *Plants (Basel, Switzerland)* 11, 15. DOI: 10.3390/plants11151993.

Damasceno, G.A. de B., Ferrari, M. and Giordani, R.B. (2016) *Prosopis juliflora* (SW) D.C., an invasive species at the Brazilian Caatinga: Phytochemical, pharmacological, toxicological and technological overview. *Phytochemistry Reviews* 16(2), 309–331. DOI: 10.1007/s11101-016-9476-y.

Effah, E., Barrett, D.P., Peterson, P.G., Potter, M.A., Holopainen, J.K. *et al.* (2020) Effects of two invasive weeds on arthropod community structure on the Central Plateau of New Zealand. *Plants (Basel, Switzerland)* 9(7), 919. DOI: 10.3390/plants9070919.

El Bouzidi, L., Abbad, A., Hassani, L., Fattarsi, K., Leach, D. *et al.* (2012) Essential oil composition and antimicrobial activity of wild and cultivated moroccan *Achillea ageratum* L.: A rare and threatened medicinal species. *Chemistry & Biodiversity* 9(3), 598–605. DOI: 10.1002/cbdv.201100115.

Elshafie, H.S., Gruľová, D., Baranová, B., Caputo, L., Martino, L. *et al.* (2019) Antimicrobial activity and chemical composition of essential oil extracted from *Solidago canadensis* L. growing wild in Slovakia. *Molecules (Basel, Switzerland)* (7), 24. DOI: 10.3390/molecules24071206.

Elshafie, H.S., Camele, I. and Mohamed, A.A. (2023) A comprehensive review on the biological, agricultural and pharmaceutical properties of secondary metabolites based-plant origin. *International Journal of Molecular Sciences* 24(4), 3266. DOI: 10.3390/ijms24043266.

Fatur, K. and Kreft, S. (2020) Common anticholinergic solanaceaous plants of temperate Europe - A review of intoxications from the literature (1966–2018). *Toxicon* 177, 52–88. DOI: 10.1016/j.toxicon.2020.02.005.

Fiorito, S., Epifano, F., Palmisano, R., Genovese, S. and Taddeo, V.A. (2017) A re-investigation of the phytochemical composition of the edible herb *Amaranthus retroflexus* L. *Journal of Pharmaceutical and Biomedical Analysis* 143, 183–187. DOI: 10.1016/j.jpba.2017.05.051.

Flade, J., Beschow, H., Wensch-Dorendorf, M., Plescher, A. and Wätjen, W. (2019) Occurrence of nine pyrrolizidine alkaloids in *Senecio vulgaris* L. depending on developmental stage and season. *Plants (Basel, Switzerland)* 8(3), 54. DOI: 10.3390/plants8030054.

Frick, K.M., Kamphuis, L.G., Siddique, K.H.M., Singh, K.B. and Foley, R.C. (2017) Quinolizidine alkaloid biosynthesis in lupins and prospects for grain quality improvement. *Frontiers in Plant Science* 8, 87. DOI: 10.3389/fpls.2017.00087.

Galappaththi, H.S.S.D., de Silva, W.A.P.P. and Clavijo Mccormick, A. (2022) A mini-review on the impact of common gorse in its introduced ranges. *Tropical Ecology* 64(1), 1–25. DOI: 10.1007/s42965-022-00239-9.

Gallegos-Saucedo, R., Barrios-García, T., Valdez-Morales, E.E., Cabañas-García, E., Barajas-Espinosa, A. *et al.* (2024) Cytotoxic activity of *Lepidium virginicum* L. methanolic extract on human colorectal cancer cells, Caco-2, through p53-mediated apoptosis. *Molecules (Basel, Switzerland)* 29(16), 3920. DOI: 10.3390/molecules29163920.

Gerber, E., Krebs, C., Murrell, C., Moretti, M., Rocklin, R. *et al.* (2008) Exotic invasive knotweeds (*Fallopia* spp.) negatively affect native plant and invertebrate assemblages in European riparian habitats. *Biological Conservation* 141(3), 646–654. DOI: 10.1016/j.biocon.2007.12.009.

Ghosh, R.K., Price, A.J. and Maity, A. (2025) Allelopathic effects of horseweed (*Erigeron canadensis*) on germination and growth of seven common weeds of the southern United States. *Weed Science* 73(1), e63. DOI: 10.1017/wsc.2025.10034.

Górska-Drabik, E., Golan, K., Sempruch, C., Chrzanowski, G., Dybowski, M.P. *et al.* (2025) Chemical composition and toxicity of *Achillea millefolium* L. essential oil against *Acrobasis advenella* (Lepidoptera,

Pyralidae) under laboratory conditions. *Molecules (Basel, Switzerland)* 30(9), 1927. DOI: 10.3390/molecules30091927.

Grandi, R., Marchesini, A., Pagnoni, U.M. and Trave, R. (1972) β-Elemen-9β-ol from *Achillea ageratum*. *Phytochemistry* 11(11), 3363–3365. DOI: 10.1016/S0031-9422(00)86409-2.

Hajmoradi, F., Fathi, H. and Moghadami, F. (2025) Phytochemical components and antibacterial activity of two populations of *Senecio vulgaris* L. essential oils as traditional medicine plant. *Iranian Journal of Microbiology* 17(1), 171–179. DOI: 10.18502/ijm.v17i1.17815.

Hartley, M.K., Rogers, W.E. and Siemann, E. (2010) Comparisons of arthropod assemblages on an invasive and native trees: Abundance, diversity and damage. *Arthropod-Plant Interactions* 4(4), 237–245. DOI: 10.1007/s11829-010-9105-4.

Hassan, G., Rashid, H.U., Amin, A., Khan, I.A. and Shehzad, N. (2018) Allelopathic effect of *Parthenium hysterophorus* on germination and growth of some important crops and weeds of economic importance. *Planta Daninha* 36, e018176372. DOI: 10.1590/s0100-83582018360100132.

Horn, P.J. and Vaughan, J.G. (1983) Seed glucosinolates of fourteen wild *Brassica* species. *Phytochemistry* 22(2), 465–470. DOI: 10.1016/0031-9422(83)83025-8.

Hussien, M.S. and Al-Hamashi, A.A. (2024) Polyphenolic compounds from *Lactuca serriola* aerial parts growing in Iraq exhibit anticancer properties and concurrent HPLC-DAD profiling. *Journal of Research in Medical and Dental Science* 12(10), 12–18.

Ilgün, S., Küpeli Akkol, E., Ilhan, M., Çiçek Polat, D., Baldemir Kılıç, A. *et al.* (2020) Sedative effects of latexes obtained from some *Lactuca* L. species growing in Turkey. *Molecules (Basel, Switzerland)* 25(7), 1587. DOI: 10.3390/molecules25071587.

John, M.G.St., Wall, D.H. and Hunt, H.W. (2006) Are soil mite assemblages structured by the identity of native and invasive alien grasses? *Ecology* 87(5), 1314–1324. DOI: 10.1890/0012-9658(2006)87[1314:ASMASB]2.0.CO;2.

Kato-Noguchi, H. and Kato, M. (2023) Evolution of the secondary metabolites in invasive plant species *Chromolaena odorata* for the defense and allelopathic functions. *Plants (Basel, Switzerland)* 12(3), 521. DOI: 10.3390/plants12030521.

Kato-Noguchi, H. and Kurniadie, D. (2024) The invasive mechanisms of the noxious alien plant species *Bidens pilosa*. *Plants (Basel, Switzerland)* 13(3), 356. DOI: 10.3390/plants13030356.

Kawahara, A.Y., Storer, C., Carvalho, A.P.S., Plotkin, D.M., Condamine, F.L. *et al.* (2023) A global phylogeny of butterflies reveals their evolutionary history, ancestral hosts and biogeographic origins. *Nature Ecology & Evolution* 7(6), 903–913. DOI: 10.1038/s41559-023-02041-9.

Keilwagen, J., Lehnert, H., Berner, T., Budahn, H., Nothnagel, T. *et al.* (2017) The terpene synthase gene family of carrot (*Daucus carota* L.): Identification of QTLs and candidate genes associated with terpenoid volatile compounds. *Frontiers in Plant Science* 8, 1930. DOI: 10.3389/fpls.2017.01930.

Khan, M.K., Karnpanit, W., Nasar-Abbas, S.M., Huma, Z. and Jayasena, V. (2015) Phytochemical composition and bioactivities of lupin: A review. *International Journal of Food Science & Technology* 50(9), 2004–2012. DOI: 10.1111/ijfs.12796.

Khare, C.P. (2007) *Erigeron canadensis* Linn. In: *Indian Medicinal Plants*. Springer Verlag, New York. DOI: 10.1007/978-0-387-70638-2_572.

Kim, J.M., Lee, Y.M., Lee, G.Y., Jang, D.S. and Bae, K.H. (2006) Constituents of the roots of *Pueraria lobata* inhibit formation of advanced glycation end products (AGEs). *Archives of Pharmacal Research* 29(10), 821–825. DOI: 10.1007/BF02973900.

Kolombar, T.M., Komlyk, V.O., Berezovskyi, V.O. and Brygadyrenko, V.V. (2025) The effect of spices, essential oils and extracts on the locomotor activity of *Porcellio laevis* (Isopoda, Porcellionidae). *Biosystems Diversity* 33(1), e2517. DOI: 10.15421/012517.

Konarska, A., Weryszko-Chmielewska, E., Materska, M., Sulborska-Różycka, A., Dmitruk, M. *et al.* (2025) Phenolic compounds in flowers and herb of *Achillea millefolium* L.: Histochemical and phytochemical studies. *Molecules (Basel, Switzerland)* 30(9), 2084. DOI: 10.3390/molecules30092084.

Kozuharova, E., Pasdaran, A., Al Tawaha, A.R., Todorova, T., Naychov, Z. *et al.* (2022) Assessment of the potential of the invasive arboreal plant *Ailanthus altissima* (Simaroubaceae) as an economically prospective source of natural pesticides. *Diversity* 14(8), 680. DOI: 10.3390/d14080680.

Kozuharova, E., Malfa, G.A., Aquaviva, R., Zarev, Y. and Ionkova, I. (2024) Utility potential of the aggressive invasive *Solanum elaeagnifolium* Cav. (Solanaceae) as an option to control its expansion: Bioactive compounds and medicinal properties. *Euro-Mediterranean Journal for Environmental Integration* 9(4), 1987–1992. DOI: 10.1007/s41207-024-00533-z.

Kozyra, M., Kukula-Koch, W. and Szymański, M. (2022) Phenolic composition of inflorescences of *Carduus nutans* L. *Chemistry & Biodiversity* 19(3), e202100827. DOI: 10.1002/cbdv.202100827.

Kuddus, M.F.H., Sukmawati, I.H. and Inayah, H. (2023) Utilization of Cacao fruit peels waste (*Theobroma cacao* L.) as a means of invasive pest control *Spodoptera frugiperda*. *AIP Conference Proceedings* 2596(1), 100004. DOI: 10.1063/5.0136845.

Kumar, R.P. and Singh, J.S. (2020) Invasive alien plant species: Their impact on environment, ecosystem services and human health. *Ecological Indicators* 111, 106020. DOI: 10.1016/j.ecolind.2019.106020.

Kumar, R., Guleria, N., Deeksha, M.G., Kumari, N., Kumar, R. *et al.* (2024) From an invasive weed to an insecticidal agent: Exploring the potential of *Lantana camara* in insect management strategies—a review. *International Journal of Molecular Sciences* 25(23), 12788. DOI: 10.3390/ijms252312788.

Kumari, S., Nazir, F., Maheshwari, C., Kaur, H., Gupta, R. *et al.* (2024) Plant hormones and secondary metabolites under environmental stresses: Enlightening defense molecules. *Plant Physiology and Biochemistry* 206, 108238. DOI: 10.1016/j.plaphy.2023.108238.

Lachowicz, S. and Oszmiański, J. (2019) Profile of bioactive compounds in the morphological parts of wild *Fallopia japonica* (Houtt) and *Fallopia sachalinensis* (F. Schmidt) and their antioxidative activity. *Molecules* 24(7), 1436. DOI: 10.3390/molecules24071436.

Lalk, S., Hartshorn, J. and Coyle, D.R. (2021) Invasive woody plants and their effects on arthropods in the United States: Challenges and opportunities. *Annals of the Entomological Society of America* 114(2), 192–205. DOI: 10.1093/aesa/saaa054.

Lau, J.A. (2012) Trophic consequences of a biological invasion: Do plant invasions increase predator abundance? *Oikos* 122(3), 474–480. DOI: 10.1111/j.1600-0706.2012.20774.x.

Lee, M.R., Flory, S.L. and Phillips, R.P. (2012) Positive feedbacks to growth of an invasive grass through alteration of nitrogen cycling. *Oecologia* 170(2), 457–465. DOI: 10.1007/s00442-012-2309-9.

Levin, L.A., Neira, C. and Grosholz, E.D. (2006) Invasive cordgrass modifies wetland trophic function. *Ecology* 87(2), 419–432. DOI: 10.1890/04-1752.

Lindsay, E.A. and French, K. (2006) The impact of the weed *Chrysanthemoides monilifera* ssp. rotundata on coastal leaf litter invertebrates. *Biological Invasions* 8(2), 177–192. DOI: 10.1007/s10530-004-5856-0.

Litt, A.R., Cord, E.E., Fulbright, T.E. and Schuster, G.L. (2014) Effects of invasive plants on arthropods. *Conservation Biology* 28(6), 1532–1549. DOI: 10.1111/cobi.12350.

Liu, C.-W. and Murray, J.D. (2016) The role of flavonoids in nodulation host-range specificity: An update. *Plants (Basel, Switzerland)* 5(3), 33. DOI: 10.3390/plants5030033.

Liu, Z., Yu, H., Sun, X. and Ding, J. (2021) Effects of elevated temperature on chemistry of an invasive plant, its native congener and their herbivores. *Journal of Plant Ecology* 15(3), 450–460. DOI: 10.1093/jpe/rtab013.

López-Rodríguez, A., Hernández, M., Carrillo-Galvez, A., Becerra, J. and Hernández, V. (2022) Phytotoxic activity of *Ulex europaeus*, an invasive plant on Chilean ecosystems: Separation and identification of potential allelochemicals. *Natural Product Research* 37(5), 769–775. DOI: 10.1080/14786419.2022.2081851.

Lowry, E., Rollinson, E.J., Laybourn, A.J., Scott, T.E., Aiello-Lammens, M.E. *et al.* (2013) Biological invasions: A field synopsis, systematic review, and database of the literature. *Ecology and Evolution* 3(1), 182–196. DOI: 10.1002/ece3.431.

Massadeh, R.K., El-Elimat, T., Al-Gharaibeh, M., Tawaha, K. and Alali, F.Q. (2021) UPLC-HRESI-MS and GC-MS analysis of the leaves of *Nicotiana glauca*. *Acta Pharmaceutica (Zagreb, Croatia)* 72(1), 97–108. DOI: 10.2478/acph-2022-0001.

McCary, M.A. and Wise, D.H. (2019) Plant invader alters soil food web via changes to fungal resources. *Oecologia* 191(3), 587–599. DOI: 10.1007/s00442-019-04510-0.

McCary, M.A., Mores, R., Farfan, M.A. and Wise, D.H. (2016) Invasive plants have different effects on trophic structure of green and brown food webs in terrestrial ecosystems: A meta-analysis. *Ecology Letters* 19(3), 328–335. DOI: 10.1111/ele.12562.

McCormick, C.A., Effah, E. and Najar-Rodriguez, A. (2023) Ecological aspects of volatile organic compounds emitted by exotic invasive plants. *Frontiers in Ecology and Evolution* 11. DOI: 10.3389/fevo.2023.1059125.

McLeod, M.L., Cleveland, C.C., Lekberg, Y., Maron, J.L., Philippot, L. *et al.* (2016) Exotic invasive plants increase productivity, abundance of ammonia-oxidizing bacteria and nitrogen availability in intermountain grasslands. *Journal of Ecology* 104(4), 994–1002. DOI: 10.1111/1365-2745.12584.

Mgobozi, M.P., Somers, M.J. and Dippenaar-Schoeman, A.S. (2008) Spider responses to alien plant invasion: The effect of short- and long-term *Chromolaena odorata* invasion and management. *Journal of Applied Ecology* 45(4), 1189–1197. DOI: 10.1111/j.1365-2664.2008.01486.x.

Morriën, E., Duyts, H. and Van der Putten, W.H. (2011) Effects of native and exotic range-expanding plant species on taxonomic and functional composition of nematodes in the soil food web. *Oikos* 121(2), 181–190. DOI: 10.1111/j.1600-0706.2011.19773.x.

Motard, E., Dusz, S., Geslin, B., Akpa-Vinceslas, M., Hignard, C. *et al.* (2015) How invasion by *Ailanthus altissima* transforms soil and litter communities in a temperate forest ecosystem. *Biological Invasions* 17(6), 1817–1832. DOI: 10.1007/s10530-014-0838-3.

Murrell, C., Gerber, E., Krebs, C., Parepa, M., Schaffner, U. *et al.* (2010) Invasive knotweed affects native plants through allelopathy. *American Journal of Botany* 98(1), 38–43. DOI: 10.3732/ajb.1000135.

Muthusamy, M. and Lee, S.I. (2024) Abiotic stress-induced secondary metabolite production in *Brassica*: Opportunities and challenges. *Frontiers in Plant Science* 14. DOI: 10.3389/fpls.2023.1323085.

Nakano, H., Nakajima, E., Fujii, Y., Shigemori, H. and Hasegawa, K. (2004) Structure-activity relationships of alkaloids from mesquite (*Prosopis juliflora* (Sw.) DC.). *Plant Growth Regulation* 44(3), 207–210. DOI: 10.1007/s10725-004-5574-8.

Nwauche, K.T., Enwereuzo, O.C., Agonmuo, E.N. and Ugoh, A.I. (2024) Proximate and phytochemical profile of *Cirsium arvense* leaves. *Asian Plant Research Journal* 12(6), 78–85. DOI: 10.9734/aprj/2024/v12i6283.

Özçelik, B., Gürbüz, I., Karaoglu, T. and Yeşilada, E. (2009) Antiviral and antimicrobial activities of three sesquiterpene lactones from *Centaurea solstitialis* L. ssp. solstitialis. *Microbiological Research* 164(5), 545–552. DOI: 10.1016/j.micres.2007.05.006.

Özenver, N., Efferth, M. and Efferth, T. (2022) Ethnopharmacology, phytochemistry, chemical ecology and invasion biology of *Acanthus mollis* L. *Journal of Ethnopharmacology* 285, 114833. DOI: 10.1016/j.jep.2021.114833.

Pälike, H. (2005) EARTH. Orbital variation (including Milankovitch Cycles). *Encyclopedia of Geology* 410–421. DOI: 10.1016/B0-12-369396-9/00123-4.

Parhomenko, O.V., Ostapchuk, V.V., Komlyk, V.O. and Brygadyrenko, V.V. (2024) Influence of medicinal plants on *Blaberus craniifer* cockroaches and their parasites, gregarines and nematodes. *Biosystems Diversity* 32(3), 398–405. DOI: 10.15421/012443.

Parkhomenko, A., Oganesyan, E.T., Andreeva, O.A., Dorkina, E.G., Paukova, E.O. *et al.* (2006) Pharmacologically active substances from *Ambrosia artemisiifolia*. Part 2. *Pharmaceutical Chemistry Journal* 40(11), 627–632. DOI: 10.1007/s11094-006-0208-2.

Passos, I.D. and Mironidou-Tzouveleki, M. (2016) Hallucinogenic plants in the mediterranean countries. In: Preedy, V.R. (ed.) *Neuropathology of Drug Addictions and Substance Misuse*. Academic Press, Cambridge, MA, pp. 761–772. DOI: 10.1016/b978-0-12-800212-4.00071-6.

Piwowarski, J.P., Waltenberger, B., Stuppner, H., Kiss, A.K. and Granica, S. (2018) The analysis of phenolic compounds from the aerial parts of *Eupatorium cannabinum* L. subsp. cannabinum. *Biochemical Systematics and Ecology* 79, subsp. DOI: 10.1016/j.bse.2018.05.001.

Poljuha, D., Sladonja, B., Uzelac Božac, M., Šola, I., Damijanić, D. *et al.* (2024) The invasive alien plant *Solidago canadensis*: Phytochemical composition, ecosystem service potential, and application in bioeconomy. *Plants (Basel, Switzerland)* 13(13), 1745. DOI: 10.3390/plants13131745.

Porter, T.H. and Mabry, T.J. (1969) Sesquiterpene lactones. Constituents of *Ambrosia artemisiifolia* L. (Compositae). *Phytochemistry* 8(4), 793–794. DOI: 10.1016/S0031-9422(00)85858-6.

Quintana, N., El Kassis, E.G., Stermitz, F.R. and Vivanco, J.M. (2009) Phytotoxic compounds from roots of *Centaurea diffusa* Lam. *Plant Signaling & Behavior* 4(1), 9–14. DOI: 10.4161/psb.4.1.7487.

Remezok, M., Kolombar, T.M., Parhomenko, O.V. and Brygadyrenko, V.V. (2022) Influence of aromatic substances on locomotor activity of *Deroceras agreste* slugs. *Regulatory Mechanisms in Biosystems* 13(3), 247–256. DOI: 10.15421/022232.

Ribeiro-Varandas, E., Ressurreição, F., Viegas, W. and Delgado, M. (2014) Cytotoxicity of *Eupatorium cannabinum* L. ethanolic extract against colon cancer cells and interactions with Bisphenol A and Doxorubicin. *BMC Complementary and Alternative Medicine* 14(1). DOI: 10.1186/1472-6882-14-264.

Roberge, G., Handa, I.T., Juette, T. and Kneeshaw, D. (2024) Invasive glossy buckthorn (*Frangula alnus*) has weak impact on native understory plant and saprophagous macroarthropod communities. *Diversity* 16(9), 584. DOI: 10.3390/d16090584.

Rodriguez, L.F. (2006) Can invasive species facilitate native species? Evidence of how, when, and why these impacts occur. *Biological Invasions* 8(4), 927–939. DOI: 10.1007/s10530-005-5103-3.

Sagar, L., Sehgal, R. and Ojha, S. (2005) Evaluation of antimotility effect of *Lantana camara* L. var. acuelata constituents on neostigmine induced gastrointestinal transit in mice. *BMC Complementary and Alternative Medicine* 5(1), 18. DOI: 10.1186/1472-6882-5-18.

Salem, A., Szumacher-Strabel, M., López, S., Khalil, M., Mendoza, G. *et al.* (2012) *In situ* degradability of soyabean meal treated with *Acacia saligna* and *Atriplex halimus* extracts in sheep. *Journal of Animal and Feed Sciences* 21(3), 447–457. DOI: 10.22358/jafs/66113/2012.

Sanz, M., Simón, B., Esteruelas, E., Muñoz, Á.M., Cadahía, E. *et al.* (2011) Effect of toasting intensity at cooperage on phenolic compounds in acacia (*Robinia pseudoacacia*) heartwood. *Journal of Agricultural and Food Chemistry* 59(7), 3135–3145. DOI: 10.1021/jf1042932.

Savadi, S., Vazifedoost, M., Didar, Z., Nematshahi, M.M. and Jahed, E. (2020) Phytochemical analysis and antimicrobial/antioxidant activity of *Cynodon dactylon* (L.) Pers. rhizome methanolic extract. *Journal of Food Quality* 1–10. DOI: 10.1155/2020/5946541.

Šćepanović, M., Košćak, L., Šoštarčić, V., Pismarović, L., Milanović-Litre, A. *et al.* (2022) Selected phenolic acids inhibit the initial growth of *Ambrosia artemisiifolia* L. *Biology* 11(4), 482. DOI: 10.3390/biology11040482.

Schädel, M., Yavorskaya, M. and Beutel, R. (2022) The earliest beetle *Coleopsis archaica* (Insecta: Coleoptera) – morphological reevaluation using Reflectance Transformation Imaging (RTI) and phylogenetic assessment. *Arthropod Systematics & Phylogeny* 80, 495–510. DOI: 10.3897/asp.80.e86582.

Sengupta, T., Vinayagam, J., Nagashayana, N., Gowda, B., Jaisankar, P. *et al.* (2010) Antiparkinsonian effects of aqueous methanolic extract of *Hyoscyamus niger* seeds result from its monoamine oxidase inhibitory and hydroxyl radical scavenging potency. *Neurochemical Research* 36(1), 177–186. DOI: 10.1007/s11064-010-0289-x.

Sessa, R.A., Bennett, M.H., Lewis, M.J., Mansfield, J.W. and Beale, M.H. (2000) Metabolite profiling of sesquiterpene lactones from *Lactuca species*. *Journal of Biological Chemistry* 275(35), 26877–26884. DOI: 10.1016/S0021-9258(19)61456-0.

Seybold, S.J., Huber, D.P.W., Lee, J.C., Graves, A.D. and Bohlmann, J. (2006) Pine monoterpenes and pine bark beetles: A marriage of convenience for defense and chemical communication. *Phytochemistry Reviews* 5(1), 143–178. DOI: 10.1007/s11101-006-9002-8.

Simao, M.C.M., Flory, S.L. and Rudgers, J.A. (2010) Experimental plant invasion reduces arthropod abundance and richness across multiple trophic levels. *Oikos* 119(10), 1553–1562. DOI: 10.1111/j.1600-0706.2010.18382.x.

Skliar, M.I., Toribio, M.S. and Oriani, D.S. (2005) Antimicrobial activity of *Centaurea diffusa*. *Fitoterapia* 76(7–8), 737–739. DOI: 10.1016/j.fitote.2005.08.006.

Soni, P., Siddiqui, A.A., Dwivedi, J. and Soni, V. (2012) Pharmacological properties of *Datura stramonium* L. as a potential medicinal tree: An overview. *Asian Pacific Journal of Tropical Biomedicine* 2(12), 1002–1008. DOI: 10.1016/S2221-1691(13)60014-3.

Speranza, J., Miceli, N., Taviano, M.F., Ragusa, S., Kwiecień, I. *et al.* (2020) *Isatis tinctoria* L. (Woad): A review of its botany, ethnobotanical uses, phytochemistry, biological activities, and biotechnological studies. *Plants (Basel, Switzerland)* 9(3), 298. DOI: 10.3390/plants9030298.

Stefanowicz, A.M., Frąc, M., Oszust, K. and Stanek, M. (2022) Contrasting effects of extracts from invasive *Reynoutria japonica* on soil microbial biomass, activity, and community structure. *Biological Invasions* 24(10), 3233–3247. DOI: 10.1007/s10530-022-02842-2.

Strzępek-Gomółka, M., Gaweł-Bęben, K. and Kukula-Koch, W. (2021) *Achillea* species as sources of active phytochemicals for dermatological and cosmetic applications. *Oxidative Medicine and Cellular Longevity* 2021(1), 6643827. DOI: 10.1155/2021/6643827.

Suprun, A.R., Kiselev, K.V., Aleynova, O.A., Manyakhin, A.Y. and Ananev, A.A. (2024) Analysis of phenolic compounds of *Reynoutria sachalinensis* and *Reynoutria japonica* growing in the Russian Far East. *Plants (Basel, Switzerland)* 13(23), 3330. DOI: 10.3390/plants13233330.

Szewczyk, K., Kalemba, D., Komsta, Ł. and Nowak, R. (2016) Comparison of the essential oil composition of selected impatiens species and its antioxidant activities. *Molecules (Basel, Switzerland)* 21(9), 1162. DOI: 10.3390/molecules21091162.

Tapera, R.F., Siwe-Noundou, X., Shai, L.J. and Mokhele, S. (2024) Exploring the therapeutic potential, ethnomedicinal values, and phytochemistry of *Helianthus tuberosus* L.: A review. *Pharmaceuticals (Basel, Switzerland)* 17(12), 1672. DOI: 10.3390/ph17121672.

Tian, Y. and Zhang, Z. (2023) Insecticidal activities of *Sophora flavescens* Alt. towards red imported fire ants (*Solenopsis invicta* Buren). *Toxins* 15(2), 105. DOI: 10.3390/toxins15020105.

Tun, K.M., Clavijo McCormick, A., Jones, T., Garbuz, S. and Minor, M. (2020) Honeydew deposition by the giant willow aphid (*Tuberolachnus salignus*) affects soil biota and soil biochemical properties. *Insects* 11(8), 460. DOI: 10.3390/insects11080460.

Turbelin, A.J., Malamud, B.D. and Francis, R.A. (2016) Mapping the global state of invasive alien species: Patterns of invasion and policy responses. *Global Ecology and Biogeography* 26(1), 78–92. DOI: 10.1111/geb.12517.

Umiljendić, J.G., Sarić-Krsmanović, M., Radivojević, L., Šantrić, L., Šćepanović, M. *et al.* (2025) Phytochemical analysis of *Asclepias syriaca* L. leaf extract and its potential phytotoxic effect on some invasive weeds. *Scientific Reports* 15(1). DOI: 10.1038/s41598-025-90940-1.

Uzelac, M., Sladonja, B., Šola, I., Dudaš, S., Bilić, J. *et al.* (2023) Invasive alien species as a potential source of Phytopharmaceuticals: Phenolic composition and antimicrobial and cytotoxic activity of *Robinia pseudoacacia* L. leaf and flower extracts. *Plants* 12(14), 2715. DOI: 10.3390/plants12142715.

Vasquez, A., Belsky, J., Khanal, N., Puri, H., Balakrishnan, D. *et al.* (2024) *Melanaphis sacchari/sorghi* complex: Current status, challenges and integrated strategies for managing the invasive sap-feeding insect pest of sorghum. *Pest Management Science* 81(5), 2427–2441. DOI: 10.1002/ps.8291.

Vaughn, S.F. and Berhow, M.A. (1999) Allelochemicals isolated from tissues of the invasive weed garlic mustard (*Alliaria petiolata*). *Journal of Chemical Ecology* 25(11), 2495–2504. DOI: 10.1023/A:1020874124645.

Verma, R.S., Joshi, N., Padalia, R.C., Goswami, P., Singh, V.R. *et al.* (2017) Chemical composition and allelopathic, antibacterial, antifungal and *in vitro* acetylcholinesterase inhibitory activities of yarrow (*Achillea millefolium* L.) native to India. *Industrial Crops and Products* 104, 144–155. DOI: 10.1016/j.indcrop.2017.04.046.

Whaley, A.K., Lukashov, R.I., Whaley, A.O., Zhokhova, E.V., Gurina, N.S. *et al.* (2023) Flavonoids from *Impatiens grandulifera* and their antioxidant activity. *Drug Development & Registration* 12(4), 223–230. DOI: 10.33380/2305-2066-2023-12-4-1587.

Woerdenbag, H.J. (1986) *Eupatorium cannabinum* L. *Pharmaceutisch Weekblad Scientific Edition* 8(5), 245–251. DOI: 10.1007/BF01960068.

Woodworth, G.R., Ward, J.N. and Carr, D.E. (2020) Exotic tree and shrub invasions alter leaf-litter microflora and arthropod communities. *Oecologia* 193(1), 177–187. DOI: 10.1007/s00442-020-04657-1.

Yelenik, S.G., Stock, W.D. and Richardson, D.M. (2004) Ecosystem level impacts of invasive *Acacia saligna* in the South African fynbos. *Restoration Ecology* 12(1), 44–51. DOI: 10.1111/j.1061-2971.2004.00289.x.

Zazharskyi, V.V., Brygadyrenko, V.V., Zazharska, N.M., Borovik, I.V., Boyko, O.O. *et al.* (2024) Antibacterial and nematicidal activities of extracts from plants of the asteraceae family. *Regulatory Mechanisms in Biosystems* 15(3), 587–593. DOI: 10.15421/022482.

Zeni, V., Benelli, G., Campolo, O., Giunti, G., Palmeri, V. *et al.* (2021) Toxics or Lures? Biological and behavioral effects of plant essential oils on Tephritidae fruit flies. *Molecules* 26(19), 5898. DOI: 10.3390/molecules26195898.

Zhai, J., Hou, B., Hu, F., Yu, G., Li, Z. *et al.* (2024) Active defense strategies for invasive plants may alter the distribution pattern of pests in the invaded area. *Frontiers in Plant Science* 15, 1428752. DOI: 10.3389/fpls.2024.1428752.

Zhang, J.-P., Tian, X.-H., Yang, Y.-X., Liu, Q.-X., Wang, Q. *et al.* (2016) Gleditsia species: An ethnomedical, phytochemical and pharmacological review. *Journal of Ethnopharmacology* 178, 155–171. DOI: 10.1016/j.jep.2015.11.044.

Zhang, S.-P., Bai, B., Chen, G.-M., Wang, Y.-Q., Hu, C. *et al.* (2024) Secondary metabolites in host pears defense against two fruit borers and cytochrome-P450-mediated counter-defense. *iScience* 27(4), 109518. DOI: 10.1016/j.isci.2024.109518.

Index